CONCERNING BUILDINGS

CONCERNING BUILDINGS

Studies in honour of Sir Bernard Feilden

Edited by

Stephen Marks

Architectural Press
An imprint of Butterworth Heinemann

Butterworth-Heinemann
Linacre House, Jordan Hill, Oxford OX2 8DP
A division of Reed Educational and Professional Publishing Ltd

A member of the Reed Elsevier plc group

OXFORD BOSTON JOHANNESBURG
MELBOURNE NEW DELHI SINGAPORE

First published 1996

British Library Cataloguing in Publication Data
A catalogue record for this book is available from the British Library.

ISBN 0 7506 2350 0

Library of Congress Cataloguing in Publication Data
A catalogue record for this book is available from the Library of Congress.

Composition by Scribe Design, Gillingham, Kent, UK
Printed in Great Britain by The Bath Press, Bath

CONTENTS

Advisory Committee

Sherban Cantacuzino
Henry Cleere
Derek Linstrum
John Warren

CONTRIBUTORS

Dame Jennifer Jenkins D.B.E.
Chairman: Historic Buildings Council for England (1975–84); National Trust (1986–90); Royal Parks Review Group (1991–96); Architectural Heritage Fund since 1994. Co-author of *From Acorn to Oak Tree: The Growth of the National Trust, 1895–1994* (1994).

Sherban Cantacuzino C.B.E., F.S.A., F.R.I.B.A.
Architect, consultant and lecturer in architectural conservation, writer on architecture; Executive Editor of *The Architectural Review* (1973–79); Secretary of the Royal Fine Art Commission (1979–94); member of Executive Committee of ICOMOS since 1990, and President of ICOMOS U.K. Publications include *European Domestic Architecture* (1969), *New Uses for Old Buildings* (1975), and *What makes a Good Building?* (1995).

Henry Cleere O.B.E., Ph.D., D.Litt., F.S.A.
World Heritage Coordinator for the International Council on Monuments and Sites (ICOMOS), based in Paris; Director of the Council for British Archaeology (1974–91); Fellow of University College London; first Secretary of the European Association of Archaeologists. Editor of *Approaches to the Archaeological Heritage* (1984) and *Archaeological Heritage Management in the Modern World* (1989).

Jo Cox B.A., Ph.D.
Partner in Keystone Historic Buildings Consultants since 1988; Hon. Research Fellow in School of English, Exeter University; worked on re-survey of listed buildings for Devon and Cornwall (1984–88); has sat on main committees of SPAB, Exmoor National Park Planning Committee, Council of Management of Devon Historic Buildings Trust, Devon Gardens Trust; founder member of Devon Nineteenth-Century Churches Project; founder member and committee member of the Devon Buildings Group; occasional lecturer on conservation studies, Plymouth University.

Peter Goodchild B.Sc., Dip.Land.Des., Dip.Cons.Studies
Lecturer in Conservation Studies, Institute of Advanced Architectural Studies, University of York; Research Fellow at IoAAS in 1979, work led to setting up of Centre for the Conservation of Historic Parks and Gardens, now part of Landscapes and Gardens unit; corresponding member of ICOMOS–IFLA International Committee on Historic Gardens and Sites, and coordinator of its working group on education.

Jukka Jokilehto D.Phil.(York), Architect (Finland)
Assistant to the Director General of ICCROM; President of the International Training Committee of ICOMOS (CIF); architect and town-planner in Finland; conservation architect at ICCROM since 1972; responsible for international training programmes in architectural conservation, international advisory missions; representative of ICCROM for the World Heritage Convention. Publications include *Management Guidelines for World Cultural Heritage Sites* (1993) (with Sir Bernard Feilden).

Derek Linstrum Dip.Arch., Ph.D., F.S.A., A.R.I.B.A.
Practiced as architect for eighteen years before taking up academic career: Senior Lecturer in Leeds School of Architecture (1966–71); Radcliffe Reader in Architectural History and Conservation at University of York (1971–92); Hoffman Wood Professor of Architecture at University of Leeds (1991–93); now Honorary Fellow of Institute of Advanced Architectural Studies, University of York, and of Bretton Hall College, University of Leeds; has travelled and lectured extensively, including teaching at ICCROM regularly from 1973 onwards.

Publications include *Sir Jeffrey Wyatville: Architect to the King* (1972), *Catalogue of the RIBA Drawings Collection: The Wyatt Family* (1974), and *West Yorkshire Architects and Architecture* (1978).

David McLaughlin Dip.Arch., D.A.(Manc.), Registered Architect
Principal Conservation Officer, Bath and North East Somerset Council, successor of Bath City Council (since 1975); on Bath and Wells Diocesan Advisory Committee for the Care of Churches since 1983; appointed (1995) Visiting Fellow for Life at Centre for Conservation Studies, York University; Cathedral Fabric Committee of St Woolos, Newport; author and lecturer on conservation.

Stephen Marks M.A., F.S.A., R.I.B.A.
For last twenty years, since spells in local government, a planning inspector taking public inquiries for the Government, including major inquiries about listed buildings and city-centre development. Publications include *The Map of Mid-sixteenth-century London* (1964) and annual *Transactions of the Association for Studies in the Conservation of Historic Buildings* (editor since 1977); other publication work, including facsimiles and new editions, mainly for local, literary and topographical societies.

Richard J. Morrice B.A., Ph.D.
Inspector of Historic Buildings, English Heritage, dealing with control and grant matters; Hon. Secretary of the Society of Architectural Historians of Great Britain; teaches architectural history module on RICS Building Conservation Course, College of Estate Management, Reading.

Patrick Nuttgens C.B.E., M.A., Ph.D., A.R.I.B.A.
Professor of Architecture and first Director of Leeds Polytechnic (1970–86); Honorary Professor of University of York; President of York Georgian Society. Publications include *The Landscape of Ideas* (1972), *York the Continuing City* (1976, 1980), *Leeds Old and New* (1979), *Pocket Guide to World Architecture* (1980), *Understanding Modern Architecture* (1988); frequent broadcaster and contributor to journals.

Keith V. Parker B.A., A.L.A.
Librarian, Institute of Advanced Architectural Studies, The University of York, since 1964.

Matthew Saunders M.A.(Cantab.), F.S.A.
Conservationist, lecturer, writer; Secretary, SAVE Britain's Heritage (1976–77); Secretary, Ancient Monuments Society since 1977; Secretary, Joint Committee of the National Amenity Societies since 1982; Honorary Director of The Friends of Friendless Churches since 1993; member, Churches Advisory Committee, National Heritage Memorial Fund, since 1995; author of biography of S. S. Teulon. Publications include *The Historic Homeowner's Companion* (1987).

James Simpson B.Arch., F.R.I.A.S., R.I.B.A., F.S.A.Scot.
Architect, conservator, writer, lecturer; set up firm of Simpson & Brown in 1977; trained at Edinburgh College of Art, worked for Sir Bernard Feilden in 1972–74 on St Paul's and Norwich Cathedrals, has been architect to St Giles Cathedral, Edinburgh, and, briefly, Surveyor of the Fabric of York Minster; curated the William Adam Exhibition in 1989; member of the Ancient Monuments Board for Scotland (1983–95).

John Warren M.Litt., F.S.A.
Architect, planner, conservator, writer; founding partner of APP, architects and planners; founding member and Hon. Architect of The Weald and Downland Open Air Museum; founder and past Chairman of The Amberley Museum of Industrial Archaeology; consultant to The Conservation Practice; Chairman of World Heritage Committee of ICOMOS UK; Fellow of Institute of Advanced Architectural Studies, University of York.

David Yeomans B.Sc.(Eng.), Ph.D., A.I.W.Sc.
Engineer and architectural historian; teaches construction and history at Manchester School of Architecture; consultant to UNESCO for the Kathmandu World Heritage Site. Publications include *The Trussed Roof: Its History and Development* (1992) and *The Architect and the Carpenter* (1992).

FOREWORD

The conservation of historic buildings and areas in Britain is now an accepted object of national policy, attracting strong popular support and engaging the attention of architects and planners. Yet fifty years ago, at the end of the Second World War, the subject concerned few beyond a handful of architectural historians and organizations like the Society for the Protection of Ancient Buildings. They lacked the influence to stem the tide which in the 1960s swept away many historic town centres to make room for new roads and shopping centres and removed large swathes of Victorian housing in the name of slum clearance on a scale far more destructive than wartime bombing.

This collection of essays explains how during the last thirty years, in a reaction against the continuing devastation Britain has developed a system for protecting historic buildings more comprehensive than in any other country. The essays fall into two groups. First are a group dealing with the legislative, administrative and financial framework for the conservation of buildings and landscapes, accompanied by a couple assessing the impact of international organizations. In the second group are several essays about specialized aspects of building conservation practice.

Bernard Feilden, in whose honour this *Festschrift* is most appropriately assembled, not surprisingly features frequently. But he does so not only as the dedicatee, but also naturally, as it were, because he was one of the few architects of his generation to specialize in historic buildings, pioneering the use of modern technology and showing how new elements could be designed in an unmistakably contemporary idiom yet fit the existing context. He was also an inspiring educator at English universities and at international centres. I knew him in both capacities. I first met him in Chesterfield, Derbyshire, when the Borough Council were proposing to build a new shopping centre on the ancient market place, and to demolish the quirky Victorian town hall and older buildings around it. The ensuing uproar enabled the Historic Buildings Council to broker a solution: Feilden was commissioned to design a scheme for reinstating the Town Hall and some other buildings, and for providing several larger shops behind existing façades. Some years later we met as trustees supporting the Indian National Trust when he took infinite pains to select the best projects for grant-aid and to choose the most suitable students for an exchange scheme.

Matthew Saunders sets the scene for these essays in his authoritative survey of the measures taken to conserve buildings in Britain since 1945. He ascribes the greater comprehensiveness of the British system compared with that in other countries to grass-roots support, effectively articulated through the SPAB and its successor organizations which specialize in buildings up to the present day. He also points to the nearly three million members of the English and Scottish National Trusts as evidence of the interest in the conservation of landscape as well as buildings.

In response to public pressure statutory controls over individual historic buildings have been put in place by successive governments. The extension of controls over historic areas as a whole was, however, the result of independent initiatives. The concept of conservation areas was originated in an Act of Parliament, the Civic Amenities Act 1967, which was introduced as a Private Member's Bill by Duncan Sandys, and control over demolition within their boundaries was brought in by the Town and Country Amenities Act 1974, another private initiative. I was involved in the latter and in the course of my negotiations was told by a senior civil servant that the Bill had no chance of success. In fact, it went through with acclamation, despite having to be introduced for a second time because of an unexpected general election. Conservation areas have proved extraordinarily popular (over 8,000 have been designated by local authorities in England alone) but I do not believe that they would have been introduced or made effective, at least without prolonged delay, had the amenity societies not taken action.

Saunders points out that the protective régime is only as good as the protectors. With the establishment of English Heritage their power has strengthened, although the system would not work as effectively as it does without the formal involvement of the national amenity societies in the planning process. At a local level the hundreds of local civic societies and similar groups act as watchdogs over the local authorities who are responsible for deciding on alterations to most listed buildings.

The protection of historic buildings and areas has been increasingly tied into the planning system. The weakest point is the failure of the taxation system to facilitate conservation, in contrast with the position in France and Germany. The result is that in Britain, almost alone in the European Union, 17.5 per cent VAT is imposed on the repair of listed buildings, while demolition, new buildings and alterations are zero-rated. At a time when grants from English Heritage are being cut, though in some cases the Heritage Lottery Fund may come to the rescue, the burden of VAT is a major obstacle to proper maintenance, particularly for churches, charities and private owners.

The conservation of parks and gardens attracted attention later than buildings, both from the voluntary movement and from the Government. The Garden History Society was not founded until 1965, a century later than the SPAB, despite the traditional British love of gardening. The first legislative recognition came nine years later when the Town and Country Amenities Act provided for parks and gardens to be eligible for grant-aid. However, the disadvantage caused by the lack of official support quickly became apparent. Officers were not appointed to identify and list historic gardens, an essential requisite for their protection, and obstacles were put in the way of the Historic Buildings Council proceeding on its own. Not until English Heritage was established, with a greater degree of independence, was the Register of Parks and Gardens of Special Historic Interest published and this carries no additional statutory protection. Grant-aid has been minimal, though with the announcement of the Heritage Lottery Fund's programme for the restoration of urban parks, these will qualify for substantial funds.

Peter Goodchild's essay ranges from the relatively small geographical area covered by designated parks and gardens to wider historical landscapes, a subject which has been the focus of a number of international conferences. The analysis by Jukka Jokilehto and Henry Cleere of the work of the international organizations concerned with historic conservation concluded that they have proved more influential than some of their critics might suppose. International conventions and charters have over the years set clear objectives and defined the measures needed for the protection of the world's monuments and landscapes. These have been adopted by an increasing number of countries and, where not adopted in principle or implemented in practice, have formed a convincing agenda for voluntary organizations to follow.

The international organizations have a number of specific achievements to their credit. Particularly creative has been the International Centre for the Study and Preservation and Restoration of Cultural Property (ICCROM), the training and research centre in Rome of which Feilden was director from 1977 to 1981. Other high-profile projects are the compilation and monitoring of the list of World Heritage Sites, the identification of World Heritage Sites in danger from neglect, development or natural catastrophe, and the establishment of a World Heritage Fund. Inclusion on the list of World Heritage Sites is eagerly sought but not easily gained owing to the high standards of management and protection insisted on: the Lake District, for example, has not been accepted.

Perhaps the pivotal essay in the book is that of Derek Linstrum on the education of a conservation architect. The architect, in Feilden's words, 'is in a central position from which he can see the historic building as a whole and become its "curator". He is able to understand how it was conceived, how it was built, how it has responded to its environment, and how it has been used ot abused.' Such responsibilities cannot be assumed without a proper training, yet in Britain, unlike France, architectural schools after the Second World War abandoned the teaching of architectural history and traditional building techniques, concentrating instead on modern design and methods. The inevitable result was that by the 1960s, when older architects who had had a broader training were approaching retirement, there was a shortage of conservation architects.

Since then a number of specialized part-time and short postgraduate courses have been provided, but conservation still does not form an integral part of the professional training. It is difficult for a layman to conceive how an architect can be thought qualified without having studied the work of his predecessors, quite apart from the need to prepare for the many practices where the re-use of old buildings now accounts for a significant proportion of the workload. In the interest of the 'general practitioner' Linstrum makes the case for conservation to be included in the training of all architects.

The group of essays concerned with the techniques of building conservation covers the whole process from the initial historical assessment and the preliminary archaeological investigation for major schemes through the actual repair works organized by the architect or surveyor with whatever help is needed from engineers and other experts. As the contributors make clear, the involvement of different disciplines and the results of research into the performance of materials and techniques have led to an increasingly sophisticated approach to the restoration of old buildings.

Conservation raises ethical and aesthetic issues as well as technical problems. These are not simply matters of philosophical debate but require practical decisions as to the authenticity of repairs, how far new structures should be added to existing buildings and what degree of loss or damage to the fabric is acceptable. Such dilemmas are most acute in dealing with buildings which have become redundant and can no longer be used for their original purpose: churches left empty because of declining faith, barns because of modern agricultural practices, textile mills because industry has moved away. Sherban Cantacuzino shows how redundancy can affect not just a single building but also a whole historic centre or neighbourhood as in Venice, San Francisco and parts of London. In solving the problems of dereliction, conservation has to become an integral part of planning policy which includes the management of open spaces and traffic as well as the repair of old buildings and the construction of new ones where necessary.

Cantacuzino takes Urbino in Italy as an example of how a historic town can retain its character while revitalizing its economic and social life. In order that the University and tourism might replace agriculture and the church as sources of livelihood redundant convents and other buildings have been converted for academic faculties, community arts centres and tourist facilities. Underpinning the changes was the decision to exclude traffic from the historic centre, a revolutionary idea when applied in the 1960s.

The future of Urbino as a living city depended upon the introduction of new activities. Similar changes have been brought about in limited areas of many British cities – in Covent Garden, London, and Albert Dock, Liverpool, for example. However, only in a few cities such as York and Chester has a coherent overall strategy been adopted and based on the twin objectives of conserving the historic fabric and finding new uses for redundant buildings. These have proved so successful that the problem has become one of limiting rather than encouraging new activities.

The conclusion to emerge from the *Festschrift* is that during the last half century there has been a sea change in attitudes to conservation. Among architectural and related professions the conservation of old buildings is no longer seen as negative, restrictive and dull, but as stimulating and calling for creative skills no less than those required for new building. Among the general public conservation is seen as economically sound and socially beneficial as well as being essential for a sustainable environment.

Jennifer Jenkins

INTRODUCTION

If ancient fabrics nod, and threat to fall,
To patch their flaws, and buttress up the wall,
Thus far 'tis duty: but here fix the mark;
For all beyond it is to touch the ark.
To change foundations, cast the frame anew,
Is work for rebels, who base ends pursue.

John Dryden (1631–1700)

Dryden lived almost exactly two hundred years before William Morris, to whom we almost daily attribute the most forceful and influential expression of our responsibility for our heritage and to our successors. Yet here we have the sentiments encapsulated in six pithy lines from a poet of another age, when there was much political turbulence and also much destruction of the heritage at the hands of the Puritans. Dryden may, of course, have been writing metaphorically and cryptically, but the concept of the duty to preserve, patched if need be, is clear. In his day, as in ours, there were no doubt those for whom old was bad because it was old!

A reader might think, then, that, by quoting so extreme a view of non–intervention, I did not approve of the work of those who saved York Minster, for example, from dire catastrophe, and have done much more than merely hold it up. It is not so, of course, but I do think that one must always start from this extremity to understand how far one may go: if the instinctive approach is to find out how little one needs to do to achieve the true preservation of our heritage, of whatever kind, then there is a good chance that what is valuable will indeed be altered as little as may be, passing on to others, as we must, that heritage, although taking advantage of opportunities which present themselves.

In editing this volume, I have learned so much about the continuing efforts all over the world in more than a century to reach an understanding of the rules and guidelines for conservation, preservation, maintenance, repair – call it what you will – that I now have a very good idea of the difficulty of reaching any certain conclusion or agreement. That we have a whole bevy of Charters, Commissions, Recommendations, Committees, Conventions, is evidence of that difficulty. I have to confess that it was not long ago that I was first made aware in detail of the problem of the need for different guidance for different cultures and traditions; this was the outcome of an article by Dr Wim Denslagen (of the

Dutch Rijksdienst voor de Monumentenzorg), which he worked up for me for publication in ASCHB *Transactions*.[1]

I was very pleased to discover recently that my grandfather, A. R. Powys (Secretary of the Society for the Protection of Ancient Buildings from 1911 till his death in 1936) was one of the two British delegates to the Congress in Athens in 1931 which gave us that foundation stone, the Charter of Athens of the same year. It is evident that the influence of Morris, of Philip Webb, of the SPAB Manifesto, and of the English way, which was already following such lines, had a direct conduit to those deliberations. One recent writer has suggested that the Charter contains characteristically Powysian phrases.[2] I have therefore had an extra incentive to see published the two contributions which flow from this. I have taken some labour to make ample cross-referencing in the Index to help in this confusing world of international charters.

The other contributors have most excellently covered their fields, though some at greater length, some at less, than requested (I feel, as an editor, that authors should have freedom to express things in the measure that suits them). I do not at all mind that you will find differences of opinion or approach between authors: differences will encourage readers to think for themselves. Nor do I mind that there is a degree of duplication, where one chapter overlaps with another, but with a different emphasis. I hope you will take your time to look at each chapter for its own message.

I wish to thank the publisher, Mr Neil Warnock-Smith, and the four members of my advisory committee (all contributors to the volume) for helping to compile a co-ordinated choice of subjects and to bring together the whole team of writers, all experts in their fields, and for their encouragement. I am also most grateful to the contributors who have all put up with my demands and questions with a good grace, and who have, more or less, met deadlines.

The volume is in honour of Sir Bernard, but it is not intended, except for the tribute written by Professor Patrick Nuttgens and the bibliography, to be about him. We set out to cover a co-ordinated and wide range of conservation issues, theoretical and practical, and wanted the book to be regarded as a reference work for our position and our achievements in the art of conservation today; it was not, however, to be a technical treatise, but to deal with the *approach* to conservation. I pray that we are successful.

Stephen Marks

NOTES

1. W. Denslagen, 'Restoration theories, East and West', *Association for Studies in the Conservation of Historic Buildings Transactions*, xviii (1993), 3–7.

2. P. Burman, '"Obstinate Sincerity": A reading of A. R. Powys', *The Powys Journal*, vi (1996), 125–51; see particularly pages 126–7 and 139–40.

1

A TRIBUTE TO SIR BERNARD

Patrick Nuttgens

I first met Bernard Feilden in 1962 in the early days of the new University of York. But I did not really know him as a person until, with my wife and children, I stayed, with his wife and children, in his house, a former coastguard's house, at Morston in Norfolk.

It was a memorable and utterly exhausting episode. He pulled boats and people down creeks to the sea, tipped us all into an inflatable motorboat, pumped it up when it looked like sinking in the sea, carried it home, fed us on warm Guinness and cockles, talking all the time, and then when I was ready to drop into bed, took me for a four-mile walk over the sands. It was not the only time he astonished people with his remorseless energy. He chugged in that inflatable Zodiac up the North Sea from Norfolk, then along the Humber, deflated it on the quayside in the middle of York and walked up to the Minster to do a day's work.

Bernard was one of that magical generation of students who studied at the Architectural Association after the Second World War. He was a mature student (he was born in 1919), but I suspect he was always mature, even in his childhood. His record and personality confirm what I have always known about him; he reassures us just by being there, the very symbol of competence and reliability.

He was born in London, worked briefly in an architect's office and attended classes at the Bartlett School of Architecture before the war. Then, during the war, he served in the Royal Engineers, in India, the Middle East and Italy. He is still, in my opinion, an engineer; he brings to architecture an engineer's capacity for taking pains and a rational analysis of the practical problems that have to be solved. But of course there is far more to his architectural life than that. Like Kent who 'jumped the fence and found all nature a garden' he looked around and found all architecture was a great adventure; and he was going to be part of it.

He was one of the first architects to attend the early courses in the 1950s on the Protection and Repair of Historic Buildings at what was then the York Institute of Architectural Study. That soon became the Institute of Advanced Architectural Studies and was the first institute to join the new University of York in 1962.

Bernard later became one of its authoritative lecturers. But he did not only start as a specialist in conservation. He founded, with David Mawson, his own architectural practice in Norwich. Their work included (and still includes) a wide conspectus of social architecture – houses, factories, schools, universities, hotels and the design of central urban areas.

But he will always be known as the outstanding practitioner and wisest adviser on the protection and repair – the conservation – of historic buildings. In 1956 he started work on Norwich Cathedral, carrying out an impeccable survey and applying engineering techniques to its decaying architecture. That part of his practice quickly spread; within a few years he was inspecting and reporting on several hundred parish churches. In 1965 he was appointed Surveyor to the Fabric of York Minster. And that was another turning point.

Despite the eminence and fame of his predecessors he found a great building that had not been properly surveyed and analysed for a hundred years. He explored every bit of it. You could see him in the evening walking along the ridge of the roof and noting every crack or movement. His detailed inspection took, he later recorded, two years – 2000 man-hours. It was the first time for many years that the Minster had been inspected stone by stone. All the cracks that had been found were marked on a perspex model of the Minster. Bernard then sat down to study them in three dimensions.

He had, by that time, already restored and adapted two major buildings for the new University of York. Heslington Hall, dating mostly from the sixteenth and nineteenth centuries, was in deplorable condition and had not been occupied for several years. He did not just restore it. He made some drastic alterations and additions, including a new main stair in the Great Hall – a change which would probably not be allowed nowadays. The King's Manor in the centre of York was even more disparate and decayed. Parts of it date from the fifteenth and sixteenth centuries, and other parts from the nineteenth and twentieth centuries. Again it was a thorough job, with historic areas brought to life and new elements unmistakably of their own date. The Manor became a virtual textbook of architectural history and imaginative conservation.

Both there and at the Minster, the study and discovery of the history of the place were fundamental to Bernard's understanding of what had to be done. For a historic building is not a dead relic; it is a living building whose life depends on the maintenance of the old elements and the introduction of new ones. The conservation work itself, he says, should be invisible; the total living building should, on the other hand, be read and understood and used. The additions he made there added a positive chapter to the story of its years of change.

But to understand how Bernard's mind works there is no better illustration than the Minster. I emphasize his work there for several reasons. One is that I saw it from the start and was already familiar with its problems; another is that it tells more than any other job about Bernard Feilden's approach to architectural problems. From the start, for example, he backed the archaeological investigations. So keen were the archaeologists that rumour had it in York that they wanted the Minster to fall down so that they could find out still more about the underlying ruins of the Roman legionary fortress. In fact, the architect and the archaeologists worked closely together and the results were different from what they might otherwise have been.

For Bernard, the detailed survey and the study of its history are partners in the revelation of a historic building and its future. That is very clear if you look at the excavation below the central tower of York Minster. The ground had to be dug away to reveal the lower levels of the piers. But the excavation also revealed the lower walls of the Roman praetorium which had occupied the site before the Minster was even thought of. Then, instead of putting back the ground once the structural work had been completed, Bernard and his team kept it open and *created* a crypt which had never been there before and which could be used to show the Roman walls, an exhibition of the repair work, and a treasury for the Minster's jewellery and plate. It was astonishing for most of us that the preservation of an ancient cathedral tower should result in a *new* crypt. The Minster, like the King's Manor, had become a living exhibition of architectural history.

Bernard Feilden is always the first to insist that his work is not just his own but the product of a team of experts. That includes structural engineers, archaeologists, art historians, mechanical and electrical engineers, quantity surveyors and building contractors, among whom he singles out the managers, foremen and the skilled labourers. The team as a whole is greater than the sum of its participants. I used to think of Bernard as the conductor of a great orchestra of skilled performers, to all of whom he would bow at the completion of the work. In the years of watching the saving of York Minster, I was assured by several individuals that they themselves were responsible for all the work. And they were proud of it. One of the workmen even ticked Bernard off for using a bit of mildly bad language in a sacred building.

What I found (and still find) definitive about the Minster story is what I imagine happened when Bernard studied that perspex model. It is still there in the new crypt so that you can see the lines of cracks marked on it in colour, and work out the movement of the structure by viewing it from every side and angle. It was from that analytical study (over whatever period was necessary) that Bernard and the engineers were able to diagnose the sickness and understand what was happening. The piers of the central tower were sinking differentially. There was movement also at the east and west ends, much of which may have occurred when the Minster was being built. The study of history as well as the record of its present state again went together.

But in the Feilden view of a monument, there is still more to be considered. It must be possible to note movements not only in the past but also in the future. The Minster must be left in good condition and ready for future maintenance. Any survey needs a firm datum; so a datum level was established eighty feet (twenty-four metres) below the floor near the west end and marked with a plate on the floor. Small plates on the columns could then be read from a level and movement noted. At the east end a vertical plumb was inserted. Bernard left the Minster in workable condition. There should be no excuse for ignorance of its condition ever again.

If the investigation was original, as well as thorough, the cure for the disease was even more original. Discarding any idea of underpinning, the architect and engineer devised a remarkable solution which, of course, you can see in the new crypt. The four major piers of the central tower were wrapped in massive concrete collars linked together, and stainless steel rods inserted in the holes drilled through the collars and the piers and post-stressed. You can see and stroke the massive bolts at the end of the rods which are holding the building up. Further necklaces of stainless steel were inserted above the arches and at the ends of the building. I used to explain to visitors that the Minster was now a sophisticated steel structure within a stone structure – hidden away but essential to its survival.

Then Bernard's expertise took him further. From 1969 to 1977 he was surveyor to St Paul's Cathedral in London. From 1977 to 1981 he was Director of the International Centre for the Study and the Preservation and Restoration of Cultural Property in Rome, and then became Chairman of the UK National Committee of the International Council on Monuments and Sites. His work and influence which had started in Norwich was now extended to an international role.

I have written about Bernard as an architect – not a special kind of architect but an architect in the fullest sense. There is of course another side to him which I have space only to mention. His family is central: his wife Ruth (who died recently while this chapter was being composed), endlessly hospitable and critically observant; his children; his second wife Tina, Secretary of Icomo UK; and a vast array of friends. He has played a major part in the work of national and local bodies and amenity societies. And he is a gifted teacher; his lectures and seminars are packed with information and insights.

What, in conclusion, has he contributed lastingly to his profession? He sees architecture as one activity and in his own practice has worked on modern buildings as well as conservation. He sees buildings as 'unified spatial environmental systems'. So, again in his words, 'the conservation of historic buildings and modern practice go hand in hand'. In an old building the new work should not be too strong for the old work; judgement is vital. And what is that environment of which he speaks? It is made of space, light and structure. How do you react to it? How do you learn about it? His answer is clear and calm: 'Let the building speak to you'.

The unity of old and new – in architecture – can be known by all architects and should be an essential part of their training. He has insisted many times that all architects should understand the causes of decay in buildings and the importance of designing for maintenance. Many modern buildings are seriously flawed, he wrote in a letter to me, 'because nature has been flouted'. If all architects understood the causes of decay in buildings, they would design better buildings. And it is to that aim that Bernard Feilden has devoted his professional life.

2

THE CONSERVATION OF BUILDINGS IN BRITAIN SINCE THE SECOND WORLD WAR

Matthew Saunders

Since the guns ceased in 1945 there has been a sea change in the conservation of historic buildings; a revolution so spectacular that the principal emotion provoked in retrospect is the unattractive and treacherous one of smugness.

1945–95: EVOLUTIONARY CHANGE BUT REVOLUTIONARY IMPACT

In 1945 there were no buildings under national statutory protection, apart from a handful of scheduled ancient monuments. Now the total of listed buildings in England, Scotland, Wales and Northern Ireland nudges 520 000, and there are well over 9 000 conservation areas in England alone. Britain has the most comprehensive system of protection of any country in the world, by a factor of three. The only equivalent, and that rivalry is not posed by a nation state, is Bavaria where 110 000 structures are protected. The National Trust (for England alone) has nearly 2 500 000 members, more than those subscribing to all political parties of whatever persuasion. More than 300 000 people belong to English Heritage, 30 000 to The Friends of Historic Scotland, 13 000 to Cadw (Welsh Historic Monuments), a further 365 000 to the 930–odd local civic societies, 100 000 to local archaeological organizations and some 20 000 to the National Amenity Societies (see page 11). From a base point in 1953, when centralized public grants first began to flow, English Heritage now distributes nearly £50 000 000 a year, Historic Scotland £11 000 000, Cadw £3 500 000 and the Department of the Environment, Northern Ireland, £1 700 000. The National Heritage Memorial Fund, purveyor of the 'Heritage' monies raised through the National Lottery has been able to draw upon £250 000 000 in its first full year of operation.

Of course, the smugness would be unjustified in several important respects. On English Heritage's figures of 1992, some 37 000 listed buildings in England were then in varying degrees of disrepair.[1] Listing establishes a presumption against

Figure 2.1

Losses: the Octagon Chapel, Wisbech, Cambridgeshire (1826). The loss of the Octagon Chapel in the 1970s was the sort of demolition which would not be countenanced in today's climate. (Reproduced from an old postcard.)

Figure 2.2

Losses: Baynards Park, Cranleigh, Surrey. The Gatehouse still stands, but the mansion beyond, built in 1587 and largely rebuilt by Rickman in 1834, was demolished in 1979. Its fate was sealed after a full-scale public inquiry. It was among the losses which confirmed SAVE and others in the belief that the panoply of legislative protection was only as good as those enforcing it.

demolition and insensitive change but a presumption is just that and can be set aside. In the years before the formal introduction of listed building consent in 1968, supposedly protected buildings were demolished in their hundreds even if, as Lord Kennet (1972) records, central government kept no statistics so that precise totals are unknown. Even thereafter the protective régime was only as good as the protectors. In 1977 applications were lodged to totally demolish 693 English listed buildings, and the annual figure[2] consistently ran to between 500 and 600 until the late 1980s when the recession and the increasing strength of English Heritage, as policeman and as donor, began to emerge.

The purpose of this chapter is to mix the plaudits with the brickbats, to note a generally linear progress upset by failings in detail.

EARLY STIRRINGS

In 1945 the country emerged from the most terrible war humanity had ever seen, and the framework of protection that has sprung up in the last two generations

Figure 2.3

Losses: the Great Barn, Butchers Farm, Worsthorne, Lancashire, threatened by application to demolish in the late 1970s. Since the Second World War, few building types have been so under threat as the historic barn: in some years in the 1980s, barns were the subject of applications for listed building consent for demolition at the rate of one every fortnight.

was in its infancy. The new régime benefited immediately from two overarching post-war emotions, the natural desire to cherish what had survived an orgy of largely indiscriminate destruction and the interventionist reforming mood of the incoming Labour government.

Hitler's supposed use of consciously discriminate destruction in the 'Baedeker Raids' on Exeter, Bath, Norwich, York and Canterbury (in answer to the Allies' attack on Lübeck) does seem, in part, to have been a half-truth manufactured by wartime propaganda, at least in the definition of intent. Nevertheless, a German Foreign Office official had described the new policy of 1942 as one where 'the Luftwaffe will go for every building which is marked with three stars in Baedeker' (Rothnie 1992, 131). Whatever the motive, the impact on the ground had been real enough. Street upon street of historic properties had gone up in flames. Even relatively small towns like Dover and Great Yarmouth were badly hit. Some of the country's greatest works of architecture, whether St Paul's Cathedral or the Houses of Parliament, had gaping holes in their roofs or fire-gutted interiors. Sir James Richard's *The Bombed Buildings of Britain* reads like a survey of the greatest products of the nation's collective architectural genius (Richards 1946). The rhetoric of resistance had glamourized the language of Armageddon. Churchill declared he would rather see London in ruins than 'tamely and abjectively enslaved'. However, as the smoke cleared and it was no longer necessary to sacrifice historic buildings in order to secure freedom, the wish to conserve the survivors became resurgent.

Even as the bombs were falling, (Sir) John Summerson and Walter Godfrey were among those who established the National Buildings Record in 1941, to record buildings that fell, and the associated 'salvage scheme' to identify structures which should be safeguarded if damaged by bombing. In 1944, with victory on the horizon, the Council for British Archaeology was founded to co-ordinate and to educate, and in the same year an impressive seven country houses passed into the hands of the National Trust, almost two a quarter. Perhaps as conscious of his ego as of his country, George Bernard Shaw also chose the penultimate year of conflict to leave to the Trust his architecturally indifferent Rectory at Ayot St Lawrence in Hertfordshire.

The Town and Country Planning Act 1944 first introduced the concept of the listing of historic buildings, even though the exigencies of war precluded all but desultory survey work.

The Act had been prefigured by its identically named equivalent of 1932 which gave local authorities the right to serve Preservation Orders, even if these were rarely activated because of the attendant rights to compensation. Individual cities had also taken action: The Bath Corporation Act 1937 scheduled the front, and in some cases side, elevations of 1 251 buildings in that city erected before 1820 for protection on the grounds of 'historic interest' or 'architectural interest or beauty'. (The presence of the last named is significant for it has not appeared in any subsequent legislation. Buildings are listed for their objective 'special interest' not for the more subjective perception of 'beauty'.) Any alterations required the consent of the Corporation. There was a right of appeal to an independent tribunal. A concomitant ability granted to the Corporation to offer loans had not been taken up at all by 1945 (Abercrombie *et al.* 1945).

The voluntary and urgent human wish to husband resources accompanied the ideological defeat of *laissez-faire* as Attlee legislated and nationalized. Chancellor Hugh Dalton's establishment in 1946 of the National Land Fund with a core endowment of £50 000 000, derived from the sale of surplus military stock, was nationalization in the sense that it was based expressly on the wish to commemorate the war dead and improve the lives of the living by securing the future of beautiful countryside and historic buildings in communal hands. The fact that that Fund was emasculated in just over ten years, when its capital was reduced to £10 000 000, is one of the great post-war tragedies (see page 26). In the following year, 1947, Lewis Silkin's Town and Country Planning Act expanded on that of 1944, and introduced the régime of listing that would be recognizable today.

The continuation of wartime restrictions and rationing for four or five years beyond 1945, together with the new tax régime, took a severe toll, both on country-house owners, as James Lees-Milne discovered on his journeys around the countryside, as recorded in his racily indiscreet diaries (Lees-Milne 1970, 1986, 1987, 1992), and on churches. The seminal report of the committee chaired by Ivor Bulmer-Thomas *The Preservation of our Churches* (Bulmer-Thomas 1952) highlighted the disastrous backlog in repairs caused by the virtual ban on 'inessential' building activity enforced under the new licenses introduced in 1940. It was that report which launched the Historic Churches Preservation Trust (now so strong that its county trusts raised £900 000 in 1994 in a single day's sponsored bicycle ride) and the Inspection of Churches Measure, under which five-yearly or 'quinquennial' structural assessments became mandatory.

For the country house, the equivalent was the Gowers Report of 1950, which led to the establishment in 1953 of the Historic Building Councils (in England, Wales and Scotland) with their attendant grant-giving powers (the money being channelled through the then Ministers and latterly through the Secretaries of State of the relevant departments). It was a measure of the Socialist resolve to follow up Gowers with tough legislation that a Labour bill, which pre-figured the 1953 Conservative legislation, proposed not only the compulsory purchase of listed buildings if 'in danger of damage or destruction through neglect or injudicious treatment' but also the listing of moveable fittings comprising historic houses collections, the latter a move which has never been subsequently implemented (Jenkins and James 1994, 155). The government subsidy to houses did not extend to churches even though the principle of State help for churches and chapels of all religious denominations had been established by the War Damages Act of 1943 under which hundreds of thousands of pounds were spent on repairs and rebuilding.

The post-war reconstruction debate espoused both conservative repair and the radical new architecture of the Modern Movement. War damage was generally repaired in the dreary new language of the late 1940s and early 1950s and Communist Poland's faithful rebuilding of the heart of historic Warsaw as if the War had never happened was a precedent generally eschewed by the parliamentary Left of the West, except in the Low Countries and some towns in Germany.

GROWTH OF THE VOLUNTARY MOVEMENT

The major difference between Britain and the rest of the world in her espousal of conservation is not just the comprehensiveness of protection summarized in the opening section but the fact that the legislative framework has grown out of grass-roots support. Generally speaking it has not been imposed from above. The Society for the Protection of Ancient Buildings, founded in 1877, although not the first conservation society (being more than pipped at the post internationally by a society in Norway and locally by another christened in delicious ambiguity as the Sid Vale Association in South Devon, both set up in 1840), was nevertheless in 1945 (and even more so now) regarded as the 'Grand Old Man of Conservation'. Its earlier associations, through William Morris, with political radicalism had soon lapsed and the overarching figure of the last four decades has been the Duke of Grafton (initially as Lord Euston), whose wife is Lady-in-Waiting to Her Royal Highness the Queen Mother (as well as being sister to Sir John Smith, see page 17).

From a modest membership base (standing now at 6 000) it has exerted considerable authority, particularly through its Technical Panel and its sheltering of the guttering flame of truth in its Manifesto which inveighs against shamming as well as destruction. In 1924 it had spawned, as much by reaction as by direct insemination, the Ancient Monuments Society (now combined with The Friends of Friendless Churches) which retained its base in Manchester until 1953 and regarded itself as the informal equivalent in the north of Morris's creation.

Incontestable offspring included the Georgian Group, set up in 1937 by a diverse band including Robert Lutyens and the architect Duke of Wellington, which moved into its own freehold headquarters in a Robert Adam house in Fitzroy Square in London's West End in 1995, and the Victorian Society, first established in 1958. The latter's founders were an illustrious, if occasionally ill-digested, group including (Sir) John Betjeman, (Sir) James Richards and (Sir) Nikolaus Pevsner. It has always been an indication of the eclectic nature of the conservation movement in this country that its leaders have included men, like Richards and Pevsner, who have made their reputation as apostles of the Modern Movement.

No man was better placed than Sir Nikolaus to understand the full range of England's inheritance. Between 1951 and 1974 he described practically every county for Penguin's 'Buildings of England' series (revised before and after his death by others and now being expanded into Ireland, Wales and Scotland). His selective championing of the nineteenth century was especially important, for the task facing the Victorian Society has always been greater than for the others. Even Arthur Oswald, writing in *Old Towns Revisited* (Oswald 1952) referred to the 'material benefits but artistic bankruptcy of the great Victorian age'. At roughly the same time the more functional products of the eighteenth and nineteenth centuries were defended by the Association for Industrial Archaeology.

In the years after the Victorian Society the organizations tumbled forth. The last of those dealing with a set chronological period, the Thirties Society, emerged in 1980. Despite its name, later changed to the Twentieth Century Society, it has always dealt with the period after 1914. (The decade-and-a-half at the beginning of the twentieth century was then, and still is, regarded as the preserve of the Victorian Society.) There has on occasion been tension within that organisation between the various strands of architectural thought, although there is a refreshing freedom from ideological intolerance. On the one hand, there are the traditionalists for whom Lutyens is beyond criticism, and on the other, the flat-roofers who respond to the white walls and rectilinearity of the Modern and the New Brutalists, whose chief delight is the bush-hammered concrete of more recent years. DoCoMoMo UK, founded in 1990, has the narrower brief of defending only the pure products of the Modern Movement.

The Garden History Society burst into flower in 1965 and has, like nearly all the societies, seen its brief as dual: to research and educate; and to campaign and conserve. Its younger and more obviously Epicurean sister the Folly Fellowship came together in 1988 but it too marries the pursuit of pleasure with a journal that is essential for anyone interested in the field.

SAVE (full title, SAVE Britain's Heritage) burst upon the scene in 1975, its principal mover and shaker being then, as now, Marcus Binney. It saw its role initially as one of blowing the occasional raspberry after what it saw as the self-congratulation of European Architectural Heritage Year (even though acknowledging that the latter did mark a critical watershed in formal attitudes). Press releases on individual cases were its staple but these have been largely superseded by highly respected strategic and polemical reports on 'buildings at risk' or on particular towns, problems and building types whether railway stations, barracks, hospitals, churches, or country houses. It also stepped in where others feared to tread, as Binney (1984) chronicles in his summary of the heroic age of SAVE and its major cases; the Lyceum, Liverpool, All Souls, Haley Hill, Halifax, and Barlaston Hall, Staffordshire.

Its ultimate stand against a Goliath was the contesting through the courts, mostly in 1990, of the consent Lord Palumbo received to demolish nine listed buildings on the 'Mansion House Square' or 'Number One Poultry' site in the City of London. It won the first stages but lost the last and the knowledge of defeat was further embittered by a legal bill of £90 000. Thanks to its friends this was paid off and a strategic victory was obtained, even as the nine were being pulled down, by the Government's declaration in 1994 in Planning Policy Guidance (PPG) 15 that 'subjective claims for the merits of replacement buildings should not in themselves be held to justify the demolition of any listed building' (PPG 15, 1994).

The late 1970s and 1980s saw the Tiles and Architectural Ceramics Society and the Historic Farm Buildings Group launched in 1985, both of them the unrivalled focus of their particular discipline. The Fortress Study Group took under its protective and scholarly wing military establishments both medieval and modern, while even relatively limited building types, such as piers, acquired their own stalwart band of defenders. In 1995 came the launch of 'Save Our Parsonages', with the aim of stemming the now well-advanced programme of disposal of historic rectories and parsonages by the church authorities. It seems extraordinary that the Association of Conservation Officers (ACO), founded in 1981 by Michael Pearce and representatives from thirty counties, is barely into its second decade, so authoritative has become its voice, that of its unrivalled magazine *Context*, and its ability to defend conservation activity primarily at local government level.

This coming together of the like-minded is unparalleled in the world. Other countries have seen the formation of conservation organizations but nearly all are generalists and without the specialist brief. Absolutely unique has been the way that Her Majesty's Government draws upon this reservoir of commitment and knowledge. From the late 1960s the national conservation organizations, christened for these purposes the National Amenity Societies, have been formally drawn into the planning process under the requirement that every application to

Figure 2.4

Each of the national amenity societies was spurred into life by particular cases: the restoration of Tewkesbury Abbey in 1877 in the case of the SPAB; the demolition of the Adam brothers' Adelphi, off the Strand, London, which brought together the founders of the Georgian Group in 1937; the gratuitous destruction of the Euston Arch which led to the foundation of the Victorian Society; and the demolition of Sir Edwin Cooper's Lloyds building in the City of London (shown here) in 1979 which spawned the Thirties Society (now the Twentieth Century Society).

(a)

(b)

Figure 2.5

The fury of local people at the threat to much-loved landmarks gave strength and being to many action groups and local civic societies, of which there are almost a thousand. Great battles raged over the threats to demolish E. B. Lamb's Town Hall, Berkhamsted, Hertfordshire (a) (reproduced from a postcard of 1904) and a modest group of eighteenth-century properties in the High Street, Stroud, Gloucestershire (b). Threats of demolition provoked sit-ins and demonstrations, and led, eventually, to repair.

demolish a listed building, in whole or in part, in England and in Wales has to be notified, to the Ancient Monuments Society, the Society for the Protection of Ancient Buildings (SPAB), the Georgian Group, the Victorian Society, the Council for British Archaeology and, indirectly, the Twentieth Century Society, to allow them to comment. Under the aegis of the Joint Committee of the National Amenity Societies they are also able to make nominations to each of the forty-two Anglican Diocesan Advisory Committees and the non-Anglican equivalents, set up after 1st October 1994 (see page 23).

There is a similar arrangement north of the border that involves the Scottish Civic Trust and the Scottish Architectural Heritage Society (formerly the Scottish Georgian Society). In Northern Ireland the highly respected Ulster Architectural Heritage Society is not drawn in by law to the planning process but has had a greater say in the identification of listed buildings than its mainland equivalents, a number of its draft lists being adopted with few alterations. The latest statutory consultee, the Garden History Society, was accorded that status in 1995 (see page 22).

The final full member of the Joint Committee, the Civic Trust (which really can be said to have been founded (in 1957) by one man, Duncan Sandys, rather than by a group) does not have an exactly equivalent statutory role but nevertheless enjoys direct and considerable government subsidy to facilitate study of particular problems and to pay for locally based regeneration task forces in settlements as diverse as Wirksworth in Derbyshire, Ilfracombe in Devon and Ulverston in Cumbria. Commercial and charitable sponsorship underlies the very influential Civic Trust Awards. It has always regarded its brief as broader than that of historic buildings even though it was the principal author of the concept of the Conservation Area in 1967 (see page 22). It coordinates the many local civic societies which have provided the bedrock for the communal pride and appreciation without which conservation policies would be blowing into the wind. Indeed no account of the growth of conservation sentiment since the War can or should ignore this explosion of interest. In 1939 there were 101 local amenity societies. By 1964 the Trust had registered 300, by 1972 760, and the figure is now 930. A handful have over 2 000 members apiece.

The expansion of architectural history

Campaigning, to be successful, needs to depend not just on raw sentiment but on facts. Since the War, the voluntary movement in its broader sense, alongside the Royal Commission on Historical Monuments, has been the principal agent in the spectacular growth of information in architectural history, voluntary both in the corporate sense in light of the rôle played in this by the Society of Architectural Historians of Great Britain and the Vernacular Architecture Group, but also individually in that the leading works have been written and researched by people who were not paid to do so. Even the help Pevsner received from the Gulbenkian Foundation and others met his expenses alone. Only his research assistants were paid. A number of those presently revising the Pevsners receive no salary.

In 1945 there were still no biographies at all on giants like Smythson, Hawksmoor or Smirke, while the only nineteenth-century architects to be taken seriously had been captured in early, largely hagiographical, pieces such as Benjamin Ferrey on Pugin and Winifride de L'Hôpital on J. F. Bentley. John Harvey's great voyage of discovery into the Gothic Age had only just begun and to many the Puginian fallacy, that polite medieval architecture was largely a communal effort with only anonymous architectural conductors, still held sway. There were occasional regional studies of the vernacular, as with Fox and Raglan in Monmouthshire, but only in the last five decades has the extent of that legacy been appreciated. Many are still coming to terms with the studies of Cecil Hewett which have shown that much more in the way of Romanesque carpentry of the twelfth and thirteenth centuries survives than hitherto suspected.

Architectural history still has its inexplicable lacunae: there are five full biographies of Nash but still none on Street, Bodley, Cachemaille-Day or Giles Gilbert Scott. And yet there is hardly an area which has not been worked on or where scholarly analysis is entirely absent, three-quarters of it the product of the last fifty years. Conservationists are thus immeasurably better briefed than they were in 1945.

The National Trust

Neither the Society of Architectural Historians nor the Vernacular Architecture Group campaigns for conservation. Neither, strictly speaking, does the greatest post-war success, the National Trust.[3] Its role has been as a property owner rather than as a campaigner. In 1945 it held 112 000 acres with 40 000 more covenanted and had a staff of 15. By 1990 the acreage owned was 569 000 with 78 000 more covenanted and the staff level 2 747 (Jenkins and James 1994). Its total expenditure in 1992 was nearly £128 000 000. Expenditure on properties was approaching £100 000 000 of which the cost of capital works, as opposed to routine maintenance, was £35 400 000. It is now the single largest private landowner in Britain. The National Trust for Scotland, founded in 1931 and constitutionally separate, spent £12 000 000 in 1992–3 out of an income of £14 900 000 and either owned or protected 56 366 acres.

The corresponding rise in membership has been equally spectacular. The total in England and Wales in 1939 stood at 7 100. Even in 1948 the membership in Wales was a meagre 500. And yet by 1970 the figure had soared to 226 000, passing 1 000 000 in 1981 and 2 000 000 in 1990. It now skirts 2 500 000. With the ownership of 600 000 acres, 531 miles of coastline and more than 230 houses and gardens it is the most successful conservation organization in the world, whatever Paula Weideger (1994) might say about its nature as a guided democracy or what she sees as a distortion in favour of the architectural as opposed to the horticultural.

Building preservation trusts

Almost entirely post-war, again, is the growth of that other arm of the property-owning conservation movement, the Building Preservation Trust (BPT).[4] Some, such as the Blackheath Preservation Society, which was established even as the war-clouds loomed, preceded 1939, but nearly all the 130-or-so have grown from the establishment of the Architectural Heritage Fund, which offers them low interest loans and which was one of the most unambiguously welcome products of the European Architectural Heritage Year of 1975.

An earlier concentration on England has been broken in recent years by increased growth in Wales and Scotland and indeed one of the most adventurous of them all is now the Scottish BPT which has just taken on the remarkable challenge of conserving Jack Coia's St Peter's College, Cardross, an icon of 1960s Scottish architecture, abandoned barely two decades after construction. The standard BPT case is, however, normally much more humble, the sort of house, chapel or mill that is the stuff of the British historic townscape. A number of buildings saved have not even been listed. By operating on the revolving-fund principle, limited amounts of money can go a long way, allowing a structure to be purchased, repaired and sold and the profits then ploughed back into subsequent projects.

THE CRITICAL ROLE OF THE INDIVIDUAL

It is an obvious truism, but a very important one, that most conservation activity is not corporate but individual. The great majority of historic houses are the responsibility of their owner-occupier, even if each displays varying degrees of commitment to repair and maintenance. The national grant structure (see pages 26–8) still leaves the privately owned Grade II building outside a conservation area largely to its own financial devices and without the enlightened self-interest of the owner the rate of disrepair and dereliction would be inestimably higher. This huge constituency, remembering that some 350 000 homes are presently listed, remains very amorphous and those organizations which have been set up in recent years to defend its interests tend to be little more than vehicles for the purveying of commercial products, many of them unsuitable. The only really successful 'trade union' for owners, and it would greatly resent the term, is the Historic Houses Association which fights for the interests of the greater and middling country house, whether it be for tax relief or freedom from meddling bureaucracy.

Even the marketing potential of the constituency seems to be unpredictable and it is one of the more depressing aspects of the post-war story that several magazines founded to educate and inform such people have failed, including the first of the lot, *Period Home*, and Marcus Binney's *Landscape*. *Traditional Homes*, undoubtedly the best, was killed off by a shotgun marriage with the gruesomely named *Period Living*. *Perspectives*, first issued early in 1994 under the aegis of the Prince of Wales's Institute of Architecture, is a very professional attempt to interest the educated man in the street in architecture both old and new. It is also early days for *The Historic Garden*, launched in 1991.

If the average man in the historic street is not an enthusiastic reader this is probably not through meanness. All but a minute percentage of appeals for historic buildings since the War have met or exceeded their target. This is true of the greatest consumers of them all, the Cathedrals, Abbeys and Minsters. Even in post-war austerity the National Trust's Golden Jubilee Appeal, which for the first time sought money for core funding rather than a particular project, raised £61 000 000 from the private sector (doubled, on a pound-for-pound basis, by the government). The ability of even the smallest parish to raise thousands of pounds for their church remains truly astonishing.

The largesse of the very richest may not be as critical as in the days when there was no listing and no government help – Lord Curzon's action in buying Tattershall Castle in Lincolnshire, for example, did directly save it from demolition – but they remain important players in the field. It was J. Paul Getty Jr's generosity which permitted the Dean and Chapter of Ely to carry out repairs without an enabling development on land close to the cathedral and that of Hereford to do the same without selling the Mappa Mundi. Ernest Cook's bountiful generosity, begun in 1931 with the saving of Montacute and the Bath Assembly Rooms for the National Trust, continued until 1949 by which time this shy man with a passion for the English countryside had purchased for the purposes of preservation some 37 000 acres with attendant historic buildings (Collins *et al.* 1989). One of the most spectacular re-creations of an eighteenth-century interior has been carried out not by the National Trust but by Lord Rothschild at Spencer House in the heart of London's West End. He acquired the lease in 1985 with the express aim of restoring the much-abused and concealed interiors and repatriating the furniture, dispersed when the building was let to commercial tenants in 1927.

The man voted in the *Observer* as the 'Conservationist's Conservationist', Sir John Smith, has directed most of his personal wealth through the Manifold Trust and above all the Landmark Trust[5] which, since its establishment in 1965, has acquired and rescued over 200 historic buildings, many from states of extreme dereliction. Letting as holiday homes brings only a partial return on capital expended, and as Sir John has never sought assistance from English Heritage it would not have flourished but for his willingness to dip deeply into his own pocket.

MAINTENANCE AND REVIVAL OF BUILDING SKILLS

Derek Linstrum deals in Chapter 6 with the development of training. It does seem valid, however, to trespass slightly on his territory to stress again the importance of voluntary effort, alongside the commercial and government-led, to revive and maintain standards and skills. The most coveted bursaries for young architects, surveyors and engineers and latterly craftsmen remain those offered by the SPAB. The Building Conservation Trust may recently have changed its name to Upkeep and decided to devote itself to training housing officers alone, but it began as an initiative to spread the conservation message among house-owners.

The Lime Centre, founded at Morestead near Winchester in 1990 to champion the use of building lime, has been supported in its first critical years by its founder Bob Bennett, the celebrated masonry consultant who lives on the site. A similar centre for conservation training has been established by the Woodchester Mansion Trust centred on the eponymous Grade I listed building abandoned before final completion in the mid-nineteenth century, and acquired by Stroud Council in 1987. The National Stone Centre, mooted in 1983 but first opened in Porter Lane, Wirksworth in the Derbyshire Dales in 1990, is a further private initiative with bold plans for future expansion including, in a touch of what the eighteenth century would have called 'the sublime', a *son et lumière* simulation of a quarry explosion. The arrival or survival of the vernacular tradition of construction in cob or earth depends a great deal upon the recent launch of the Devon Earth Building Working Group. The first new apprentice cob mason for many years was employed in 1992 through the Devon Rural Skills Trust.

Fortunately, some products have been re-launched commercially including cleft oak shingles, which have been available since 1989 through the Carpenter Oak & Woodland Co. and crown glass, which was reintroduced in 1992 by Historic Buildings Services of Leeds and the London Crown Glass Co. of Gerrards Cross, having not been produced commercially since the 1870s. The Ironbridge Museum in Shropshire has pioneered the production (as opposed to the recycling) of wrought iron, which had been unavailable since the closure of Walmesley's of Bolton in the early 1970s.

THE VANDAL, THE THIEF AND THE CRITIC!

No panorama of public opinion or grass-roots activity would be believed if it did not touch upon the 'opposing camp'. The occasional facelessness and intolerable delays of bureaucracy have made enemies even among sympathetic owners. Many young people have vented their spleen on historic buildings that happen to be symbols of authority or wealth. The post-war growth in vandalism is such that it is taken for granted that buildings in disuse will be set upon. In certain parts of the country more listed buildings have been lost to the pickaxe of the vandal and the fire of the arsonist (and the occasional owner) than have succumbed through the legitimate channels of the listed building consent procedures.

The thief is as old as the civilization that he raids but the very success of the conservation movement since the war has nurtured a substantial sub-group which has exploited the constant demand for authentic building materials by purloining these from other buildings and selling them on. The legitimate architectural salvage industry is now so large that it has its own representative organization, SALVO, its own news-sheet and a handbook of practitioners running to 700 domestic firms. The Open Churches Trust which got under way in 1995 following a £1 000 000 donation from its founder, Sir Andrew Lloyd-Webber, aims to allow more churches to deter the thief but encourage the visitor.

There are those who reject the whole notion of conservation, especially where it is enforced by others. But those articulating strong and rounded objections are relatively few. One of the longest-standing, most consistent and the most intellectually plausible is the journalist Martin Pawley. Robert Hewison is the acknowledged scourge of 'the heritage industry' (Hewison 1987) but is joined in his attack by those within the conservation movement for whom the sickly anaesthetized re-creations in some 'heritage centres' are also fabrications and falsehoods. The occasional occupant of the Presidency of the Royal Institute of British Architects has cursed the name of conservation, most notably Michael Manser, and some architects have on occasion lamented the loss of a brave new scheme at the hands of a timid or conservative Planning Committee. And yet most work with both new and existing structures and the RIBA recently recognized such realities in the spring of 1995 by allowing a Conservation Group to be set up within it. This Group may help to reverse the decline in traditional architectural education, and in particular re-emphasize the values of history and measured drawing which have diminished since the 1960s.

THE PUBLIC LEADS, THE LAWYERS FOLLOW

There are three forms of protection: scheduling principally for archaeological sites, buried remains and ruins; listing for the individual historic structure; and designation as a conservation area for historic districts with definable character worth conserving or enhancing.

Scheduling

Scheduling, first introduced for a handful of sites in the Ancient Monuments Protection Act 1882 and extended more broadly by the Act of 1913, probably did begin before the bulk of the public was ready for it, hence the long delay between the Acts. And yet listing on a national basis, when it came in 1947 (see page 8), seemed to attune very much with the public mood and soon became accepted as the mainstay of any régime of protection. Even now, the number of scheduled sites in England is a mere 13 000 compared with the 450 000 listings, albeit that the expansion of the scheduling programme which began in the early 'nineties is likely to bring the grand tally up to some 60 000.

Listing

Listing in 1947 was a complicated, indeed, in the words of Lord Kennet (1972, 76), a 'Byzantine' affair. The instrument of protection was a Building Preservation Order (in theory possible from 1932), which remained provisional until the Minister had confirmed it. Once confirmed this entitled the applicant to Development Value Compensation and placed the onus on the local planning authority to refuse consent for the works desired within two months. Failure to do so could be construed as acquiescence. Again, as Lord Kennet has picturesquely termed it, the 'Sick Typist Syndrome' meant that buildings were coming down merely because the two-month deadline was being missed by the indisposition of a secretary.

Kennet's own Town and Country Planning Act of 1968, promoted by him as Minister, introduced greater clarity and firmness of purpose, with the express need for the applicant to seek and obtain 'listed building consent'. Controls have been since intensified by the introduction of: the two methods by which local authorities can enforce repairs (1971); a five-year life for listed building consents (1980); and a potentially seminal, if little noticed, amendment to the General Development Order requiring consent for the demolition of any house, whether listed or not (1992).

As the number of listings grew, central government felt that a fully centralized system could not cope and in 1977 it was agreed that alterations to Grade IIs need not be referred to Whitehall. In the late summer of 1978 the leaking of an internal Department of the Environment memorandum proposing either the abolition of statutory protection over Grade II listed buildings at a time when their

Figure 2.6

Listing has not prevented applications being made, albeit unsuccessfully, for consent to demolish: Highfield United Reformed Church, Huddersfield, W. Yorkshire (1843). This richly proud essay in late Neo-Classicism came under threat of demolition in the early 1980s at the hands of the local authority (Kirklees) which is now one of the most conservation-conscious in the North, an indication not just of a localized change of heart but of the transformation of local authorities, from being the authors of one-in-three applications to demolish to a presently infinitesimal proportion.

number was 230 000, or the complete delegation of control over them to local planning authorities was swiftly rejected by the Historic Buildings Council and, following an outcry, buried by Ministers. In 1994 only 1 in 33 applications to demolish came from a local authority. In 1976 the figure had been 1 in 5 and in 1977, 1 in 3 (236 out of 693). Faced with such a record SAVE and the Amenity Societies were able to argue irrefutably that the conservation ethos was still so fragile within local government that to give officers and councillors the final say would be to consign hundreds of listed buildings to the demolition man. It remains one of the most rewarding of post-war statistics that local authorities have developed from the biggest single agencies for demolition to, with some notable exceptions, the greatest bulwarks against it. Admittedly this is partly a factor of local government emasculation as well as a change of heart, but the effect remains benign.

The same doubts within the Department of the Environment at the régime they had created took their toll, too, on the number of Investigators expanding the lists. These fell from 17 in 1974 to 4 in 1980. It was clear that the re-survey launched in the late 1960s to revise the first lists and to give better protection to hitherto neglected interiors, nineteenth-century buildings, vernacular structures and items of group value would take well into the twenty-first century to complete. Even so, the total had risen from 120 000 by 1969 to 273 000 by the end of 1980. The reaction of Michael Heseltine when Environment Secretary was to require the then fledgling English Heritage to complete the re-survey within four or five years in a new campaign using 90 personnel either employed directly or, in the case of the great majority, drawn from the county councils or the private sector, normally architectural practices. For many of those participating in this latter-day Domesday Book it constituted the most intensive examination, principally of the rural architecture of this country, ever attempted (Robertson 1993; Walker 1994).

By the end of 1994 the total listings in England were 447 000, in Scotland 39 000, in Wales 15 200 and in Northern Ireland 8 200. The magic threshold of 500 000 would have been crossed some years ago had not those compiling these statistics decided to change the counting procedure and described terraces, for example, as one item. Michael Pearce (1994, 8) estimates that a third is earlier than 1700, another third is eighteenth-century and the remainder, nineteenth-century and after. Two-thirds of the total are occupied as dwellings. Even with this total, the number of listed buildings in relation to the total building stock is only between 2 per cent and 3 per cent.

The catholicity of the lists is astonishing. Structures capable of protection through listing have by precedent included headstones, lamp-posts, stiles, ha-ha's, cricket pavilions, lychgates, boundary walls, wellhouses, garden terracing and aircraft hangars. If stone has been placed purposely on stone, or brick on brick, the structure could qualify. The younger it is, the tougher are the criteria applied, but even the Thirty Year Rule (introduced in England in 1987, having already been in force in Wales and Scotland), does not preclude the listing of a building more than ten-years-old in cases where the building is of national importance.

Some monuments of the twentieth century are acquiring iconic status. In April 1991 Sir Norman Foster's Willis Faber building in Ipswich, constructed in 1975, was listed in the virtually inviolate Grade I whilst Berthold Lubetkin's Finsbury Health Centre was upgraded to I in 1994. The number of Grade Is, probably nearer 10 000 but normally quoted as slightly over 6 000, is 2 per cent of the total. Historical association is prized too. Charles Dickens' Gads Hill in Kent is Grade I, as is the house at 36 Craven Street, immediately west of Charing Cross station, that was the London home of Benjamin Franklin. Even in his lifetime the country cottage of the painter and stained-glass artist John Piper in Dyfed was listed, while 1992 brought protection for the Lichfield home of the composer Clementi, 'the father of piano playing', and in 1994 for the house at 28 Mallord Street in Kensington built in 1913 for Augustus John.

Constant updating and re-surveying has uncovered the rarefied and the unique; the Iron House at Hopton Wafers in south Shropshire, designed with a remarkable iron vault in accordance with a patent of 1809; the earliest known covered ride for racehorses (*c.* 1830 in Newmarket); the Singing Theatre at Larmer Grounds, Tollard Royal in Wiltshire of 1895 constructed by the archaeologist General Pitt-Rivers who firmly believed that education should be linked to entertainment; the medieval rabbit enclosure walls at Nassington in Northamptonshire;

the monument in Horseferry Road, Westminster, to Christopher Cass the Elder, the master mason responsible for St Johns, Smith Square and Blenheim Palace, and said to be the earliest granite monument in England; and the dovecote at Rougham Hall in Norfolk of about 1690 built for the Hon. Roger North who described it at length in the critically important treatise *Of Building*.

Conservation areas

The third method of control, the conservation area, a concept introduced under the Civic Amenities Act of 1967, is wholly the preserve of the district council, although English Heritage retains the old Greater London Council power to designate in London alone. The Secretary of State has the theoretical power to designate as well, but Central Government only exercises that right in Northern Ireland given the eccentricity of the Province's political landscape at local level. Acts of 1972 and 1974 brought demolition of buildings under statutory control and required notification before the felling of trees more than 3 in. (7.6 cm) in diameter within such areas. In the summer of 1995 local authorities earned the further right to restrict damaging change when the previous need for Department of the Environment permission to limit Development Rights through an Article 4 Direction was abolished.

Conservation areas have proved popular as a means of protecting areas of definable character, both for the purposes of protection and the channelling of grants. In England, by 1988, only two district councils did not have at least one conservation area, and now they all do. The total is well over 8 000 in England, some 600 in Scotland and 400 in Wales. Michael Pearce has estimated (1994, 9) that one-third of English villages have now been designated. For some authorities it has become a means of concentrating minds in areas where it is planned landscape rather than buildings which dominate, whether the estate to a country house or a Victorian municipal park, this despite only half-hearted encouragement to do so in PPG 15 (1994, paragraph 4.6).

Historic parks and gardens

Indeed, of all aspects of the historic environment it is only the planned park and garden where protection relies upon a code among gentlemen rather than the full weight of the law (Roberts 1995). Registers of Parks and Gardens of Special Historic Interest have been compiled in England, Wales and Scotland, that in England identifying some 1 200 sites between its launch, in 1983, and 1994. As with the similar Register of Battlefields published by English Heritage in 1995, the lists are non-statutory but intended to guide decision makers. In Wales, its advisory status is emphasized by the veto which owners enjoy over whether their land should be included. From June 1995 any application affecting a Registered site in England has to be notified to the Garden History Society for its comment and any graded I or II* (some 40 per cent of the total) also has to go to English

Figure 2.7

Conservation areas were designated also to encourage the sort of good neighbourliness that is so clearly absent in the Telephone Exchange which jostles the Grade I Ladybellegate House in Gloucester.

Figure 2.8

One of the great revolutions in thinking pioneered very largely by SAVE Britain's Heritage is the belief that demolition is not just distasteful but also wasteful, that conservation has strong economic, as well as aesthetic grounds. The re-use of buildings, as with the Corn Exchange, Leeds (shown here) which is now the location for many small shops, transformed the attitude of many. Conservation became not just a rarefied luxury but essential to any sensible economic amd planning policy.

Heritage. Scotland already had a similar system of consultations and one was suggested for Wales, in June 1996. The same openness of mind to the works of the twentieth century shown in the listing programme also guides these compilers. Ten compositions by Lanning Roper are included.

The ecclesiastical exemption

In the 1990s, after years of controversy, undoubtedly the most glaring exclusion from listed building consent and conservation area consent procedures was ended with the considerable trimming back of the ecclesiastical exemption on 1st October 1994 (DNH/Cadw 1994). Hitherto, Government had regarded all religious bodies as indivisible and the exemption granted to the Church of England in 1913 had passed by extension to all places of worship. It could be argued that the Anglican church deserved its exemption as a result of the faculty jurisdiction over churches in use, and the Advisory Board for Redundant Churches and the Redundant Churches Fund (see page 29) set up in 1969 to deal with churches out of use. However, with most other denominations internal controls were either lax or non-existent.

It was Peter Brooke as Secretary of State for the National Heritage in 1994 who trod where other angels had feared to tread and carried through the major change of that autumn. Only five denominations were allowed to maintain the exemption, the Anglicans, all chapels affiliated to the Baptist Union Corporation, the Methodists, the United Reformed Church and the Roman Catholics. The same régime extended to Wales through that country's Secretary of State but no changes have yet been introduced in Scotland. All other denominations, both Christian and non-Christian, lost the exemption across the board over both external and internal work contrary to earlier suggestions by less audacious ministers that arguments of 'freedom of worship' should be allowed to permit even Grade I interiors to be gutted without secular hindrance. Institutional chapels and Peculiars are likely to retain the exemption until the end of the decade to allow them to decide whether to pass into secular or ecclesiastical control and to permit time for the passage of any attendant legislation. It seems highly likely that any decisions on these will post-date the two-year review of the new régime planned for the 1st October 1996.

THE MACHINERY OF CONTROL

Listed building control

This substantial legislative framework does not seek to prevent change but rather to control it. At the time of European Architectural Heritage Year 1975, listed buildings were coming down at the rate of one a day. Michael Pearce's infinitely useful analysis (1994, 9) has taken government statistics to show that in the seven years between 1984 and 1991 approval was given for the demolition of 1 077 listed buildings in England, and that in the same period 2 721 were removed from the lists ('de-listed'). Some suffered demotion after a renewed reassessment of their worth but the majority of cases followed very serious collapse or fire or inappropriate alterations permitted or ignored by indulgent planning authorities.

Some of the losses in the fifty years since the Second World War still shame a civilized society. Many more have been saved by the skin of their proverbial teeth, and even though rejected, applications to demolish have been lodged in respect of buildings as important as Hellaby Hall near Rotherham (*c.* 1680, Grade II*); the Memorial Hall, Manchester, the more modest but important counterpart of 1864 to Waterhouse's Town Hall in the same Albert Square; the Grade I listed Woolton Hall, Liverpool; John Dobson's magnificent Court House at Morpeth in Northumberland; James Gibbs' Grade I listed house in Arlington Street in the West End; Eden Nesfield's seminal Queen Anne house at Kinmel in Clwyd; and James Cubitt's Grade II* listed Union Chapel in Islington. All were reprieved but not before they had been brought to the brink. The law precluded demolition but it

Figure 2.9

The Maltings, Sleaford, Lincolnshire (1892–1905), at 1 000 000 sq ft the largest grouping of listing maltings in the country, with three of them now gutted by fire. Listed building consent to demolish was refused in the late 1970s, an indication of the increasing seriousness with which industrial archaeology was being viewed.

was only private and public action which subsequently gave them a new lease of life.

In the 1980s, the most significant legislative change was not in the protective framework but in the creation of new organizations. Least important were the Urban Development Corporations (UDCs) by which the Thatcher Government hoped to tackle urban dereliction and yet they do deserve a mention. It was the UDC for Merseyside which rescued the country's largest agglomeration of Grade I buildings, the Albert Docks in Liverpool, and that for the London Docklands which married the same mixture of flagship conservation schemes with modern development.

National Heritage Memorial Fund

Vastly more influential for the country as a whole has been the National Heritage Memorial Fund, the 1980 reincarnation of the National Land Fund (see page 8) and English Heritage (formal name, soon to be abolished, The Historic Buildings and Monuments Commission for England) which followed three years later. Between them they are responsible for over 80 per cent of all historic buildings grants from the public sector (local authorities in England disburse £8 000 000 per annum, half of it as contributions to the Town Schemes operated jointly with English Heritage) (Pearce 1994, 11).

As already mentioned, the National Land Fund was intended to use the proceeds of war to brighten the lives of those who had survived. Even after its emasculation in the late 1950s[6] (page 8) it continued to be indispensable in allowing great houses to be purchased and endowed, nearly always in the hands of the National Trust or the National Trust for Scotland. The feathers in its cap in this regard include Cotehele in Cornwall, Penrhyn Castle near Bangor, Saltram House in Devon, Ickworth in Suffolk, the ill-fated Heveningham in the same county (which was later to be sold on the orders of Michael Heseltine) and, as virtually its last gesture, Norman Shaw's Cragside in Northumberland.

It was the Government's momentous refusal to buy another nineteenth-century house, Mentmore in Buckinghamshire, from the proceeds of the Fund that prompted the outcry which led to its transformation into the National Heritage Memorial Fund (the 'Memorial' was introduced during the passage of the Bill). This has independent trustees with sole control over their own capacity to be generous. From the outset, their interpretation of the 'Heritage' was broad. Grant-aided activities embraced the acquisition of natural and planned landscapes, the conservation of historic film, archives, church monuments and wall paintings, and the purchase, and endowment for the nation, of houses like Belton in Lincolnshire, Calke and Kedleston in Derbyshire, and Fyvie in Aberdeenshire, as well as the less conventionally beautiful manifestations of industrial archaeology. The £40 000 000 capital raided by Government in the late 1950s was never fully restored and its income was irregular until it was chosen to disburse the Heritage monies raised from the National Lottery.

The Heritage Lottery Fund, which offered its first grants in 1995, faced the occasionally ambiguous pleasure of an income for grants of £250 000 000 a year in 1995 and £300 000 000 in 1996. This is set to rise at the end of the decade, century and millennium to some £340 000 000 if, as is presently envisaged, the allowance of the Millennium Commission (which will then close) is divided between the four other good causes (Heritage, Sport, the Arts and Charities). The Lottery has been such a success, with almost two-thirds of the adult population participating, that the National Heritage Memorial Fund is now far and away the greatest source of money for historic buildings, with grant levels at its disposal which are six times those of English Heritage.

English Heritage and grants from taxation

It is English Heritage, established in 1983 under the initiative of Michael Heseltine, which is still the most influential organization to emerge in the period under review. Under three Chief Executives, Peter Rumble, Jennie Page and, from 1995,

Christopher Green, and two Chairmen, Lord Montagu and the more turbulent Jocelyn Stevens, English Heritage has rapidly found its feet. Heseltine wanted the new agency to bring 'a greater commercial and entrepreneurial flair' to the promotion of the 400 monuments transferred to its care from the DoE, and generally speaking this has been done. A considerable increase in income from membership and gate receipts has helped it to promise the final completion by the end of the century of the backlog in repairs at its own properties, estimated in 1992 as £56 000 000. In 1994–5 it expected to spend £34 400 000 on its own properties, £19 600 000 on education, development and promotion and £61 300 000 on conservation grants, advice and statutory casework, that is the defence it offers for those listed buildings which are the subject of applications to demolish or alter. In that respect its arrival has transformed the scene.

Assisted, of course, by the recession and the drop in the number of applications to demolish (page 6), English Heritage Inspectors (and other conservationists) have helped to render the chances of winning consent to demolish relatively remote. The number of public inquiries convened to consider proposals to destroy listed buildings called in by the Secretary of State at the suggestion of English Heritage has plummetted from thirty-three in 1987 to three in 1994. They have been greatly helped in the capital by inheriting the London Division of the Greater London Council when that was abolished in 1984, and with it the full executive powers to determine all applications for listed building consent within all thirty-three London boroughs. Under English Heritage's Forward Strategy for the 1990s some limited powers are being handed back to the London Boroughs under 'flexible authorization' but all important decisions will remain in-house.

It has helped too that the grant régime has steadily broadened to cover practically all building types, the principal exceptions dictated by the ecclesiastical exemption, places of worship in use and cathedrals becoming eligible in 1979 and 1990 respectively. The Great Storms of 1987 precipitated the first grants towards historic gardens which have continued, modestly, since. Eligibility extends too to the monuments of the Industrial Revolution: in 1993–4 £500 000 was offered for the repair of the remarkable Anderton Boat Lift in Cheshire which transfers barges from the Trent and Mersey Canal to the River Weaver 50 feet (15 metres) below (see Figure 13.2 on page 289). The 'Buildings at Risk' grants, introduced at the beginning of the 1990s, were designed to fund emergency measures of protection although at present these are limited to conservation areas. The grant régime may be free of prejudices in respect of types, but Grade II buildings outside conservation areas remain ineligible for English Heritage assistance even though they represent half the listed buildings in the country.[7] In Wales the situation is marginally better in that the Historic Buildings Council does not limit the definition of 'outstanding' for the purpose of Section 3A grants to those listed Grade II* and Grade I, which English Heritage now does (even though reserving the right to upgrade from Grade II to II* in parallel with the offer of a grant).

All grants to, and indeed the upkeep of, all listed buildings are overshadowed and burdened by Value Added Tax (VAT) which was levied on all building repairs, listed or otherwise, from 1979 at the then standard rate of 15 per cent (now raised to 17.5 per cent). Concessions of sorts followed in 1984 and 1989. These combined to zero-rate works of alteration and extension on residential properties (and, from 1995, non-residential properties converted to dwellings) but retained the excise at full rate on repairs and maintenance. Alterations won tax relief, careful conservation and self-effacing repairs attracted an impost of 17.5 per cent. The system is illogical and indefensible but the prospects of change seem remote unless the Government can be persuaded to reduce the rate for listed buildings as it already has done for domestic fuel.

For those who can register, the burden can be passed on but for charities and the private owner there is little or no prospect of such a transfer. This distortion of the market against listed buildings and in favour of new buildings is the more serious because, as Pearce (1994) has demonstrated in a diligent comparison of English Heritage Annual Reports, far fewer buildings are now grant-aided than ten years ago. He has shown that, apart from Town Scheme grants and those for rescue archaeology, English Heritage now rejects around a third of all grant applications, 56 per cent of these for financial reasons. Again, as he has shown, the value of grants offered by the Historic Buildings Council for England in its last year, 1983–4, was nearly £21 000 000. In 1993–4 the amount distributed by English Heritage was just under £42 000 000 and in 1994–5 £48 000 000. In real terms this is an increase, for to match the value of the original £21 000 000 required £35 000 000 ten years later. There are now more listed buildings and there is slightly more available in central government subsidy in real terms, but the trend is towards larger grants going to a smaller number of buildings; and this at a time when local government finances are so stretched.

The list of 'Grants 1985–93' published in 1995 by Cadw, the equivalent of English Heritage in the Principality, tabulates the £15 700 000 offered by that agency and taken up by the owners of 100 ancient monuments and 486 historic buildings. North of the border, Historic Scotland, established in 1991, spent £10 700 000 in grants in 1992–3 which went to 145 buildings and monuments, and £11 400 000 in 1994–5 which went to 140. The equivalent figure in Northern Ireland for the two years 1991–3 was £3 500 000.

Department of National Heritage

The body which, by its name, seems to accord the ultimate accolade to historic conservation is the Department of National Heritage (for England), established without any prior consultation with outside parties, immediately after the 1992 General Election. However, for many professionals in the field it is a hybrid, and a weak one at that. The divorce of historic buildings from the mainstream interests of Town and Country Planning defended by the super-ministry of the Department of the Environment makes them appear dangerously close to a rarified extra. The new Department distorts the term 'heritage' to meaninglessness when it embraces not just historic buildings but also Arts, Sport, Broadcasting and, from the spring of 1996, the supervision of charities. The fact that the system of identification, listing, has been transferred to the Department of National Heritage but decisions over what happens to the same building subsequently, the listed building consent machinery, remains with the Department of Environment offers the prospect too of tension and dissent. That it has not arisen so far is due in no small measure to the strong conservationist sensibilities of John Gummer, Secretary of State at the Department of the Environment, and the fact that the longest-serving Secretary of State at the Department of National Heritage, Peter Brooke, brought to the post a personal commitment to the cause. His father had purchased Wordsworth's Dove Cottage to save it from despoliation.

TRIUMPH OF THE LETTER OVER THE SPIRIT

As we approach the end of the twentieth century the most glaring irony is that many people profess themselves believers but then clearly have only half-understood the message. Now that the rates of demolition remain serious but are appreciably lower than they were in the 1980s the greater, and more insidious, risk comes from misdirected affection, from the owner who believes he is doing the right thing by exposing stone or timber-framed elevations that were never meant to be revealed, by replacing genuine doors and windows with standardized substitutes, many of them spuriously described as 'Georgian', or by applying sham stick-on 'leaded lights'. The half-digested message that stone is a good thing led to a complete industry in the 1970s and 1980s, now largely dead, of 'stone cladding' where brick surfaces were covered in crudely sham 'sticky stone'.

The most tricky opponent is one who combines the appeal of 'home improvement' with supposedly historical detailing, none more so than the plastic window. Fenestration in PVCu (formerly uPVC) is entirely a product of the last twenty-five years, being first imported from West Germany in the late 1970s. It is destroying the character of villages, towns and cities the length and breadth of the country as original, reparable and serviceable timber casements and sashes are ripped out. English Heritage is fighting back through its 'Framing Opinions' campaign, particularly on the merits of the comparatively invisible 'secondary glazing' installed against the inner face, as are the SPAB, the Georgian Group and the Victorian Society in their own publishing programmes, but the efforts appear largely ineffective against the systematic marketing of an industry that now employs thousands.

Looking towards the next millennium some existing problems will no doubt fade, to be succeeded by others. Building types already in retreat will decline further. None more so perhaps than the formal place of worship as the financial and demographic problems mount, and as non-conformist denominations talk further about unity, a process which would inevitably lead to further 'rationalization' of building stock. Greater redundancy among Anglican churches has not so far shown an increase over recent years, despite the well-publicised reduction in the paper value of the holdings of the Church Commissioners, although some effect seems unavoidable. The Churches Conservation Trust (formerly the Redundant Churches Fund) already owns some 300 Anglican churches too important to be demolished or converted, while its younger brother, the Historic Chapels Trust, established in 1992 which has five properties in its care, is meant to act as an equivalent holder of last resort for non-Anglican properties. But the

Figure 2.10

Those who designed and constructed, and indeed those who gave permission for, this new hotel in Cambridge, built about 1990, no doubt thought that Classicism, however ill-digested, would be safer than more challenging Post-Modern or 'frankly modern'. The Conservation Movement should not become the Conservative Movement: where good adventurous design is rejected or toned down in favour of the anodyne neo-historical, it cannot escape some responsibility.

greatest risk remains that posed by the explosion of interest in 'home improvements'. In a very real sense there is a fate worse than death or demolition. Always beware of those that speak your language but do not understand the grammar.

I began with a warning over the dangers of smugness. And yet there clearly are reasons for genuine satisfaction. The conservation of historic buildings has support across most segments of society from St James's Palace down and within all political parties. It is enshrined in European Charters and is a major plank in the Green Movement which now dominates so much policy making. It is widely regarded as 'a good thing' in the language of *1066 and All That.* Controversy is not stilled, it never will be nor should it be, but this is normally provoked by arguments over the means not the end.

And yet the present situation has been reached in barely two generations. Is it the swing of a pendulum or, as I ventured earlier, a sea change? I do believe that the central principle is now entrenched. It is fundamental to the legislative planning framework and an express goal in every Local, Structure and Unitary Plan. The dismantling of the régime of protection would not get through Parliament, short of a revolution. Even that most committed opponent of consensus and theoretical apostle of *laissez-faire*, Margaret Thatcher, left the planning system broadly the same as when she came to power.

The Government's latest pronouncement on policy, Planning Policy Guidance 15 on *Planning and the Historic Environment* issued in September 1994, begins with language that is almost biblical in tone: 'It is fundamental to the Government's policies for environmental stewardship that there should be effective protection for all aspects of the historic environment. The physical survivals of our past are to be valued and protected for their own sake, as a central part of our cultural heritage and our sense of national identity.' The mealy-mouthed is rejected too in its statement (at paragraph 3.17) that 'there are many outstanding buildings for which it is in practice almost inconceivable that consent for demolition would ever be granted. The demolition of any Grade I or Grade II* building should be wholly exceptional and should require the strongest justification'.

Smugness, No. Not as long as 200 listed buildings come under threat of demolition each year, some 30 000 languish in disrepair and crude modernization mutilates the character of thousands more. And neither, too, in the face of an unthinking and occasionally asinine tax structure and a clear need to broaden the membership base of conservation societies, both local and national, beyond the middle classes.

However, even with all these caveats we have travelled further forward in consciousness in these fifty years than in all the preceding centuries put together. And to that extent a feeling that missionary zeal has been rewarded is warranted.

BIBLIOGRAPHY

Abercrombie, P. (Sir Patrick), Owens, J., and Mealand, H. A. (1945). *A Plan for Bath.* Bath Corporation.

Baker, D. (1983). *Living With the Past, the Historic Environment.* Published privately by the author who was then, and remains, the Chief Conservation Officer for Bedfordshire County Council. (An excellent analysis of the State of the Art in 1983 with a strong historical perspective.)

Binney, M. (1984). *Our Vanishing Heritage.* Arlington Books. (A personal account of the first ten years of campaigning by SAVE Britain's Heritage.)

Binney, M., and Burman, P. (1977). *Chapels and Churches: Who Cares.* British Tourist Authority in association with Country Life, London. (A very thorough examination, with many statistics, of the problems facing places of worship. The appendices include notable losses since 1945.)

Binney, M., and Hanna, M. (1978). *Preservation Pays.* SAVE Britain's Heritage. (One of the first systematic efforts to explain the economic benefits of conservation, an initiative taken up much later by joint research in the 1990s by English Heritage and the RICS.)

Bulmer-Thomas, I. (1952). *The Preservation of Our Churches.* Church Information Board.

Collins, Giles and Malleson (1989). *Innovation and Conservation – Ernest Edward Cook and his Country Estates.* The Institute of Agricultural History.

Cormack, P. (Sir Patrick) (1978). *Heritage in Danger.* Quartet Books. (A comprehensive survey of the problems faced by historic architecture as well as works of art.)

DNH/Cadw (1994). *Ecclesiastical Exemption, What it is and How it works.* Department of National Heritage and Cadw.

Hanna, M. (1977 and every year since). *The English Heritage Monitor.* British Tourist Authority and English Tourist Board Research Services. (The *Whitaker's Almanack* of the conservation world with annual statistics on losses, grants, new listings, buildings open to the public, etc.)

Harvey, J. (1972). *The Conservation of Buildings.* John Baker.

Haskell, A. (1993). *Caring for Our Built Heritage: Conservation in Practice.* E. & F. N. Spon. (A review of conservation schemes carried out by County Councils and National Park Authorities in England and Wales.)

Hewison, R. (1987). *The Heritage Industry: Britain in a Climate of Decline.* Methuen.

Jenkins, J., and James, P. (1994). *From Acorn to Oak Tree: The Growth of the National Trust 1895-1994.* Macmillan.

Kennet, W. (Lord Kennet) (1972). *Preservation.* Temple Smith. (History by an insider; the only time a post-war Minister engaged in the protection of historic buildings has put pen to paper.)

Lees-Milne, J. (1970). *Another Self* Hamish Hamilton;

Lees-Milne, J. (1986). *Caves of Ice.* Faber & Faber.

Lees-Milne, J. (1987). *Ancestral Voices.* Faber & Faber.

Lees-Milne, J. (1992). *People and Places: Country House Donors and the National Trust.* John Murray.

Lowenthal, D., and Binney, M. (1981). *Our Past Before Us: Why Do We Save It?.* Temple Smith. (An overview, both historic and philosophical with particular exemplar chapters on Dartmoor, Leeds, Suffolk, the London Suburb and Cathedrals.)

North, R. (1981). *Of Building: Roger North's Writings on Architecture,* ed. H. Colvin and J. Newman. Clarendon Press.

Oswald, A. (1952). *Old Towns Revisited.* Country Life.

Pearce, D. (1989). *Conservation Today.* Routledge. (A handbook to an exhibition at the Royal Academy, explaining the history of conservation in Britain since European Architectural Heritage Year (1975).)

Pearce, M. (1988). *Conservation: A Credit Account.* SAVE Britain's Heritage. (An authoritative analysis of the finances of conservation.)

Pearce, M. (1994). *The Management of the Architectural Heritage in the United Kingdom.* (Salisbury) published privately by the author.

Pearce, M. (1995). 'Paying for the past', a lecture given in a seminar to the University of Oxford Department for Continuing Education, May 1995, on PPG 15.

PPG 15 (1994). *Planning Policy Guidance: Planning and the Historic Environment.* PPG 15 (September 1994), Department of the Environment, Department of National Heritage. (The most up-to-date summary of Government policy and interpretation of the law.)

Richards, J. M. (Sir James) (1946). *The Bombed Buildings of Britain.* Architectural Press. (See also Kent, W. (1947). *The Lost Treasures of London.* Phoenix House Ltd.)

Roberts, J. (1995). 'Historic parks and gardens – Listing, awareness and the future', *Journal of Architectural Conservation* i (March 1995), 38–55. (A detailed review of the Register compiled by English Heritage.)

Robertson, M. et al. (1993). 'Listed buildings: the national resurvey of England', *Transactions of the Ancient Monuments Society* xxxvii, 21–94.

Ross, M. (1991). *Planning and the Heritage, Policy and Procedures.* E. & F. N. Spon. (Broad-brushed but informative account of the development of the present conservation régime written by a former Head of Listing at the Department of the Environment.)

Rothnie, N. (1992). *The Baedeker Blitz, Hitler's Attack on Britain's Historic Cities.* Ian Allan Publishing.

Saunders, M. (1986). 'Architectural Conservation: An Apology and Historical Perspective', *Transactions of the Ancient Monuments Society*, xxx, 219–25.

Strong, R., Binney, M., and Harris, J. (1974). *The Destruction of the English Country House*, London. (The handbook to the seminal exhibition at the Victoria and Albert Museum. An historical panorama over a full century but concentrating especially on the inordinate number of losses sustained in the 1950s.)

Walker, D. (1994). 'Listing in Scotland: Origins, survey and resurvey', *Transaction of the Ancient Monuments Society*, xxxviii, 31–96.

Weideger, P. (1994). *Gilding the Acorn: Behind the Façade of the National Trust.* Simon and Schuster.

NOTES

1 *Buildings at Risk, London, a sample survey* (English Heritage, 1992) found that 36 700 of England's listed buildings were 'at risk' and 72 850 were 'vulnerable'. Further information from the Buildings at Risk Survey, English Heritage, Keysign House, 429 Oxford Street, London W1R 2HD.

2 The Ancient Monuments Society has compiled annual lists of applications to demolish listed buildings in England and Wales since 1978. Statistics, but not full lists, of consents granted have been published in the same period by the Department of the Environment.

3 The centenary of the National Trust was celebrated by four books: *From Acorn to Oak Tree: The Growth of the National Trust 1895–1994*, by J. Jenkins and P. James, Macmillan, London, 1994 (the principal author, Dame Jennifer Jenkins, is a former Chairman of the National Trust); *The National Trust: The Next 100 Years*, by H. Newby, The National Trust, 1995; *The National Trust: The First 100 Years*, by M. Waterson, The National Trust and the BBC, 1994; *Gilding the Acorn, Behind the Façade of the National Trust*, by P. Weideger, Simon & Schuster, New York, 1994 (a view by an expatriate American which sees the Trust as secretive and overly burdened by buildings as opposed to gardens.) Preceding the centenary was *Figures in a Landscape: A History of the National Trust*, by J. Gaze, Barrie & Jenkins in association with the National Trust, 1988.

4 The Architectural Heritage Fund publishes an extremely useful *Annual Report* and these read together give an illustrated history of the growth of the Building Preservation Trust movement.

5 The work of the Trust is summarized in the extremely full and illustrated *Landmark Handbooks* published by the Trust on a regular basis.

6 The Third Report from the Expenditure Committee of the House of Commons, Session 1977–78, offers the raw material for a definitive history of The National Land Fund. The activities of its successor body, the National Heritage Memorial Fund, is easily obtained from its *Annual Reports* and, in the case of the National Heritage Lottery Fund, from the regular 'Updates' which it issues.

7 English Heritage has recently adopted the policy of publishing full lists of its grants. Those for the period 1984–92 were issued in February 1994, and there have been regular annual updates thereafter. The lists cover Historic Buildings (Section 3A) grants, Church and Cathedral grants and Ancient Monument grants only.

3

PRINCIPLES AND PROBLEMS: ETHICS AND AESTHETICS

John Warren

If we stand back far enough, and if our sense of detachment is great enough, we may see mankind as no more than the most remarkable of the animals to have evolved in the thin and fragile life-bearing shell which surrounds this planet, its biosphere: and much of what we see man doing is, in essence and in greater form, what other animals are doing also: living, congregating, mating, fighting, hunting, building colonies, harvesting, dominating the environment – and working to achieve objectives, curious as many of them may be. Man is not the only creature to use tools, nor to tend and cultivate; nor the only animal to build, to create colonies and common structures; or the only one to decorate and restore, to remodel and repair, to re-use and to conserve. The ant, the beaver and the swallow rebuild with equal energy structures that they make home; and while security and the sense of well-being that is the reward of such activity may be sensed by each creature in its special and different way there is probably much that is common to each. The beaver in his lodge repaired, and man in his castle buttressed, may share a sense of security deep in their subconscious. Any disturbance to their lodges and castles may be followed by a flurry of restoration activity, but there the similarity probably ends.

Man is set apart from his fellow creatures by an ability to perceive himself, that is to view himself from without. This is not so much a consequence of brain size itself; it is a consequence of a brain structure which is radically different. Biologists conjecture that it is the double layer in the human brain, the inner and outer cortex, which enables him to look inward and so to know himself. He finds that he is an inhabitant of a minute planet on which he has built social structures of extreme complexity to facilitate his existence and safeguard his environment. He knows himself to be the minutest dot in an unimaginable immensity of expanding space; yet, to be vast by comparison with most of his fellow creatures. Knowing himself, he can analyse his activities and construct principles. Thus he can argue the ethics and the purposes of his manifold activities, not least among them conservation. This capacity to perceive and rationalize leads to evaluation and the setting of standards. Rules are formulated and laws written; principles emerge to control his perceptions, and absorbed by the conservator they are the precepts which guide his individual actions.

Having created in – and through – their mind these manifold concepts, they have also unwittingly set up interactions between these entirely metaphysical entities. As a basis for structural arguments and coherent progress, conservation must share a common framework of mutual presumptions if it is to be carried forward as a coherent social phenomenon.

Conservation is one of many social activities. This social dimension is profoundly encapsulated in the concept that man holds his stock of buildings (and the land itself) in trust for future generations. The idea was not entirely new when William Morris famously enunciated it. Generations of thinkers had moved towards justifying the animal instinct to restore familiar surroundings while adding the social rationale of benefit to present and future communities.

Although conservation is often undertaken with intense private fervour, the holding on to and putting back of the built fabric is not demanded of man as his private benefit but as a duty to society at large in response to the deep need to recover the familiar and loved shape of things. The city dwellers and the merchants of London who suffered the trauma of the great fire in 1666 were unable to grasp the opportunity to rebuild to Christopher Wren's new and rational plan. In part this was because of the complexities of negotiations over land ownership: in part it seems to have flowed from the need to re-establish the comfortable urban structure of inconvenient lanes, alleys and irregular land parcels of medieval London – to remake the familiar; and so the pattern of the medieval city survived.

The phenomenon is well-known to archaeology as the 'persistence of place'. A flattened city is rebuilt with the same ill-placed gateways on the same irrational street pattern, despite the opportunity to achieve a more logical plan; which logic these same builders would employ if they were founding a new colony. The old plan persists with walls winding around long vanished obstacles for the same reason that the ants' nest and the beavers' lodge are repaired in their original haphazard form. Familiarity, the comfort of the known and well-used, re-establishes in collective memories tracks and tunnels ingrained in the twenty-four neurons of the ant or the millions of neurons of a human: and the swallow returning to her nest, refurbishing the mud cup in which she brooded last year; is she merely being practical, economizing on labour and regaining a place already proven safe? Or is she also finding pleasure and satisfaction in the security of familiar surroundings? Or is that pleasure and the sense of security one and the same thing?

Only to humanity is it given to ponder such satisfactions and legislate over the processes of conservation in which the most fundamental urge may be a basically animal instinct, neither understood nor contemplated by the very large majority, for whom it is sufficient that conservation is a 'good thing'. The rationale of building conservation is based on the best interests of society, and from that base is built a complex pyramid of social activity which protects and controls the built (and by extension the natural) environment.

PERSPECTIVES

The whole philosophical basis of conservation depends upon a social imperative in which individuals play active and passive parts, the results being meaningful only in the context of society at large, over long spans of time, viewed both forward and backward. For the individual any understanding must relate to the point from which it is viewed. No metaphysical concept is absolute because it is not isolated. In social terms there is an inevitable relativity between observed and observer. In recognizing these positions we introduce the concept of *perception*. Any view or judgement is inherently related to the standpoint of the observer or judge. Each is seen in a *perspective* of time or space in both physical and social dimensions and no perception can be defined without, by implication, defining the perspective with which it is seen. To illustrate: a single town house of a type once common now survives in isolation, the only such example. In contemporary terms a perception of its social importance, of rarity and of loss akin to extinction becomes the key to its defence. It is unique. It must survive. But by that argument the same house surviving among hundreds of similar houses could disappear without loss, one of many to have gone; and so, if it is one among many, the arguments for its conservation might seem to diminish, even to disappear.

But a residual argument of intrinsic merit then predominates. The building is of a particular type and age. All buildings of such type and age are of merit. Therefore it must survive. The perspective upon the house has changed and the argument has shifted. The perspective may change again if this one house is part of an urban plan. It, and any of its fellows if removed, leaves an unacceptable gap in the townscape. Its retention then becomes essential to the completeness of the scheme. The entity must survive entire. The argument has shifted again. The perspective now is of its importance, not for itself but as a component. If the townscape is an undamaged formal urban design, this argument of completeness rises to great importance. The loss would scar the perfection of the whole. So the arguments for the retention of the one single house move from extinction of species, through intrinsic merit to completeness of the ensemble, depending not upon itself but on the surrounding circumstance. We might, therefore, call this a *circumstantial perspective*. The circumstances or context bear upon the conservation of the building. The same problem introduces a perspective of scale. Thus, if the object of conservation is the retention of architectural harmony in an urban composition it will matter little, *in that context*, if the sculpture in the pediment on the façade were removed for conservation in a protected environment and replaced by a replica. Even less so the replacement of a door knob. However, if the wide view of the townscape is replaced by a critical view of the façade of the house alone the importance of the sculpture in the pediment is different. The question of whether it is a replica, of the quality of the replica and its level of maintenance becomes important. If the perception focuses on the door itself, the loss of the knob is critical. So there is a *perspective of scale* directly related to the size and importance of the object in its context.

There is also a *perspective of time*; as the dividing line, which we call the present, moves forward – as the future becomes the past – generations change.

The future of one generation is the immediate past of the next. The past is loved, the future feared: some part in this process is fear of the unfamiliar and the reassurance of the familiar. Communities frequently exhibit a resistance to change, particularly if it is a change in their surroundings. Antagonism can arise as change approaches, but having been accepted the new condition becomes familiar, even loved. With growing acceptance fear gives way to relish, justifying defence and retention. Antipathy has become conservation, a paradox explained by the simple fact that we adhere to and love the surroundings in which we grow up; and this, in itself, may be deeply primeval.

This one aspect of the perspective of time is complemented by another: the evolving view or change in perception which has less of the primeval and more of the intellectual about it. Medieval man conserved even if he saw the reasons for doing so very differently from his successors. He might have retained out of affection, out of a sense of familiarity or usefulness but primarily he retained because in so doing he saved labour. Thus he made good the damage to his cathedral and repaired the castle. But he applied no intellectual rationalization to the process, neither declaring himself a trustee for the next generation nor retaining the work of his forefathers as significant evidence of historical evolution. These concepts had however already become apparent in the nineteenth century although ruthless destruction of earlier fabrics was unquestioned by most communities, while on the other hand the concept of the museum was well established. In that climate a few pioneers became a conservation movement. A century or more has further refined and added to these concepts, improving some and amending others quite radically. There is, therefore, a *perspective of time* which derives from the standpoint of the observer in history. Time moves forward and with increasing distance the perspective is continually changing, enhancing the receding past.

There is also a *perspective of quality*. Quality is a value of merit deriving from intrinsic factors such as materials and workmanship, but merging into values acceptable to society at large. A fine artefact is held to be more meritorious than one less fine; one in good condition more desirable than one damaged; one of better material more worthy than one of poor stuff. The sculpture in the pediment of the town house is more deserving of protection if it is a superbly cut piece of marble worked by a famous sculptor and in good condition than if it is a decaying piece of friable stone carved by an unknown. The market tells us that it is worth more. Popular response to its potential destruction is more powerful but social merit also counts, confusing the issue further. Our ordinary town house will be more worth keeping, whatever the decay of the sculpture in its pediment, if a prince kept a lady there or a great painter made it his home. The quality of the fabric is then enhanced or replaced by a social value and these matters may interlock. A church wall of no apparent importance may be found to carry wall paintings beneath its limewash. The quality of the fabric itself has not changed but the wall paintings are an added value and show that the structure was significant, so that in addition to aesthetic merit and historical importance the wall acquires a social dimension fitting it into the wider evolution of society. This dimension then has a worth and this worth can be tested. If, for instance, the wall paintings are removed their visual and historic value goes with them, but the wall itself retains its importance by reason of its survival as an historic artefact, made the more valuable by paintings which no longer exist. The paintings proved something. Their quality has rubbed off on the wall itself, although to different degrees according to the observer. The perspective of quality and the social perspective may then interlock with the *perspective of scarcity*, the view which made the isolated town house important as the survivor of a type or of a generation of buildings.

These perspectives of quality and social value also fluctuate with time because values change with each succeeding generation. 'Queen Elizabeth slept here' is an argument to justify the retention of any structure, whatever its visual merit, while social values remain static. But in some circumstances a change in political, social or intellectual viewpoint may completely invert the values on which retention is based. Generally the change will be expressed in neglect and consequent decay, but it may take the form of the levelling of statues of overthrown leaders and the burning of fine buildings because they were the headquarters of hated régimes or the dwellings of a hated class of society. Thus social values may overtake visual merit. Conversely a revolutionary turned statesman may find his glory transferred to the otherwise meritless house of his birth, unwittingly achieving its conservation. The changing views of society, the changing status of individuals, the evolving structures of social power, the changes in taste and perception of significance and even loss of status may all produce circumstances of conservation. Even abandonment may justify conservation. Among the many arguments for the most careful preservation of the ruins of Herculaneum and Pompeii are their encapsulation of social history by having been frozen in time through the eruption of Vesuvius. In Norway, a World Heritage Site consists of the ruins of an ill-built mining town, abandoned on the cessation of work. For neither site could conservators claim architectural merit to justify conservation but each qualifies by rarity and the encapsulation of the knowledge of a moment in the history of man's evolution on this planet.

One perception is more intangible than any other, the *perspective of place*. In one aspect this might almost be called a perspective of spirituality or sacredness. To the Australian aborigine and the bushman a patch of bare earth may carry this special intangible quality that makes it meaningful, sometimes with great intensity. It is a quality within the place itself, separate from its historic merit, rarity, quality and circumstance. It may be found in the bare stone-bordered praying space of the Muslim nomad. It is felt by the Japanese contemplating newly raked gravel at an ancient shrine where every piece of timber in the building has been replaced at regular intervals over long centuries of use. It is felt in places long hallowed by worship, and even perhaps abandoned. It is felt in places such as the Holy Sepulchre in Jerusalem where the whole original native rock has been removed to be replaced by an entirely new fabric. Although this spirituality is locked to one particular spot it is perhaps a quality in the mind of the observer as much as of the place itself. It is enough that the effect upon the observer is palpable. Being intangible it cannot be evaluated. It simply pervades and serves to heighten other perceptions, being meaningful in varying degrees to different people. It is a quality so nebulous that to justify conservation it is never cited of itself as a reason. Never stated and never scrutinized, it is a quality that underlies the relationship of building and site. To express it as perspective of place is to underline the quality intrinsic to the site; to express it as a perspective of spiritual value is to underline its effect upon the observer.

PRECEPTS AND PRINCIPLES

The perceptions which govern the art cannot be absolute. They subsist on the basis of judgements made by the observer whose assessment must be that of his standpoint at any particular time. These factors derive from his background, training and personality.[1] The sensitive conservator in his work will interpret the views of his society. With his own individual precepts he develops the rules of conduct under which he operates. These rules control him whether consciously or subconsciously and whether or not they have been subjected to intellectual analysis. They can vary widely while respecting the generally accepted principles of conservation, reflecting character and conviction. A conservator may, for instance, stand by his sense of responsibility to the future, to the nation, to his religion or to his employer to a greater or lesser degree. He may hold strong views on reversibility, traditional techniques, new technology and contemporaneity in design. His personal or religious sympathies may debar a conservator from some activities such as the conversion of chapels to houses or working on military buildings. They may encourage him in others. These *precepts* become characteristic of his personal behaviour. They derive from his training, understanding of his art, its place in society and his entire persona. Hence they are personal to him; they are his *vade mecum*.

They are not to be confused with the *principles* that govern his art. Principles are the established dicta which parcel up a problem. They tend to become a short cut or formula, to be written down in conventions and charters and in the most certain of cases to be carried into law. Principles emerge by way of respectable practice. Being collective they must be the product of social thought and activity. So it is that the increasingly coherent conservation movement has laid down principles, defining and redefining them by charter and recommendation. Principles have an essentially philosophical background and practical application. By consensus the most general are these:

- careful recording and research before intervention;
- minimum alteration of historic fabrics;
- minimal risk of significant loss, damage or uncertainty in performance through intervention, selection of materials or through the calibre of operatives;
- reversibility of interventions;
- retention of a maximum of the original structure;
- distinctive or distinguishable use of new and of additional material;
- sympathy in interpretation and sympathy in use;
- respect for the quality of place;
- preference for original materials and workmanship;
- longevity in the finished work.

These ten principles share one common factor to the extent that they may be said to be founded upon it. This is *integrity*. Integrity, sometimes expressed as honesty, underlies each of these subdivisions of activity in building conservation for the fundamental reason that the work of conservation is concerned with artefactual history. Every historian faces the issue of interpretation. Interpretation betrays the integrity of the historian. Ulterior motives, quality of judgement, and quality of craftsmanship, all become apparent in the quality of the finished work. Integrity, therefore, is vital.

Principles may also be expressed inversely. The most informative of negative injunctions censures improvements and misleading alterations. Fakery (work done with intention to deceive) and deliberate destruction are fundamental errors in conservation and restoration and yet they have an equivocal side.

The restoration of the walls of Carcassonne by the indefatigible and brilliant nineteenth-century French architect and scholar, Viollet-le-Duc, was as much aimed at beautification as at conservation. Such was his success that his creations are now protected as works of art in themselves. They have become historic artefacts; likewise the towering nineteenth-century walls, turrets and battlements on solid Norman foundations at Arundel Castle (Figure 3.1) and on many another stout tower made grander to meet the aspirations of potentates or to reflect the self-importance that can go with wealth. The walls of the City of Muscat were thrown down as redundant and old-fashioned only a generation ago to be rebuilt shortly afterwards as heritage. Such events have happened many times in history. The act of rebuilding did not offend against the principles of place but the new, functionless walls with their now-grander gates raise great issues of integrity. Is the reversal of one destructive act by a succeeding generation more, or less, meritorious than the same reversal by the original destroyer following a change of heart? Are purposeless and non-original walls on an original alignment less acceptable than the repair of decaying original walls? Does the act of rebuilding become less significant as time passes until the rebuilt structures acquire a certain quality of place of their own? Are the recreated squares of Warsaw and the Town Hall of Ypres invalidated by being entirely new above their basements? Are they maturing into respectability? And, if so, do we allow that re-creation is a legitimate technique of conservation activity even when no fragment of the original remains standing? Grandfather's axe, the temples of Kyoto and the city centre of Warsaw then cohere in the activity of preservation of object type as a valid purpose of conservation. Is this activity then proof that conservation consists essentially of retaining or re-creating space and place and that the action of holding on to materials is secondary?

These and similar questions may be analysed and largely answered by testing each principle against the fundamental criterion of integrity or honesty. In some circumstances conflicts arise. The conservator may have to tolerate a breach or loss of one principle, setting it against the advantages of conformity with others. The subtlety and complexity of these interactions demand that conservation is treated, taught and understood as an art on which the application of intellectual endeavour imposes refinement in providing its rationale. The process of sloughing off the irrelevant and refining the essential is therefore both selfconstricting and selfstrengthening. Thus the fundamental precept of integrity, applied to the basic principles of conservation yields a structure based on constraint:

Figure 3.1

Arundel Castle, the towering nineteenth-century walls, turrets and battlements on solid Norman foundations. (*Photo: John Warren.*)

- Recording and research before intervention requires scientific method, precision and skill, all of which demand integrity and detachment. Frequently it will demand self-discipline.
- A minimum of intervention requires restraint and the avoidance of beautification and improvement. The person who can copy out a fine poem containing one clodhopping line, and yet refrain from improving it, is at heart a conservator.
- Reversibility requires discipline in the use of materials and the rejection of some, however convenient they may be. It recognizes the future and its yet unknown methodology.
- Retention of the maximum of an existing structure imposes the burden of honesty in interpreting the building in all its phases, giving each part its own validity. Retention in preference to renewal defers to the prime evidence. Each demands analytical integrity.
- Differentiation of new from old material requires care in handling the work to give honest expression to the sequence of events. Technical methods allow 'hidden' differentiation.
- Sympathy in interpretation demands integrity in understanding the qualities of both old and new, skill in the application of techniques and a high level of judgement.
- Sympathy in use will limit possible users, possibly with financial consequences which place special demands upon integrity.

- Preference for original materials stresses the integrity of the prime process, and their visible expression.

Cumulatively man has constructed an intellectual edifice in which his behaviour as a conserving animal may be rationalized into the precepts which guide him individually and the principles by which the activity is governed. The application of integrity determines the impartiality of the precepts and the effectiveness of the principles.

STANDARDS AND VALUES

It will be clear that this edifice depends throughout on agreed values. Much philosophical ink and even more scholarly hot air have been expended on the concept of values but the subject is not avoidable as it is the very stuff of conservation where values are in constant use as tools for decision-making. Any statement on a scale of values (for example, a temperature) implicitly recognizes a zero. Any proposition involving variations in value may lead from extremes to neutral values or norms. If the quality of conservation varies this suggests extremes of value and a median standard, a neutral or norm.

The problem may be approached by analysing the effects of distortion. Just as water is made visible by the ripples which distort its surface, so bias makes visible the values used in making judgements. The more a judgement is biased the more readily we can see the values on which it is based. Conservation of a rubble wall excessively done leaves us in no doubt as to the sympathetic quality of more modest workmanship. In comparing judgements we can compare levels of bias and as a result we perceive the underlying value. Bias differs from perspective in this sense: that bias is intrinsic to the conservator whereas the variation in perspective derives from his viewpoint. Bias is therefore a distortion of precepts.

It is not to be presumed that distortion or bias is necessarily adverse. A distorting mirror does not destroy perspective. It emphasizes one aspect at the expense of others. In some guises it may be useful. The mirror which the car driver has at his elbow has a vital convexity which gives a wide-angled view of the road behind, while the convex mirror at the funfair which makes the nose stand out and the chin recede carries the same idea to excess and even absurdity.

In terms of conservation, distortion or bias may arise by way of background. Two equally capable conservators may differ in point of religion even to the extent of holding contrary views as to which historic aspect of a fabric should be conserved. Nor may it be only the conservators who are biased. The conserving society may be prepared to fund one aspect but not another for similar reasons. As a result, conservation becomes distorted and represents one aspect of history more completely than another. It might be argued, to justify such a bias, that a comprehensive exposition of one culture is more meaningful and informative than a mixed pattern of several; but a stronger similar bias might expunge a series of monuments to enhance others of competitive background.

Such action would be censured as unacceptable and the reason for its unacceptability must be historic falsification; that is, bias carried to excess.

Thus bias can be identified by degree. The greater the bias the greater the distortion until, passing a certain point, the result becomes unacceptable, falsifying the picture rather than enhancing it. Conversely, a lesser bias produces lesser distortion and is likely to be more acceptable even to the point of being too bland. Recognizing these changing values we can demonstrate that we are dealing with a genuine metaphysical phenomenon and can go on to establish a neutral value or norm which is without bias. This neutral or bias-free condition may therefore be identified and helpfully recognized to set a basic standard. The standard is important in situations where common levels of achievement are required, as for instance in the award of funds, or inclusion of a building in a specific category of merit or protection. If the standard, whatever it may be, is universally recognized or broadly accepted, the conservator has a clear basis for his work. This present argument can by-pass the actual evaluation because its purpose is to describe the mechanisms by which value judgements are reached, or compared.

THE THIRD REALM

Plato's theory of ideas introduced the 'Third Realm' – that conceptual level in which essences – good, evil, beauty, etc., may be said to dwell. These all are products of the mind, but have a commonality and a continuity beyond the mind of any one individual. Karl Popper, famous as a philosopher because of his deep involvement with theoretical physics, has expanded the basic Platonic concept to explore the way in which ideas may interact to produce further products, none of which would exist without the substantive first ideas. This is important to conservation because the Art deals in abstractions – ideas – and sets its criteria by them. Their interaction then produces values. By setting each against the other, conclusions may follow which determine actions in the practical work of the conservator.

Philosophy thus offers a rationale, a construct, by which values may be related to facts.[2] A brick wall, the activity of conserving that wall and the quality of its conservation are all facts, but they exist at different levels; the physical level, the intermediate or problem level and the idea or value level.

REALM 1	the physical level	the tangible object	brick wall
REALM 2	the intermediate level	mind	conservator
REALM 3	the idea level	concept	quality or work

A physical fact may be that its mortar was made with lime: realm 1. At the intermediate level stands the mind equipped to conserve with lime: realm 2. The principle that lime should be used in its conservation rests at the level of ideas: realm 3.

The concern here is to distinguish facts of similar calibre logically rather than intuitively in order then to compare them. A true comparison can be made only if the facts occur on the same level of analysis, that is, within the same 'realm'.

This construct could be also a computer at the physical level, a computer programme at the intermediate level and the resultant calculation at the idea level. Hardware, software and product all have their own independent existence, but Realms 1 and 3 (physical and idea) are capable of interacting only through the intermediate level. The final calculation cannot be produced by the computer without the injection of the programme! In terms of conservation the brick wall is the physical reality, the specification and the workforce are the intermediate level of activity and the calibre of the work is a final value. This process may be applied to other facets where, for instance, the building remains the physical fact, the documentation relating to it is the intermediate level and the historic interpretation is the value judgement at idea level.

The same analytical process can differentiate the mechanics of recording through the abstraction of drawn analysis to understanding the spatial values, shapes and volumes of a structure perhaps no longer extant. Here, with a projection into the past, time enters the value frame extending its usefulness. It follows that every problem can be analysed at its physical, its subjective and its idea level. Moreover, this complex level of values and ideas also becomes interactive within itself. This realm of ideas has a reality and a continuity. By interaction it breeds further concepts. Thus understanding the spatial values of a past structure now partly destroyed induces the notion of remaking the structure. By another set of values the part destruction has become a quality in itself. The concept of a pleasing ruin so delighted the eighteenth-century landowner that if he did not possess one in his park he would be tempted to simulate the craggy end of a crumbling nave on a prominent hilltop. Here the establishment of one set of values (delight in the romantic ruin) has led directly to a matching value (the idea of the simulated ruin) which led in turn to the physical fact of the new structure.

REALM 1	ruined abbey	↓	↑	simulated ruined abbey
REALM 2	landscape concept	↓	↑	landscape designed
REALM 3	romantic illusion	→		romantic vacuum

Exploration of the linkage between ideas in this way permits an analysis of value problems, demonstrating visually the position of values in relation to identifiable facts. In this way a complex intuitive process may be clarified, allowing the conservator to marshall his ideas. This disassembly and restructuring of ideas may give pause and the opportunity for rethinking, but in the end all ideas remain subjective. Their comparison can never be mathematical.

The conservator will be little interested professionally in alternative philosophical theories which postulate that all values have a perpetual existence of their own and are, like seams of coal, simply waiting to be discovered; or more convincing alternatives which hold that values are the product of problems and do not exist until created: the 'no problem – no value' causality. Philosophical discussion is relevant to the conservator's art only where an understanding of the processes can lead to more consistent methods of achieving a final judgement.

A current circumstance may illustrate the problem and give it point. Ottoman powder magazines were prone to explosion and as they were often located in the most solid monument of a subject city, such as the Parthenon, great damage was done when they went up. The Romanesque Cathedral at Bagrati in the Republic of Georgia was ruined in this way in the seventeenth century. It has long been without its crossing, dome and central steeple.

With the social change in the Republic of Georgia, the Metropolitan (bishop) has asked for advice on restoring the Great Church to its original function (Figure 3.2). The ruin has been adequately conserved as a static monument under the policies of the USSR, but now conditions at the intermediate level have changed. Ownership, desire for religious use, funding from private sources, and social support are among the many new factors. Consequently the derived values will be different. The question, not answerable here, is whether these different values support such a programme of restoration. New values arise with the consideration of new structure – restoration as opposed to conservation, and reconstruction imposed on restoration. At the level of ideas there emerge the principles of validity of use, purposefulness of structure and spatial integrity, set against imitative detail, loss of historical status and a change of user. In any such investigation all relevent background must be explored so that a single analytical process can be applied. An entire range of arguments stems from the physical condition of a damaged building through the abstract arguments of use, performance and social expression. These might lead to the value ideas (conclusions) that restoration should be undertaken in contemporary terms with contemporary materials rather than in terms of replication – or even the converse. It will be self-evident that value judgements at this level interact, often to the self-confusion of the protagonists in the argument.

Figure 3.2

Bagrati, the Romanesque Cathedral at Kutaisi in the Republic of Georgia, ruined by an explosion in the seventeeth century. (*Photo: Kakha Khimshiashvili.*)

The conservator with a rational view of the underlying philosophical processes will be better equipped to achieve judgements and validate advice in the increasingly abstruse spheres of his activity. At the intermediate or abstract level a range of established patterns in the mind translates readily into specific injunctions – methods of work, techniques, and particularities of organization. At the value level his precepts lead him to judgements within the principles of the profession. If the conservator is consistent in making these judgements the quality of the finished operation should likewise be consistent. A careful philosophical logic is the fundamental basis for such consistency.

But while this logic may seem a bastion its foundations stand on uncertain ground – the motive of affection at work in society at large. We love what we know; we conserve what we love. So fundamentally the bastion rests on values that are emotive. It is all the more important then that the work of conservation is carefully structured and reinforced through and through by stringent analysis. Principles (general rules controlling action), precepts (personal commandments) with training, and self discipline make the conservator. His environment is made by the attitudes of society, the funds it has and the rules it applies. These, if ill-conceived, may lead to absurdities. The conservator enters the picture as part of an essential process of sifting and rationalizing which involves criticism, critical analysis and the melding of conflicting ideas. By processes of debate he may establish the incompatibilities, eliminate the absurdities and harmonize apparent differences, deriving common features and balanced judgements. As an example we may look at one of the commonest of such frictions stemming from the interaction of logical and emotive pressures.

ETHICS AND AESTHETICS IN BUILDING CONSERVATION

Efforts to make time stand still, or even to roll backwards, give rise to actions based on judgement. Judgement in turn demands skill, training and the ability to be dispassionate. Where there are principles we conceive of ethics: where there is art we conceive of aesthetics. In the thought process these intangible factors lie at the level of values (Realm 3). Their interaction produces one of the more important metaphysical conflicts in conservation.

First consider definitions:

Ethics, ethical: principled; of integrity; uprightness; correctness; truthfulness – all these attributes of behaviour are matters of ethics. In short, ethics embodies the principles which underwrite good behaviour. In the converse, therefore, unethical becomes a term of censure.

Aesthetics: the qualities of artistry – the underlying quality of a work of art which ties it to its source or school, period, place or creator. There is no converse, for the obvious reason that the qualities in an artefact are, by definition, its aesthetic.

It will immediately be apparent that ethics and aesthetics are entirely different, although they are indissolubly linked. In conservation neither can exist without the other. Any work of conservation must be judged by whether it is principled or unprincipled, ethical or unethical, and its impact upon the user will be measured aesthetically according to the insight and predilections of the observer. Ethically it may be truthful or deceitful; aesthetically it may be satisfying or unsatisfactory.

It will be apparent immediately that four combinations are possible:

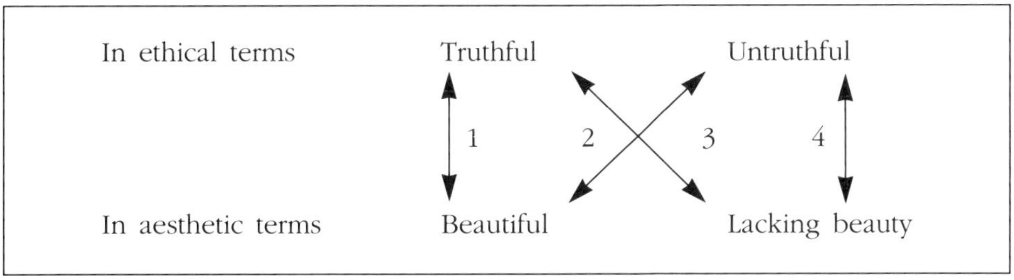

and these combinations can be a matter of degree.

The possible combinations of these attributes lead, among other conclusions, to the proposition that a work of conservation may be aesthetically pleasing but deceitful. To give the idea shape, imagine a building or feature which has been conserved or restored in such a way that it has become pleasing but has been remade so extensively that it is a very long way from the original creative work. It may delight the eye by its success aesthetically but be valueless because it is untruthful. Conversely consider fragments, which on their own are unsatisfactory aesthetically, such as a few shattered pieces of pottery. These, set in a plaster shape may reproduce with total accuracy the form of the artefact but the result is not aesthetically meaningful because the pieces themselves are scattered so that only the most scholarly can visualize the aesthetic calibre of the original. In this form visual objectives have been sacrificed in favour of total truthfulness. The result may not stir the heart but it stirs the mind.

These examples, and any other combination, will show one thing: that ethics are the dominant criterion in any serious intellectual evaluation. Where the ethics are wrong the result is intellectually unacceptable. Half cynically we might add that if ethics are the qualities that matter intellectually aesthetics are those of emotional appeal. Sometimes fate is kind and these attributes go hand in hand. Where they do the result is uncontroversial. If the pursuit of intellectual truth produces a handsome work of conservation all parties are pleased. Difficulties lie in the temptation to pursue aesthetic objectives at the expense of integrity.

We can understand the problem more thoroughly by breaking down each of these qualities to identify their key components and the methodology that goes with them. Let us pursue this quest for a moment, drawing from the general arguments key conclusions to act as guiding principles.

Initially, consider conclusions which are essentially ethical. Conservators are not free agents but are controlled by the object of conservation. The imagination of the painter or sculptor may be unconfined – a composer's or poet's even more so – but, by virtue of starting with an object of precise shape, the conservator is controlled. This control extends to the material itself and to the immaterial background of its history. The architectural conservator cannot make the right decisions unless he understands why the building is what it is. In preparation, therefore, the conservator satisfies himself, before any action is taken, that he knows sufficient about the object to make decisions correctly.

The first conclusion, therefore, is that *the object of conservation must be thoroughly researched.* It must be understood as an object created in space and in history. This in turn requires an understanding of its social background and status, the changes that have been imposed on it and the reasons why it has taken on its specific form. It becomes a matter of integrity with the conservator that no action is taken until the problem has been adequately researched.

Generally the action required is benign, that is to say non-damaging, although sometimes it may extend to destructive research such as excavation or dismantling. This problem leads directly to a second principle relating to intervention.

Conservation almost inevitably leads to intervention, that is, the addition of further, usually new, material or removal of an element which, almost inevitably, is part of its history. Before making any intervention it is essential to record all available evidence. Even evidence which may seem unimportant may ultimately prove to be significant. If an object is worthy of a conservator's attention it is so because of its historic or contextual merit. If it has such merit this, of itself, must be recorded.

The second conclusion therefore, is that *prior to any intervention or change a scrupulous record must be made of state and condition.* As it is an essential prerequisite that work is executed to a sufficient standard we can add that *intervention shall be undertaken only by those with appropriate skills and equipment.*

The concept of ethics contains an essential element of truthfulness. The essence lies in the avoidance of deceit and pretence. It is untruthful to pretend that something is old when it is not. It is untruthful to put something in a context where, although new, it appears to be original. It is neither necessary nor truthful to proclaim the newness of an intervention by making it aggressively self-evident. A non-statement may be as honest as a violent statement. In this range of opportunities the quality of sympathy arises and to this we will turn again when looking at aesthetics.

There is an ethical element also in other factors such as utilization. Almost no building or object has been created without reason or purpose. Often, too often, the reason evaporates leaving the object stranded on the shores of history, purposeless but still historically important. No one foresees a widespread return of the Roman cult of Mithras, but every Mithraic temple uncovered is a powerful candidate for conservation *in situ,* to be retained as found. Sometimes the purpose of the object can be redeemed and it can be reused in its original context. This ultimately is the most ethical of all solutions. Better to restore a mosque for use as a mosque, even if that does involve physical alterations, than turn it into a jam factory unaltered. Provided that the alterations are not excessive the ethic of retaining the religious use dominates the ethic of integrity of unaltered structure; but the decision will be one of degree, demanding a balanced judgement.

The third conclusion must be that *in any intervention or re-use truthfulness is a paramount virtue.* This truthfulness will extend beyond stating which part is old and which part is new, to fundamental honesty in handling the restored structure itself. A building may require strengthening. This might be achieved by visible insertions. If they are disguised, the disguise may of itself be a dishonest statement. The alternative of hidden reinforcement avoids the need to resort to disguise but may remain deceptive. The ethical implication is that where exposed reinforcement is required it should be handled honestly and with the least visual impact.

Ethics involve, in essence, responsible behaviour, and the principles so far enunciated derive from a sense of responsibility in holding past creations in trust for future generations. There follows, therefore, a further principle which applies in conservation as in other aspects of life – reversibility. Just as one fundamental objection to capital punishment is that the penalty is beyond recall, so the application of irreversible solutions in conservation is unsatisfactory. An original component, once destroyed, can never be called into evidence. The material it contains can never be analysed by techniques yet to be invented. There is no opportunity for better techniques to be applied in the future. A further conclusion therefore is that *every intervention shall, wherever possible, be reversible.*

We now have four conclusions governing ethics embodying the basic principles of conservation. Let us look at aesthetics, which are driven by the emotive reaction which is a part of the response to all forms of art. No aesthetic value exists without a preconditioning of the mind by comparisons and by an understanding of social and visual values. All aesthetic values derive from conditioned response; that is to say from the reaction of the mind trained to what it sees, hears or senses. These reactions run the gamut from repugnance, horror and fear to adulation, joy and pleasure. These emotive reactions are a critical part of our understanding. If our object of conservation has an aesthetic quality then our intervention should run in parallel with that quality, neither in conflict to diminish it nor exaggerating to enhance it. Aesthetically a first conclusion, therefore, can be that *the impact of an intervention should be sympathetic or neutral.*

On this basis and by this means we can ensure that the quality of the original object is seen for what it is. But there is a debatable aspect to this conclusion. In cases where the original quality of the building has been wholly or largely destroyed in the process of historical change there is a strong and sometimes irresistible argument for restoring the aesthetic quality it once had, even though the structure used in the restoration is entirely new and the material that demonstrates its aesthetic qualities may, therefore, be described as false. Replacement of a finial may pose such a problem if the shape of the original is not known. Better to have a finial than none at all, so runs the argument, leaving us with the choice of a modern design or a traditional form. We might from this derive a second conclusion that *an intervention should move towards a full appreciation of the building.*

Many buildings have been altered at different phases in their history. Intervention can heighten or diminish any one phase. The process can go so far that the conservator may eliminate one phase entirely in favour of another. They can cut out later developments to expose the original state or, in the converse, can cover the original by emphasizing the later work. They are entitled to make a judgement (indeed may be forced to do so) and that judgement must be based on knowledge which derives from comparative values, assessments of rarity, of significance, of social importance and of historical value. Many subtle questions can arise.

What, for instance, does the conservator do with the humble but very rare building which cannot be restored or exposed without destroying a much grander but relatively common structure which is totally obscuring it? Does he remove the modern but historic parts to reveal the first and most venerable foundation? From such a dilemma arises the conclusion that *the scale and nature of the intervention must respect the significance of the building and its phases of construction.*

Finally let us turn to the nature of the aesthetic. It is the fingerprint, the distinctive mark, of the creator. It is the quality apparent to the skilled observer, which tells when and by whom the building was designed, perhaps by whom it was built. It is most meaningful to those with most experience; to the newcomer it may hardly be apparent at all. It consists of nuances, inflexions and accents which betray the source and quality of the creative effort. In simple analogy it may be compared to the dialect of a language. To someone who cannot speak a language dialect is meaningless; to those well versed in the language dialect is immensely rewarding, carrying the flavour of the country, telling something of the origin of the speaker and much about his place in society. It can be ugly or beautiful; it can betray place and time.

So we have an aesthetic characteristic which may, perhaps, be encapsulated in a conclusion to the effect that *the intrinsic quality of the building must be respected, any intervention being neutral or sympathetic to the original.* This might be coupled with a further conclusion to the effect that *the building should be placed in a setting which is contextually right and meaningful.* For reasons of history or limitations of brief this may be beyond the powers of the conservator, but aesthetically it is not to be ignored.

These conclusions will suffice to allow us to discuss their rationale and to relate ethics to aesthetics.

Thorough research – the first conclusion – is likely to involve a range of special skills beyond those of the conservator and, therefore, in pursuit of high ethical standards the conservator will have a humility which allows him to seek out and sit at the feet of others who have greater and more specialized knowledge. Part of the conservation ethic is fulfilled when the conservator deploys, at the outset, the whole range of available skills to define the qualities of the building being dealt with and this, of course, will draw in judgements which may be made by others of greater specialist competence – art historians, architectural historians, archaeologists, structural specialists and archivists in their libraries may all contribute and the conservator may be actively involved in one or all of these activities. Integrity is shown in the way in which the resulting information is combined before making decisions. The skills available include many forms of physical survey, laboratory analyses of material and documentary evidence. Contextual evidence can be important and the research must extend to knowledge of the materials applicable in any intervention.

The second conclusion was that *prior to any intervention a scrupulous record must be made of state and condition.* This aspect of the work is sometimes very difficult to fund for the simple reason that most building owners see such work as relatively unimportant since no results are visible on the building itself.

With recording goes research. Physical surveys will include many aspects of measurement. Environmental conditions must be analysed while physical, planning and human circumstances are crucial to an understanding of potential and requirements, and themselves may be the subject of record. This process of recording is, of course, of small importance while the object stands unaltered but when the conservation process has gone forward a record of its earlier condition becomes increasingly significant, and at that point there is no going back to make it.

Recording is doubly important where a whole phase of a building will disappear in the work of conservation. Sometimes this is unavoidable, at other times it is simply desirable. It is part of the integrity of the conservator that information is not allowed to slip away when it can be an important resource for the future. An objective of recording and research will be to define the aesthetic value and context of the building in its original state. An object once pristine and colourful may have acquired subsequently the mellow qualities of 'pleasing decay'. Are these to be sacrificed to return to the original condition, and is this condition relevant in present circumstances? By intervening shall we have eliminated forever a special quality even though this might be an accident of history and of aesthetic value alone?

With this thought we leap forward to our final principle of ethics: that *every intervention shall wherever possible be reversible.* Reversibility is simply the ability to remove the intervention, restoring the earlier condition. Aesthetically this raises the problem of patination – the recovery of an attractive time-worn aspect.

Our third conclusion was that *in any intervention or re-use truthfulness is a paramount virtue.* Truth demands that a present intervention is not to be mistaken for old material and with care this can be achieved. Modern materials can always be identified or marked without damage to the aesthetic qualities of the work. Truthfulness carried to the extreme, however, can conflict violently with aesthetic objectives. It may be argued that because the form of a particular ancient window is unknown a modern, metal window, painted bright blue should be inserted to make a positive statement of lack of knowledge. This argument, of course, is countered by the principle of sympathy and the need for a neutral quality in any intervention. An ethical principle is thus ameliorated by aesthetic considerations.

This takes us to our essentially aesthetic conclusion: *that the impact of any intervention should be sympathetic or neutral.* Here, a conclusion driven aesthetically takes on an ethical overtone. It is hardly possible to be sympathetic to something which you do not understand, but we have already established that the conservator will make it his first duty to understand the object being conserved. The opportunity for choice may lie in several directions and the decision may depend considerably upon integrity and balance of judgement, but other pressures may enter the equation which have nothing to do with aesthetics: money, religion, ethnic loyalties and politics among them.

The conclusion that *an intervention should move towards a full appreciation of the building* is essentially aesthetic. The fullest understanding of a building is gained from an appreciation of its original state. An intervention should, therefore, enhance rather than diminish this understanding. Application of this idea tends towards the predominance of a wider characteristic over the lesser. A surface given a colour, texture or pattern on the basis of informed speculation may contribute more to understanding than one left blank for lack of detailed knowledge.

The ethic of truthfulness in an overall context may dominate that of truthfulness in minutiae. Aesthetic advantage may then become an important factor. However, aesthetic advantage is not on its own decisive and the rightness of the greater truth over the lesser becomes the determinant.

Aesthetics and ethics bear equally upon the conclusion that *the scale and nature of the intervention should respect the significance of the building and its phases of construction.* It is both responsible and visually desirable to limit the impact of the intervention upon an historic monument.

In final conclusion, *the intrinsic quality of the object must be respected, any intervention being neutral or sympathetic to the original.* This is not to say that a building of one type can only be restored using materials identical to the original. That, in itself, might sometimes confound the ethic of truthfulness. What it does mean is that the conservation should not be obtrusive. Ideally the observer should come to the object and be unaware of the detailed nature of the conservation. He should merely be aware that the object has been venerated and dealt with carefully and with skill. He should not be deceived and should not be offended; he should not be repulsed nor should be led to believe that the original object was finer, nobler, more important or more beautiful than in truth it really was. This in part depends on the surroundings and adjacent structures. For this reason it is important that *the building should be placed in a setting which is contextually right and meaningful.* This objective is often made more difficult to achieve by sequential historical events, the insertion or removal of other buildings, changes in street pattern or other circumstances outside the conservator's control. Ethical judgements will dominate over considerations of beauty. A correct setting is to be preferred to one which is made handsome irrespective of contextual proof.

THE FUTURE

Beyond these balances and checks of values there is a consideration of present and future time. The question must be asked of every act of conservation: how is this work meaningful and right for this and future generations?[3] The conservator must weigh his judgements not only ethically and aesthetically, but practically in terms of what is achievable and socially right for his generation. It is also fair that, where appropriate, the conserving generation makes its own creative statements. When it attempts to do so the reason should not be doctrinaire, or worse, the ambition of the conservator to leave his mark. A visible contemporary statement is justifiable only where a gap must be filled. But where lack of knowledge denies evidence for the original or where a new situation makes new demands upon an ancient fabric the creative opportunity changes and the benign self-effacement of the conservator must be supplemented by the skills of the restorer, providing even complete new elements of architectural design. It does not follow that the philosophies of conservation no longer apply. The nature of the debate remains the same although new premises provide variations in its starting point.

In doctoring our ancient fabrics we have in recent years made great strides technically and ethically. Society has made equal progress in regularizing and codifying the processes which control and determine conservation. It is fair to ask whether it matters that, as conservators and as citizens, we understand the philosophical procedures which govern our behaviour. Those who answer, 'No! All that matters is that we get the right results' might ponder the parallel question: does it matter that general practitioners in medicine understand the biological mechanisms which operate in response to the remedies they prescribe? To which the answer is that it is often helpful and sometimes very necessary. Perhaps it is also so in building conservation. It is the function of philosophy to give order to the processes of thinking and leave us with an understanding of the procedures in which we are engaged. Beyond this it should also enable us to see ourselves in context, to stand back far enough with a sufficient sense of detachment to see our actions in the social context of our lives and the continuous evolution of our communities and our kind.

The art of conservation is one of compromise. Essentially the skill of the conservator is self-effacing. He will never be the brightest star in the architectural firmament. His role is not to shine large and bright but to glow quietly in the galaxy of human achievement.

To conclude with a further thought of astronomic dimensions – one which can affect the judgement of the conservator in choice of method, materials and objectives. We protect our buildings by law, apparently indefinitely. How long will this be? How far can our minds be stretched? Our planet may survive as we know it for perhaps one and a half million years, maybe much longer. Some cosmologists talk in terms of a thousand million. Do we as conservators conceive of our work as extending for one thousand, or fifty thousand, or beyond a million years? Do we imagine that our principles measure up to the aeons of this vast future? Are we repairing venerable stones shaped only a thousand years ago for a future of a million years? The thought, to the mind prepared to grasp it, is awesome.

Acknowledgements

The author is grateful to those many friends with whom he has discussed the philosophies of conservation, not least to Sir Bernard Feilden; also to the Institute of Advanced Architectural Studies in the University of York for the opportunity to present lectures on the subject and to the Association for Studies in the Conservation of Historic Buildings in whose *Transactions* (vol. 18) the latter part of this material was first published.

NOTES

1 The inadequacy of such precepts may be grasped and admitted. To quote one historian, 'Totality of perception is impossible to achieve and is also probably unnecessary, but the very idea that the totality of perception may not be necessary leads one to one of the cores of art historical procedure. The perception which I (or anyone else) have of what I see is not really a neutral and, so to speak, clinical act. It is one which is already loaded either with methodological or historical prejudices. All too easily one's reading of a building becomes a search for clues to advance already-developed theories and interpretations under the guise of scientific completeness.' (Oleg Grabar (1990) *The Great Mosque of Ispahan*. The Hagop Kevorkian Lectures, New York University Press, 21.)

2 Sir Karl Popper described Plato's 'realms' as worlds. His World 1 was physical, World 2 was the mind and mental constructs, and World 3 was abstract ideas. In his World 3 he wove a fabric of great complexity. Mathematics for instance he placed in World 3, *producing of itself* numerical constructions such as irrational numbers which were not the direct product of the mind giving rise to mathematics. His concern to put such abstractions into a practical context he expressed thus: 'It is through the attempt to see objectively the work we have done – that is to see it critically – and to do it better, through the interaction between our actions and their objective results, that we can transcend our talents and ourselves.' (Sir Karl Popper (1992) *The Unended Quest*. London, 196.)

3 Edward O. Wilson, writing of the parallel sphere, the Natural Environment, says 'The stewardship of the environment is a domain on the near side of metaphysics where all reflective persons can surely find common ground. For what, in the final analysis, is morality but the command of conscience seasoned by a rational examination of consequences? And what is a fundamental precept but one which serves all generations?' (Edward O. Wilson (1992) *The Diversity of Life*, 351.)

4

INTERNATIONAL STANDARDS, PRINCIPLES AND CHARTERS OF CONSERVATION

Jukka Jokilehto

PURPOSE

The purpose of the chapter is to discuss international principles and charters in the protection and conservation of cultural heritage. An introduction will be given to the background of international concern about heritage, the question of 'universal value', the establishment of international organizations and their involvement in the preparation and implementation of international charters and standards. The trends will be discussed in their application in different cultural contexts, the broadening concept of cultural heritage, from artistic and monumental to ethnographic and vernacular, and corresponding conservation philosophies, including notes on the debate around authenticity and cultural diversity.

INTERNATIONAL CONCERN

The World Heritage Convention, adopted by the seventeenth General Assembly of the United Nations Educational, Scientific and Cultural Organization in Paris in 1972, has defined the need for 'establishing an effective system of collective protection of the cultural and natural heritage of outstanding universal value'.[1] The convention marks a milestone in the international concern for the world's heritage; it recognizes the close relationship of the built and natural environment, and lists representative examples of both 'as part of the world heritage of mankind as a whole'. The convention coincides also with the growing ecological concern for the environment, expressed in an international meeting in Stockholm in 1972, which was followed by the 'Brundtland report'[2] of the United Nations introducing the concept of sustainable development.

The roots of the modern concept of universal heritage can be seen in Antiquity; the idea of identifying the Seven Wonders of the World[3] anticipated the World

Heritage List of UNESCO, with its more than 400 sites by December 1994. However, the modern concepts of protection and conservation of cultural heritage are founded on the new historical consciousness that has developed in Europe since the eighteenth century. This is also the basis for value judgements and guidelines for their treatment. At the same time, the grand tours contributed to an increased knowledge of cultures and the recognition of great works of art as a symbol of the achievements of humanity. Observing the precarious condition of such works, travellers proposed that international action could assist in their conservation. The same concern was raised regarding the fate of cultural heritage during warfare.[4] These ideas were brought to a test when the Allied Nations took action to return to the countries of origin the collections confiscated by Napoleon in occupied territories, and the issue was again raised regarding the Elgin marbles. In 1864, Emperor Alexander II of Russia gathered the first international conference in Brussels which adopted the first project for an international declaration concerning the protection of works of art in the case of armed conflict. Culture was declared the common heritage of mankind, and artistic treasures were considered irreplaceable (Jokilehto 1986). The present Hague Convention of UNESCO results from this development.

As a result of heated debates on conservation versus restoration, various countries had initiatives concerning the protection of cultural heritage; such was the Society for the Protection of Ancient Buildings (SPAB), established in England by William Morris in 1877. The insistence on maintenance and repair instead of 'restoration' was heard widely, and the society had an impact also abroad. In Italy, the government circulated the first national charter on restoration in 1883, written by Camillo Boito, who insisted on a respect for all historic phases in a monument. In German-speaking countries, starting in 1899, there was a series of yearly seminars to debate about conservation issues, emphasizing, for example, the need for training; these continued for two decades. In 1904, the Sixth International Congress of Architects in Madrid presented recommendations for the restoration of 'dead' and 'living' monuments, and insisted on training.

The idea of a 'universal history' of art and culture is reflected in the current interest of the World Heritage Committee in preparing a global study of the world's physical cultural heritage; this is intended as a basis for the identification of sites of 'outstanding universal value', and the definition of the significance of existing sites in relation to influences and values assigned to them. The concept of 'universal value' should *not* be conceived in the sense that a site would necessarily have influenced the whole of the universe, but in the sense that the site is authentic, and that it can be interpreted as a genuine, creative contribution of the culture (or cultures) that it represents. As such, it forms a component in the 'universal heritage'. A site that is accredited 'outstanding universal value' can be referred to as an outstanding, or the most representative, example of a particular type of heritage, as a unique testimony to a civilization or cultural tradition, or as a masterpiece of the creative genius (UNESCO 1994). The idea of universal value represents the recognition of universality in authentic heritage within each culture of the world.

ESTABLISHMENT OF INTERNATIONAL ORGANIZATIONS

At the end of the First World War, the material and human losses, as well as the destruction of important cultural heritage sites, raised the question of improving international collaboration; the League of Nations was created in 1919 to settle disputes before these were to break into an armed conflict. Therein was founded the International Committee on Intellectual Co-operation that met at Geneva in 1922 for the first time. Concern for conservation was amongst the cultural initiatives of the committee, and, in 1926, the International Museums Office was established to deal with museums and works of art, to organize international meetings, and to prepare publications such as *Mouseion*. The aim was to promote international collaboration, develop an awareness of and respect for heritage, improve national legislation, provide international measures for the protection of cultural heritage, and to return illegally removed properties.[5] Many of these issues have remained in the agenda of international organizations ever since.

In 1930, an international conference was organized in Rome for the study of scientific methods for the examination and preservation of works of art. The conference proposed to establish a network of national committees for improved contacts, and recommended the development of scientific methods for the examination of works of art before restoration. A manual was proposed on the conservation of fine arts, and extracts of this were published as an 'international code' of conservation in *Mouseion*.[6] In 1931 another international meeting was organized in Athens to discuss the conservation and restoration of monuments of art and history, and to define the principles in its conclusions – the 'Charter of Athens'. These were brought to the attention of the Member States of the League of Nations in 1932, and in 1933, an International Commission on Historic Monuments was nominated to examine follow-up action, to prepare recommendations concerning education and awareness, legislation, conservation techniques and international documentation.

In 1933, there was yet another meeting in Athens, this time organized by CIAM, the International Conference on Modern Architecture, which also produced a 'Charter of Athens'. This document resulted from the contribution of distinguished architects of the Modern Movement, including Le Corbusier, and related mainly to modern architecture and city planning, but it also included observations on the protection and rehabilitation of historic urban centres.

At the end of the Second World War, several new organizations were created to facilitate international collaboration; in 1945, the old League of Nations gave way to the new United Nations Organization, and the International Committee of Intellectual Co-operation was succeeded by the United Nations Educational, Scientific and Cultural Organization (UNESCO). One of the tasks of UNESCO was to 'maintain, increase and diffuse knowledge: by assuring the conservation and protection of the world's inheritance of books, works of art and monuments of history and science, and recommending to the nations concerned the necessary international conventions'.[7] The Organization was, however, counselled to respect 'the independence, integrity and fruitful diversity of the cultures and educational systems of the States Members', and was not to intervene in matters essentially within their domestic jurisdiction.[8]

In 1946 the International Museums Office was re-established as the International Council of Museums (ICOM), a non-governmental organization to link professionals and institutions dealing with the conservation management of museums and collections. In the early days, ICOM preferred flexible arrangements in the joint coordination of international exhibitions rather than providing 'a rigid system of inter-governmental regulations'.[9] In its work the organization has given due attention to conservation, the establishment of international criteria for the treatment of collections, the definition of the profession of conservator–restorer, and the establishment of appropriate standards for training of specialists.

In 1949 UNESCO established an International Committee for Monuments and Archaeological Excavations; this had fourteen members representing different regions of the world. The committee discussed the ways to improve collaboration between experts, and criticized the tendency of some individual experts to refuse to reveal information on special techniques known only to them. There was discussion about legal issues, and about bringing up to date previously prepared drafts for international conventions from the 1930s, concerning the protection of heritage, and, for example, the return of collections to original owners in the case where these had been removed during war.[10]

In 1956, UNESCO proceeded to the creation of the International Centre for the Study of the Preservation and Restoration of Cultural Property (ICCROM), an inter-governmental organization placed in Rome to assist Member States in the protection of both movable and built cultural heritage. The statutory functions of ICCROM[11] included documentation, research, technical co-operation, and training. During the first decades, this last point has become perhaps the most important of its activities, and the statutes specify that the organization should: 'promote, develop and provide training related to the conservation and restoration of cultural property and raise the standards and practice of conservation and restoration work'.[12] In 1993, a fifth statutory function was added concerning encouragement of a better understanding in the conservation of cultural heritage.

At the invitation of the Italian Government, the Second International Congress[13] of Architects and Technicians of Historic Monuments met in Venice from 25th to 31st of May 1964 presenting a declaration on the fundamental principles of conservation and restoration of the architectural heritage, The Venice Charter. This document was considered necessary as an update since the situation in the world had changed after the writing of the conclusions of the Athens meeting in 1931. The new charter was based on the recommendations prepared by Gustavo Giovannoni after the Athens meeting, and published by the Italian government in 1932. The Venice Charter was written by a small working group, including Raymond Lemaire, Pietro Gazzola and Paul Philippot, and signed by a group of experts who were mostly European. However, the Charter has proved its validity by becoming a standard reference over a period of thirty years, and is even used in other regions of the world.

The Venice meeting also recommended the creation of a non-governmental organization, the International Council on Monuments and Sites (ICOMOS), which had its founding meeting in Warsaw in 1965. One of the statutory aims of this organization was defined as the need to 'encourage the adoption and implementation of international recommendations concerning monuments, groups of buildings and sites'.[14] The Venice Charter was later adopted by ICOMOS as its fundamental ethical guideline, and it has since been referred to in most successive documents concerning conservation principles. The national committees and the international scientific committees of ICOMOS have since contributed to the preparation of several international charters and guidelines; the subjects have included cultural tourism, historic gardens, historic towns and urban areas, archaeological heritage management, and training. Other documents have related to particular cultural or geographic areas, such as 'Las Normas de Quito', 'Burra Charter', 'Recommendation of the meeting at Castillo de Chapultepec', 'Deschambault Declaration', 'Appleton Charter' and 'Carta de Petropolis'. The latest document in the series concerns cultural diversity and authenticity, the 'Nara Document on Authenticity' results from expert meetings organized in Bergen and Nara in 1994.

DEFINITION OF CULTURAL HERITAGE

The conventions and recommendations by UNESCO, the documents prepared by ICOMOS and ICOM, as well as the work by ICCROM, form the basis for the establishment of an international framework regarding the identification, protection, conservation and restoration of cultural heritage. To this can be added the contribution of other organizations, especially in the regional context, such as the Council of Europe, the European Community, and similar organizations in Asia, Oceania, Africa, North and Latin America.

The definition of historic monuments starts with an interest in archaeological monuments and ancient works of art already in the Italian Renaissance. Although the French Revolution caused much destruction, it also promoted protection and conservation of heritage; this included coining the concept of *monument historique*, and adoption of several protective measures. In October 1790, the Commission des Monuments was given the task to care for works of art and the preparation of inventories. In 1793, a new commission, the Commission Temporaire des Arts presented a document concerning instructions for a protective inventory of cultural property, 'Instruction sur la manière d'inventorier et de conserver dans toute l'étendue de la République, tous les objects qui peuvent servir aux arts, aux sciences et à l'enseignement'. This document thus implicitly provided a definition for cultural heritage which was justified for its artistic, scientific and educational values.

The nineteenth century marked an important progress in the development of research, and consequently in the definition of cultural heritage. At the same time, the debate gave major emphasis on conservation of all historical phases, rather than restoring specific artistic features. This was emphasized by Hugo and Didron in France, Ruskin and Morris in England, as well as Boito and Giovannoni in Italy. In addition to great works of art, William Morris was interested in 'minor arts'. He himself worked in crafts and applied arts, and he also conceived architectural heritage in a broad manner, not only 'monuments' but also the simple residential houses in historic areas. In the Manifesto that he wrote for the Society for the Protection of Ancient Buildings, in 1877, Morris declared:

> If, for the rest, it be asked us to specify what kind of amount of art, style, or other interest in a building, makes it worth protecting, we answer, anything which can be looked on as artistic, picturesque, historical, antique, or substantial: any work in short, over which educated, artistic people would think it worthwhile to argue at all.

In 1931, the international conference in Athens emphasized the importance of conservation rather than restoration; it is therefore perhaps not surprising that, in the 1930s, the definition of 'monument' had been somewhat 'reduced' compared, for example, with the values in the time of the French Revolution, when emphasis had been given to artistic, scientific and educational aspects. In 1934, the International Commission on Historic Monuments suggested that the concept of 'historic monument' should be applied to a building the conservation of which was of interest to community due to its historical significance, and especially to buildings that were important to the history of art. It was also noted that some countries preferred to put an age limit to the protection of such monuments.[15] The use of the word 'monument' in English has remained more limited than in Latin languages, and, hence, misunderstandings are easily possible.

The war destruction of the 1940s had the effect of convincing the public about the importance of historic towns and neighbourhoods; in the following decades this was followed up and further developed in public debates, and led to the adoption of legal measures concerning urban conservation. In the 1970s, there was a growing concern for ecology and the environment; this enlarged the concepts even further. It is of some interest to note the different phases in this process: during the French Revolution the question was about 'objects useful to arts, sciences and education'; later the discussion emphasized the concept 'historic monument'; Morris spoke about 'ancient buildings'; the Athens Conference discussed the conservation of 'monuments of art and history'; the first UNESCO documents speak about 'cultural property', but since the 1970s, emphasis is given to the concept 'cultural heritage', involving not only the physical cultural heritage but also other fields of culture; the breadth of the concept 'cultural heritage' was reflected in its 1990–5 medium-term programme:[16]

> The cultural heritage may be defined as the entire corpus of material signs, either artistic or symbolic, handed on by the past to each culture and, therefore, to the whole of humankind. As a constituent part of the affirmation and enrichment of cultural identities, as a legacy belonging to all humankind, the cultural heritage gives each particular place its recognizable features and is the storehouse of human experience. The preservation and the presentation of the cultural heritage are therefore a corner-stone of any cultural policy.

UNESCO conventions and recommendations

The three UNESCO conventions provide a more limited scope reflecting the specific purpose of each. The first of these, The Hague Convention (1954), articulated 'cultural property' in three main categories: a) movable or immovable property of great importance to the cultural heritage of every people; b) buildings, whose main and effective purpose is to preserve or exhibit the movable cultural property; and c) centres containing a large amount of cultural property, to be known as 'centres containing monuments'. The second, the 'Convention on the means of prohibiting and preventing the illicit import, export and transfer of ownership of cultural property' (1970) concentrated on movable heritage and collections, and provided a long list of various types of properties to be considered, including rare collections, property related to history, elements of monuments and sites, antiquities more than a hundred years old, property of artistic interest. In 1972, the third convention concerning world heritage articulated cultural heritage in three categories (art. 1):

> *Monuments*: architectural works, works of monumental sculpture and painting, elements or structures of archaeological nature, inscriptions, cave dwellings and combinations of features, which are of outstanding universal value from the point of view of history, art or science;
> *Groups of buildings*: groups of separate or connected buildings which, because of their architecture, their homogeneity or their place in the landscape, are of outstanding universal value from the point of view of history, art or science;
> *Sites*: works of man or the combined works of nature and of man, and areas including archaeological sites which are of outstanding universal value from the historical, aesthetic, ethnological or anthropological points of view.

This definition was also introduced almost in the same form to the 'Recommendation concerning Protection at National Level of the Cultural and Natural Heritage', which was adopted at the same General Assembly in Nairobi, but with the difference that instead of 'outstanding universal value' the question was here about 'special value'. The articulation into monuments, groups of buildings and sites, proposed in this definition, has since been used in several other international documents.

The ten international Recommendations of UNESCO focus each on a particular issue, some on specific types of heritage, including archaeological sites, landscapes, cultural and natural heritage, historic areas, movable property, and moving images. Others discuss specific problems: rendering museums accessible, prohibiting and preventing the illicit export, import and transfer of ownership of cultural property, preservation of cultural property endangered by public works, and international exchange of cultural property. In terms of the built heritage, of particular interest is the 'Recommendation concerning the Safeguarding and Contemporary Role of Historic Areas', adopted in Nairobi in 1976; here the question is about 'historic and architectural (including vernacular) areas', as well as about 'environment'. In the general principles the document recommends that: 'historic areas and their surroundings should be regarded as forming an irreplaceable universal heritage' (para. 2); successively, fundamental importance has been given to the consideration of the needs of the existing population and of the variety of traditional types of activities in any planning scheme (para. 3):

> Every historic area and its surroundings should be considered in their totality as a coherent whole whose balance and specific nature depend on the fusion of the parts of which it is composed and which include human activities as much as the building, the spatial organization and the surroundings. All valid elements, including human activities, however modest, thus have a significance in relation to the whole which must not be disregarded.

BUILT HERITAGE

Historic urban areas have been the subject of several international or regional documents, including the 'Amsterdam Declaration', the 'European Charter of the Architectural Heritage' by the Council of Europe, both of 1975, as well as the 'Convention for the Protection of the Architectural Heritage of Europe' of 1985. The two first documents emphasized the importance of 'the groups of lesser buildings in our old towns and characteristic villages in their natural or manmade settings', the Convention spoke about monuments, groups of buildings and sites, of conspicuous historical, archaeological, artistic, scientific, social or technical interest. The ICOMOS 'Charter for the Conservation of Historic Towns and Urban Areas', of 1987, was concerned about historic urban areas, large and small, including cities, towns and historic centres or quarters, together with their natural and man-made environment. The Quebec Association for the Interpretation of the National Heritage has defined (in 1980) cultural heritage as: 'The combined creations and products of nature and of man, in their entirety, that make up the environment in which we live in space and time.' This definition was taken into the Deschambault Declaration (1982).

Several international seminars have dealt with problems related to urban conservation and management of traffic, and have proposed recommendations in this regard; these include: Oxford and Graz in 1969, Brno in 1971, and Castillo de Chapultepec in Mexico in 1972. The earliest was Las Normas de Quito, resulting from a regional conference in Peru in 1967, and discussing particularly the issue of cultural tourism and heritage in relation to social and economic development. The first Brazilian Seminar about the preservation and revitalization of historic centres, organized by the Brazilian committee of ICOMOS in Itaipava in 1987, proposed basic principles that formed the Carta de Petropolis; there an emphasis was laid on the particular conditions in large urban metropolises in rapidly developing countries. The question was here less that of restoring specific historic structures, and rather more that of managing the area 'in its totality' as an 'historical entity that is permanently undergoing a dynamic process of successive transformations'. The first two paragraphs of the Carta give the following definition of historic urban areas:

> Urban historical sites may be considered as those spaces where manifold evidences of the city's cultural production concentrate. They are to be circumscribed rather in terms of their operational value as 'critical areas' than in opposition to the city's non-historical places, since the city in its totality is a historical entity.
>
> Urban historical sites are part of a wider totality, comprising the natural and the built environment and the everyday living experience of their dwellers as well. Within this wider space, enriched with values of remote or recent origin and permanently undergoing a dynamic process of successive transformations, new urban spaces may be considered as environmental evidences in their formative stages.

Within this broad scheme, several UNESCO recommendations and ICOMOS charters focus on particular types of heritage. These include questions related to the protection of cultural property in relation to archaeological excavations, first defined by UNESCO in the Recommendation of 1956. Forty-four years later, in 1990, ICOMOS adopted the International Charter for Archaeological Heritage Management, where the archaeological heritage was defined as 'that part of the material heritage in respect of which archaeological methods provide primary information' (art. 1). While conservation of historic structures has perhaps not received sufficient international attention, some of the first initiatives by Bernard Feilden, once nominated Director of ICCROM, regarded historic structures; an international seminar of engineers was organized in Rome in 1978 with recommendations, and action was taken concerning earthquakes.[17]

Another theme has been the protection of the beauty and character of landscapes and sites which is dealt with in the UNESCO Recommendation of 1962. This was intended to supplement measures for the protection of nature, and spoke about 'the aspects of natural, rural and urban landscapes and sites, whether natural or man-made, which have a cultural or aesthetic interest or form typical natural surroundings' (para. 1). More recently, the ICOMOS 'Florence Charter', of 1982, has defined an historic garden as a 'monument', but has emphasized that 'the historic garden is an architectural composition whose constituents are primarily horticultural and therefore alive, which means that they are perishable and renewable' (art. 2). An historic garden is thus not only a 'monument', but a 'live monument' (art. 3), and treatments should therefore take into account the 'perpetual equilibrium between the cyclic movement of seasons and of the development and decay of nature and the will of the artist and artificer seeking to keep it permanently unchanged' (art. 2).

INTANGIBLE CULTURAL HERITAGE

It will be of interest, at this point, that while many cultures give importance to physical heritage, and, for example, China speaks of 'cultural relics', there are countries where conservation of cultural heritage is not limited to physical properties. The Japanese, 'Law for the Protection of Cultural Properties', of 1984, identifies several categories: tangible cultural properties, intangible cultural properties, folk-cultural properties, buried cultural properties, historic sites, places of scenic beauty and/or natural monuments. Intangible properties refer to 'the artistry and skills employed in forms of theatre, music, the applied arts, and other tangible cultural products which possess a high historical or artistic value'; the government designates notable artistry in performing arts (for example, theatre, music, dance) or mastery of skills in applied arts (for example, ceramics, textile, lacquerwork, metal, wood, doll making, paper). Such recognition can be: 1) individual recognition, that is, to a person, 2) collective, that is, to constituent members of a group, or 3) group recognition. In 1994, the Japanese State List of protected intangible properties included 76 individuals representing 52 different skills, and 23 groups representing as many skills. In the whole region of Oceania, particular emphasis has been given to indigenous cultures, and especially conservation of intangible heritage. The New Zealand, 'Charter for the Conservation of Places of Cultural Heritage Value', of 1993, states that indigenous cultural heritage is 'inseparable from identity and well-being and has particular cultural meanings'. The question is thus not so much about conserving physical objects, but rather about a way of life and about conserving the memory of traditional knowledge.

POLICY FOR PROTECTION

Interest in cultural heritage and its protection has been often awakened by its destruction; this happened in Antiquity, in the Middle Ages, during the Italian Renaissance and the French Revolution, and it also resulted from developments caused by the Industrial Revolution in the nineteenth century. The World Wars gave a motive for the establishment of international organizations in order to guarantee a broader-based protection for heritage. Such organizations have acted through international conventions, recommendations and charters, and have contributed by calling attention to threatened classes of heritage, and proposing improved standards for their treatment. International documents, especially those by UNESCO, result from a consultation of experts and Member States, and, therefore, they reflect the needs identified by the States themselves. Each time a convention or recommendation is prepared it is adopted by a representational organ of the institution concerned.

The international contribution to the protection of cultural heritage can be seen in two forms; first, through international conventions that, being ratified by the State Party, obtain legal force in the country concerned; second, through international Recommendations or Charters that remain rather moral guidelines or an invitation to implement corresponding measures in national legislation and norms. At the practical level, such documents should be taken as guidelines when defining standards for conservation treatment. The most active international measures so far have been implemented through initiatives in the framework of the World Heritage Convention of UNESCO; this is reflected also in the number of countries that have ratified the convention (about 140 at the end of 1994); this is nearly double compared with other conventions. As indicated in the Statutes of UNESCO, international organizations have limited possibility to intervene into national jurisdiction; action at this level necessarily depends on the level of care taken by each government, as well as on the political determination to react. Much depends on the cultural awareness of the population, and their possibility to influence the decision-making process. Non-governmental organizations, such as ICOMOS, Jeunesse & Patrimoine International, and Europa Nostra, can have an important role to play in this regard.

The effectiveness of international impact has been tested particularly during armed conflicts and in the case of natural disasters, but also in rapidly developing and changing societies. So far, generally speaking, culture and cultural heritage have been less successful in establishing infrastructures that could compare with the Red Cross or the World Wildlife Fund. Such problems were recognized in both World Wars and have been reiterated in the armed conflicts of the 1980s and 1990s, as well as in the numerous natural disasters, earthquakes, floods and fires that have caused extensive damage of cultural heritage in the most densely built regions of the world. Effective solutions still need to be found although various initiatives have been taken; these include the development of an improved system of communication, the creation of a network of 'Blue Shields' for the safeguard of cultural heritage at the local level, and the establishment of an international Heritage at Risk Fund as proposed by ICOMOS. Such undertakings should be carried out as a collaborative effort.

Concern has also been caused by priorities imposed through commercial interests or development; cultural heritage is rarely given due consideration. At the present, though, the trend is to include cultural heritage in the concept of sustainable development, and to promote human sustainable development. In Canada, in fact, urban conservation has been defined as 'management of change'; this was related to the Main Street Project in the 1980s that extended to the USA, and involved a large number of merchants and property owners in city centres to improve the streetscape while keeping and restoring historic features.

RECOMMENDED ACTIONS

The actions recommended in UNESCO documents mostly aim at the strengthening of conservation policies at the national level, provision and implementation of appropriate legal and administrative instruments, and development of conservation management structures. The 'Recommendation for the protection of the cultural and natural property at the national level', recommends the following types of action:

> Organization of services, including: inventory and documentation, training of specialized personnel, promotion of research, creation of laboratories, and assisting tenants and owners in carrying out necessary upkeep; multi-disciplinary collaboration amongst the various bodies involved; due balance in devising and execution of protective measures at the different administrative levels;

> Protective measures, such as: constant maintenance, thorough study of the problems, investigation of effective measures for treatments; protection of heritage against possible harmful effects of technological developments; identification of social needs of the community as part of the conservation process in historic areas; keeping abreast of advances in transportation, automatic data-processing, etc.;
> Administrative, legal and financial measures, as well as educational and cultural action, and promotion of international collaboration.

Many of these categories of actions are repeated in other recommendations with the necessary adjustment to the type of heritage concerned each time. The promotion and active implementation of such actions, in collaboration with Member States, is the statutory task of ICCROM, which has systematically based its programme activities on the guidelines proposed in international policy documents.

STANDARDS AND PRINCIPLES FOR CONSERVATION TREATMENTS

The aim of the international recommendations and charters is to provide a measuring stick for the definition of the quality standards required for treatment. Such standards are the result of the maturation of concepts and principles at the local and national level, and they reflect the results of the debate on conservation versus restoration over the past two centuries. Ever since the Athens meeting, international documents have been fairly unanimous about the priority to be given to conservation and conservative restoration. Reconstruction has been recommended to be strictly limited, or, at the most, to be carried out according to the principles of 'anastylosis'. More recently, the question of cultural diversity emphasizes that restoration should be based on a critical process, that the value judgements need to be referred to the cultural context concerned, and that the results, therefore, will vary from case to case. The standard of conservation treatments should be at the highest level, but measured against the quality and values identified with the heritage resource concerned.

Conservation and restoration

Looking at the specific treatments, there are concepts such as conservation, preservation, restoration, maintenance and repair, that are generally defined in the various documents following the lines of the Venice Charter, but taking into consideration the particular type of heritage or the cultural area concerned. While there are some differences in national terminology and between languages, in the international context, the word 'conservation' has been mostly used as the general covering concept. The declaration of Athens considered 'conservation' a general approach, and recommended the limitation of 'restoration', which was then practically interpreted as reconstruction (Section I):

> Whatever may be the variety of concrete cases, each of which are open to a different solution, the Conference noted that there predominates in the different

> countries represented a general tendency to abandon restorations in toto and to avoid the attendant dangers by initiating a system of regular and permanent maintenance calculated to ensure the preservation of the buildings.

The Venice Charter, while discussing 'conservation' through five Articles (4–8), does not really provide a specific definition. Rather, it defines the framework implying maintenance on a permanent basis and proposing an appropriate and socially useful purpose for the historic building. In addition, attention is given to keeping the relationship of the traditional setting without radical modifications. A monument should be kept *in situ*, and its integral elements should not be removed except when this is the only means to ensure their preservation. Considering that the Venice Charter was conceived and written in French, and that the English version is really a translation, terminological differences should be understood against this background. The concept 'restoration' was based on the developments in philosophy defining conservative restoration in the lines of Cesare Brandi, the first Director of the Italian State Restoration Institute (one of the first in the world). Accordingly, restoration was defined in Article 9 of the Charter:

> The process of restoration is a highly specialised operation. Its aim is to preserve and reveal the aesthetic and historic value of the monument and is based on respect for original material and authentic documents. It must stop at the point where conjecture begins, and in this case, moreover, any extra work which is indispensable must be distinct from the architectural composition and must bear a contemporary stamp. The restoration in any case must be preceded and followed by an archaeological and historical study of the monument.

The basis for this definition goes back to the development of the twentieth-century philosophy of Heidegger and Dewey, and especially of Brandi. The philosophical works of Brandi on aesthetics had been published in the 1940s and 1950s; his *Teoria del restauro* dated from 1963 and was well known to the authors of the Charter. Brandi distinguished between the restoration of a work of art, based on its recognition as a work of art each time restoration was undertaken, and the 'restoration' or repair of an industrial product that mainly aimed at re-establishing its functionality (Brandi 1945, 1963). The concept of 'restoration' has remained in the Italian language as the main term when dealing with works of art or important historic architecture, and it has been defined as a conservative treatment based on a critical process. In English, instead, due to debates in the nineteenth century and especially the attacks by John Ruskin, the concept had been replaced by the word 'conservation'.

The Burra Charter (1981), written in Australia, was based on the principles of the Venice Charter. Here 'conservation' is defined specifically as the covering concept for a combination of one or more different treatments. Conservation is understood as a multidisciplinary activity based on firm scientific basis supported by a body of experience. Appropriate conservation policy should be determined by understanding the cultural significance of the place. The concept of 'place' is used to replace 'monument' and 'site'. The aim of conservation is given in two articles:

> Article 2: The aim of conservation is to retain or recover the cultural significance of a place and must include provision for its security, its maintenance and its future.
> Article 3: Conservation is based on a respect for the existing fabric and should involve the least possible physical intervention. It should not distort the evidence provided by the fabric.

There are many similarities between the concept of 'restoration' in the Venice Charter and the concept of 'conservation' in the Burra Charter. Instead, the concept 'restoration' in the Burra Charter is defined as 'returning the existing fabric of a place to a known earlier state by removing accretions or by reassembling existing components without the introduction of new material' (art. 1). Although respecting the principles of the Venice Charter, this is more limited. Generally speaking, both charters agree that modern techniques can be used in conservation and consolidation, especially where traditional techniques prove inadequate, but with due regard to testing and proving their efficacy and durability. The essence in the treatment of historic structures is to measure the compatibility of modern technology with traditional workmanship and structural system, in order to aim at a long enough time scale, to guarantee that the treatment can be repeated in the future, and that it could even be removed if it does not prove compatible in the long run (the principle of reversibility).

Anastylosis and reconstruction

The question of 'anastylosis' was introduced in the Athens meeting where this was accepted as a possible treatment in the case of ruins (art. vi):

> In the case of ruins, scrupulous conservation is necessary, and steps should be taken to reinstate any original fragments that may be recovered (anastylosis), whenever this is possible; the new materials used for this purpose should in all cases be recognizable.

While originally meaning 'restoration' in Greek, the word has thus taken a particular meaning in the international terminology, and should be understood as a specific type of restoration. The Venice Charter defines the concept as follows:

> All reconstruction work should however be ruled out *a priori*. Only anastylosis, that is to say, the re-assembling of existing but dismembered parts can be permitted. The material used for integration should always be recognisable and its use should be the least that will ensure the conservation of a monument and the reinstatement of its form.

The Burra Charter speaks about 'reconstruction', stating that this should be undertaken only where necessary for reasons of survival, and that it should not constitute the majority of the fabric of the place; it further states (art. 19):

> Reconstruction is limited to the reproduction of fabric, the form of which is known from physical and/or documentary evidence. It should be identifiable on close inspection as being new work.

When considering such definitions, it is essential to understand them in the context of the charters or guidelines that are referred to, and not as a separate recipe. This means always to keep in mind the final objective which is to maintain the cultural heritage 'in the full richness of its authenticity'. Decisions are often made under pressure from clients, politicians, tourism operators, and also the general public, who need to be correctly informed about the issues in order to be able to participate fully in the conservation process.

Maintenance

One of the principal means to guarantee conservation is to undertake regular inspections and maintenance. ('Regular inspections and maintenance, please!' as Sir Bernard would say.) Such treatment is necessary in reference to all properties, and it is also generally recommended in all relevant international guidelines. It is of particular importance with historic gardens; the Florence Charter (art. 11) states:

> Maintenance in the case of historic gardens is an operation of paramount importance which must necessarily be continuous. Since the principal material consists of the plants, the preservation of the garden in an unchanged condition will involve both individual replacements whenever required and a long-term programme of periodic renewal (complete felling followed by replanting with sufficiently mature specimens).

In all cases, maintenance is fundamental, and this has been duly recognized in all relevant international charters and guidelines. In this regard, a useful reference will be the management guidelines by Feilden and Jokilehto (1993), which has been translated into several languages.

Management of historic urban areas

The series of conferences organized by the Council of Europe in relation to the 1975 Architectural Heritage Year produced the concept of 'integrated conservation' that is described in the 'European Charter of the Architectural Heritage'. This means that the planning process related to existing built areas should integrate conservation principles for an appropriate policy to avert the risks that this heritage is facing. The Charter states that integrated conservation is achieved by the application of sensitive restoration techniques and the correct choice of appropriate functions, and it depends on legal, administrative, financial and technical support. The process, therefore, should be based on the understanding of the existing and historical reality, the significance and the condition of the heritage resource, the specific requirements of the population, the availability of resources, and consequently an approach that will involve the population and the relevant authorities in a multidisciplinary decision-making process, in order to reach a balanced judgement.

Particularly in the American context, conservation and maintenance are often covered with the word 'preservation'. Whatever the term, the planning and management of historic urban and rural areas will require many types of treatments. The Carta de Petropolis defines the aim of such treatments as follows (para. iv):

> The main purpose of preservation is the maintenance and enhancement of reference patterns needed for the expression and consolidation of citizenship. It is through the outlook of the citizen's political appropriation of urban space that preservation may contribute to improve life quality.

This statement gives the essential scope of the work in relation to a context that needs to be understood as dynamic. The question is to define what is being conserved, planned and managed. In a broad perspective, it is certainly essential to see that 'the maintenance and enhancement of reference patterns' be given due attention, and that the traditional conservation and restoration of historic structures be seen as part of this broader perspective. It will mean development of a planning process with many points in common with the critical process required for the restoration of ancient monuments or even works of art.

CULTURAL DIVERSITY AND AUTHENTICITY

Concerning particularly sensitive fields, such as the indigenous heritage, recommendations have included specific measures to respect its integrity and authenticity. This is evident in the New Zealand Charter of 1993, which emphasizes several degrees of intervention, and includes also 'non-intervention'; attention in such areas is to be given to the fragility of the culture in the changing world, risk mitigation, and respect of the historical setting. These issues have been the topic of debates concerning authenticity, and were addressed in the Nara Document on Authenticity of 1994 (para. 4):

> In a world that is increasingly subject to the forces of globalization and homogenization, and in a world where the search for cultural identity is sometimes pursued through aggressive nationalism and the suppression of the cultures of minorities, the essential contribution made by the consideration of authenticity in conservation practice is to clarify and illuminate the collective memory of humanity.

While not attempting to define cultural heritage, the conference emphasized in the Nara Document that 'the diversity of cultures and heritage in our world is an irreplaceable source of spiritual and intellectual richness for all humankind' (para. 5), as well as that 'all cultures and societies are rooted in the particular forms and means of tangible and intangible expression which constitute their heritage, and these should be respected' (para. 7). The document further defines 'information sources', that are closely related to the concept of 'historic monument', as 'all monumental, written, oral and figurative sources which make it possible to know the nature, specificities, meaning and history of a property'.

Authenticity is one of the principal aims in the protection and conservation of cultural heritage; it has been referred to in several conservation charters since the Venice Charter, which recommends: 'It is our duty to hand them on in the full richness of their authenticity'. Authenticity in the conservation of cultural heritage can be defined as a measure of truthfulness of the internal unity of the creative process and the physical realization of the work, and the effects of its passage through historic time.

The scope of the Nara meeting was to provide a forum to discuss the concept and its definition in different cultural and heritage contexts, taking into consideration the needs of the World Heritage Convention. Accordingly, it was defined as a concept that is relative to the identification of appropriate values, as well as being 'judged within the cultural contexts to which they belong'. In practice, authenticity should be assessed according to clusters of references, such as 'form and design, materials and substance, use and function, traditions and techniques, location and setting, and spirit and feeling'. Such analysis will assist in defining the relevant artistic, historic, social and scientific dimensions of this heritage.

EDUCATION AND TRAINING

International recommendations have given particular attention to educational issues; these have been present since the beginning of the century, in the Madrid recommendation of 1904; they are given special importance in the Conclusions of the Athens meeting in 1931, and again in practically all international documents by UNESCO, ICOMOS, and other organizations, such as the Council of Europe. In 1993, the Tenth General Assembly of ICOMOS in Colombo adopted the 'Guidelines on Education and Training in the Conservation of Monuments, Ensembles and Sites' prepared by its International Training Committee. The document was first drafted by Sir Bernard Feilden. In this context, conservation is emphasized as: 'a cultural, artistic, technical and craft activity based on humanistic and scientific studies and systematic research'. It is a multidisciplinary activity requiring the specific skills of many professionals, 'with proper education and training, sound judgement and a sense of proportion with an understanding of the community's needs'.

Considering that conservation is fundamentally a cultural issue, participation of the population in the decision-making process and proper acceptance of responsibility by relevant authorities are essential requirements for successful conservation. Training has become the most important activity for ICCROM, and it has worked, as has been recommended in the Training Guidelines, in order 'to provide a reference for the institutions and bodies involved in developing training programmes, and to assist in defining and building up appropriate standards and criteria suitable to meet the specific cultural and technical requirements in each community or region'. International training strategy needs to be built up as a system of communication between national requirements and regional and international resources.

TRENDS AND CONCLUSIONS

The trend may be said to be about guaranteeing the management of heritage resources in view of sustainable human development; cultural properties are a 'non-renewable' resource, unique, and irreplaceable. This does not mean, however, that history would have stopped; rather it means that our authentic heritage is also needed to further develop our creative capacities. Only through learning and maturing in understanding qualities and values, will society be able to maintain its own cultural authenticity, and to provide its own creative contribution to heritage and the quality of life. Conservation is really not 'exportable' from one culture to another; rather, each culture requires the process for an historical consciousness and the definition of its values as a basis for heritage conservation.

International standards, charters and guidelines are developed with reference to and a critical comparison of field experiences, and proposing appropriate recommendations on this basis. Over the thirty years since the Venice Charter was written, its validity has been questioned and tested numerous times. While limited in its scope, and perhaps never intended to survive such a long period without being updated, it has however been successful in remaining a useful reference, and has been complemented literally by dozens of documents, such as the Normas de Quito and Burra Charter, which have helped to bring its message within the reach of practitioners of specific cultural regions. In 1977, also, Bernard Feilden proposed a revised edition of the charter. In the preface, he stated that international charters are necessary as a guide to legislation and individual action, and continued:[18]

> These proposed revisions are suggested against the British background of Conservation which is broadly based upon comprehensive town planning legislation and active public participation, particularly through the popular Amenity Movement which is linked to the general Conservation Movement with its aims of avoiding waste and enriching the quality of life.

The usefulness of international standards and charters is that these provide a broader framework to site managers and local decision makers. This can become an important support in debates about correct procedures, as well as being a reference to drafting guidelines or legal documents for national or local use. The limitation of international documents is in their need to apply to many different contexts; this can be best overcome if emphasis is placed on them as the indication of a critical process, and not as a model or prescription. The Venice Charter was written during the period of extensive restorations and reconstructions after the damage of the Second World War. The specific problems of that time were then emphasized, and less attention given to other concerns that may be relevant today. The Charter, however, continues to be valid as a proclamation of some of the basic principles.

The Nara Document on Authenticity recognizes that evaluation of significant cultural heritage sites should be made with full reference to the cultural diversity of the relevant area. The document is clearly referring to the Venice Charter and to the World Heritage Convention, and its aim is to make the caretakers and managers of heritage resources assume their responsibility. Such action is possible only through a coherent training process, where 'conservationists' 'know, understand and apply UNESCO conventions and recommendations, and ICOMOS and other recognized charters, regulations and guidelines', and where they are capable of making 'balanced judgements based on shared ethical principles, and accept responsibility for the long-term welfare of cultural heritage', as is expressed in the Training Guidelines of 1993 that reflect the policies sustained by Sir Bernard.

BIBLIOGRAPHY

The three conventions and ten recommendations of UNESCO are published in *Conventions and Recommendations of UNESCO* (1995). The principal ICOMOS charters are available from ICOMOS National Committees, and have been published in relevant Newsletters. A useful reference is the publication by the Canadian ICOMOS Committee for the event of the Quebec meeting on World Heritage Cities in 1992. The documents by the Council of Europe are available from the offices in Strasbourg, but some are also reprinted in the Quebec publication. The numerous charters and recommendations that have been prepared by groups of experts in various international or regional meetings are more difficult to get hold of. The best course is to contact either the ICOMOS Documentation Centre in Paris or the ICCROM Library in Rome; the Appendix to this paper contains comments on a selection of the documents.

Bekri, C. (1991). *L'UNESCO: 'Une entreprise erronée?'.* Publisud, Paris.
Brandi, C. (1945). *Carmine o della pittura.* 'Elicona I' (1945), Editori Riuniti, Roma (1992).
Brandi, C. (1956). *Arcadio o della scultura, Eliante o dell'architettura.* 'Elicona III–IV' (1958), Editori Riuniti, Roma (1992).
Brandi, C. (1963). *Teoria del restauro.* Rome.
Diaz-Berrio, F. S. (1976). *Conservacion de monumentos y zonas monumentales.* Sep/Setentas, Mexico.
Diaz-Berrio, F. S. (1990). *Conservacion del patrimonio cultural en Mexico.* Coleccion Textos Basicos y Manuales, Instituto Nacional de Antropologia y Historia: Mexico.
Feilden, B. M. (1987). *Between Two Earthquakes, Cultural Property in Seismic Zones.* ICCROM & Getty Conservation Institute, USA.
Feilden, B. M., and Jokilehto, J. (1993). *Management Guidelines for World Cultural Heritage Sites.* ICCROM, Rome.
ICOMOS Canada (1990). *Preserving our heritage: catalogue of charters and other guides.* Prepared for the International Symposium on World Heritage Towns, Quebec.

Jokilehto, J. (1986). 'History of Architectural Conservation', DPhil Thesis, University of York.

UNESCO (1985). *Conventions and Recommendations of UNESCO concerning the protection of the cultural heritage.* Paris.

UNESCO (1994). *Operational Guidelines.* World Heritage Committee, UNESCO.

World Commission on Environment and Development (1987). *Our Common Future.* Oxford University Press.

NOTES

1 Convention concerning the Protection of the World Cultural and Natural Heritage, adopted by the General Conference at its 17th session, Paris, 16 November 1972.

2 The World Commission on Environment and Development, headed by Gro Harlem Brundtland, Prime Minister of Norway, was set up in 1983 by the United Nations.

3 The Seven Wonders of the ancient World included: the Pyramids of Egypt (3rd mill. BC), the Hanging Gardens of Babylon (*c.* 600 BC), the Statue of Zeus at Olympia (*c.* 430 BC), the Temple of Artemis at Ephesus (550/356 BC), the Mausoleum of Halicarnassus (4th cent. BC), Colossus of Rhodes (*c.* 305 BC), Pharos of Alexandria (*c.* 280 BC).

4 The Swiss jurist, Emmerich de Vattel, advanced the idea of 'universal heritage' and the legal consequences in warfare in his *Le Droit des gens* ('The Law of Nations') in 1758.

5 *Mouseion*, vol. 21–22, 1–2, 1933, 274.

6 *Mouseion*, vol. 13–14, I–1, 1931, 162ff.; vol. 27–28, 3–4, 1934, 303ff.

7 *Mouseion Supplement*, No. 102, I, 46; Bekri 1991.

8 *Mouseion Supplement*, No. 102, I, 46.

9 *Museum*, (1951), IV/1, 2.

10 *Museum*, (1950), III/1, 92f.

11 At the beginning, the organization was called in short the 'Rome Centre', having been created by UNESCO; the second Director, Prof. Paul Philippot (1971–7), changed it into 'International Centre for Conservation' in order to better reflect the field; the third Director, Sir Bernard Feilden (1977–81), took the present short name 'ICCROM' that was approved by the Council in 1979. The letters ICC reflected the previous short name, I(nternational) C(entre) for C(onservation); ROM indicated Rome as the seat of the organization. Sir Bernard also designed the first flag for ICCROM.

12 ICCROM Statutes, Art. 1.

13 The first congress took place in Paris in 1957; there were recommendations but these were never formulated into an international document or charter.

14 The ICOMOS Statutes were adopted by the fifth General Assembly at Moscow in 1978.

15 *Mouseion*, vol. 25–26, 1–2, 1934, 179f.

16 UNESCO, Draft Medium Term Plan 1990–5 (UNESCO, (1989) 25 C/4, 57).

17 The recommendations are published in Feilden (1987).

18 The revised charter was dated 17 March 1977, and was prepared as a British contribution to the ICOMOS Moscow General Assembly the following year.

APPENDIX 1

COMMENTS ON A SELECTION OF CHARTERS AND RECOMMENDATIONS CONCERNING THE BUILT HERITAGE

The following documents have been selected from the numerous recommendations, charters, conclusions or declarations, that have been produced by international, regional or national seminars, conferences or workshops on conservation of the cultural heritage. Most of such documents have been written during the three decades following the Venice Charter of 1964. However, references to some earlier documents have been added in view of their importance in the development of definition of cultural heritage and conservation policies, that is, those by William Morris, Camillo Boito and Gustavo Giovannoni, and the conclusions of two international meetings in Athens in 1931 and 1933.

1877 Manifesto of SPAB, England

The Manifesto of the Society for the Protection of Ancient Buildings (SPAB) was written by William Morris in April 1877, strongly influenced by the conservation movement headed by John Ruskin. The document notes that a new interest had developed in ancient monuments over the preceding fifty years, but expresses concern for the continuing destruction that these suffer due to neglect and changes 'under the name of Restoration'. The Manifesto recognizes the significance of a broad range of historic structures, 'anything which can be looked on as artistic, picturesque, historical, antique, or substantial: any work in short, over which educated, artistic people would think it worthwhile to argue at all'. For such buildings, it recommends maintenance and minimum intervention, and proposes that one should 'stave off decay by daily care, to prop a perilous wall or mend a leaky roof by such means as are obviously meant for support or covering'. The Manifesto has become well-known through the activities of SPAB, but it has also exercised a considerable international influence.

1883 Resolution, Italy

The Resolution approved by the third National Conference of Architects and Engineers, in Rome in 1883, was written by Camillo Boito, who was well aware of policies in France and England. The document has seven paragraphs, and emphasizes that where 'architectural monuments' effectively need some work, they should be 'consolidated rather than repaired, repaired rather than restored, taking great pains to avoid any additions or renovations'. Modern work and new materials should be kept to the minimum and should differ from the historic monument but without being in conflict with its artistic appearance. The contributions of all historic periods should be respected; exception can be made and parts removed if these are manifestly of minor importance compared to forms that they cover. All new work on historic monuments should be photographed, and a fully documented report with complete measured drawings, in colour when necessary, should be deposited with the authorities. The Resolution was circulated to all government offices responsible for ancient buildings, and has been considered the first modern declaration of Italian conservation policy.

1931 'Athens Charter', International Museums Office

The conclusions of the international meeting organized by the International Museums

Office (precursor of ICOM) in Athens, in 1931, can be considered the first international recommendation to concern itself specifically with the safeguarding of architectural heritage. The document was brought to the attention of the League of Nations. It contains seven major headings dealing with general principles, and administrative and legal measures, and offering recommendations concerning aesthetic enhancement of monuments, restoration materials and techniques, education and international collaboration. It recommends that restorations be abandoned in favour of a system of regular and permanent maintenance with respect to the historic and artistic work of all periods. Surroundings of historic monuments should be protected and commercial publicity avoided. Judicious use of modern work is considered acceptable, but should be hidden in order not to distract from the artistic appearance of the monument. In the case of ruins, conservation is given priority, but steps could also be taken to reinstate any original fragments that may be recovered ('anastylosis'). The new materials used for this purpose should in all cases be recognisable. Education of the public and international collaboration are considered important.

1932 Norms for the restoration of monuments, Italy

The Italian Norms for the restoration of monuments were written after the Athens meeting by Gustavo Giovannoni, the principal conservation architect of the Superior Council for Antiquities and Fine Arts. The Norms were published by the Government in January 1932. The Norms solicit correct conservation and rehabilitation of ancient monuments, defined as artistic and historic documents that permit scientific research. Restoration should take into account 'the historic reasons whereby none of the phases which comprise the monument should be eliminated or falsified by additions which might mislead scholars, nor should the material brought to light through analytical research be lost'. The twelve paragraphs that follow emphasize the need of regular maintenance, appropriate use and rehabilitation. Historic buildings should be conserved *in situ*; any completion should be avoided especially in ancient monuments with the exception of anastylosis; alterations should be kept to the minimum, simple in form and marked; restorations should be carefully documented. Modern techniques can be valuable should traditional methods prove insufficient. In the surroundings consideration should be given not to cause isolation of single monuments or to overwhelm them with the mass of new constructions.

1933 'Athens Charter', CIAM

The recommendations of the International Conference on Modern Architecture (CIAM) organized in Athens in 1933 are often referred to as the 'Athens Charter'. The original document remained unobserved until 1941 when Le Corbusier published a book with the name *La Charte d'Athènes*, bringing the message to the attention of a broader public with his comments on the document. The charter is mainly concerned with modern city planning, but it also takes into account the question of the heritage of historic cities. The principle is to respect the architectural values of such heritage so far as this responds to the general interest and does not compromise the living conditions of the population. It recommends the removal of through traffic from residential areas, the formation of green spaces where destruction has occurred, and also the avoidance of pastiche.

1954 'The Hague Convention', UNESCO

The Convention for the Protection of Cultural Property in the Event of Armed Conflict (The Hague Convention) was first conceived at the end of the nineteenth century, and having been through several editions was re-edited and adopted by UNESCO in 1954 as its first International

Convention. The significance of the Convention is to have provided an international definition for cultural property, articulated in three categories: (1) movable or immovable property of great importance to the cultural heritage of every people, such as monuments of architecture, art or history, whether religious or secular; (2) buildings whose main and effective purpose is to preserve or exhibit movable cultural property; and (3) centres containing a large amount of cultural property. The Convention provides instructions for the marking and safeguarding of such property in the case of armed conflict, and it also proposes the possibility of establishing an international list of such properties.

1956 Recommendation on Archaeological Excavations, UNESCO

The first international recommendation, adopted by the General Conference of UNESCO in New Delhi in December 1956, concerns International Principles Applicable to Archaeological Excavations. The recommendation recalls the general interest of international cooperation concerning the study and preservation of all archaeological remains, and proposes that the criterion based on preserving all objects originating before a certain date should be abandoned, and should in any case use criteria allowing the necessary flexibility in the protection taking into account public interest and the scope of researchers. Furthermore, recommendations are given on the general principles concerning the legal and administrative system of protection, control of excavations, preservation and assignment and rights of finds, trade of antiquities, clandestine excavations and excavations in occupied territories.

1962 Recommendation on Landscapes and Sites, UNESCO

The Recommendation concerning the Safeguarding of the Beauty and Character of Landscapes and Sites was adopted by the General Conference of UNESCO in Paris in December 1962. It emphasizes the scientific and aesthetic importance of cultural and natural landscapes and sites and the fact that they form a heritage which is a major factor in the living conditions of the general public. Recommended protective measures can include urban and rural planning schemes, scheduling of isolated sites, and protection of extensive landscapes by zoning. Special attention is given to educational action addressing the public both through schools and public media.

1964 'Venice Charter'

The Second International Congress of Architects and Technicians of Historic Monuments met in Venice in May 1964 and approved an International Charter for the Conservation and Restoration of Monuments and Sites that has since been known as the 'Venice Charter'. The document was signed by twenty-three persons representing sixteen countries, UNESCO and ICCROM. The aim of the document is formulated in the preface as the common responsibility to safeguard cultural heritage for future generations in 'the full richness of their authenticity', but the Charter also emphasizes that each country is responsible for applying the plan within the framework of its own culture and traditions. The sixteen paragraphs of the Charter define some of the fundamental principles of conservation, by referring especially to the conclusions of the 1931 meeting in Athens, but in practice also to the Italian Norms by Giovannoni of 1932, and the current development of restoration policies and theory. The Venice Charter has been internationally recognized as a fundamental reference for conservation policy. The meeting recommended the foundation of ICOMOS; this was achieved in Warsaw the following year.

1968 Recommendation on Public Works, UNESCO

The Recommendation concerning the Preservation of Cultural Property

Endangered by Public or Private Works was adopted by the General Conference of UNESCO in November 1968. It resulted from a concern aroused over the danger for ancient temples caused by the construction of the High Dam in the valley of the Nile. The recommendation recognizes that it is essential to harmonise the preservation of cultural properties with the changes necessitated by social and economic development. For this purpose, there is a need to identify important heritage sites in advance to make planning and implementation of protection measures feasible. The general principle is to safeguard *in situ*, but when overriding economical and social conditions require that cultural property be transferred to another site, abandoned or destroyed, a careful study should be made with detailed records for future research.

1972 World Heritage Convention, UNESCO

As a result of the growing ecological concern for the environment, UNESCO adopted at its General Conference in Paris in 1972, an international Convention concerning the protection of the world cultural and natural heritage, the World Heritage Convention becoming the most widespread convention of UNESCO ratified by 142 States in 1995. The Convention concerns principally immovable property classified in three categories: monuments, groups of buildings (ensembles), and sites. The Convention establishes the World Heritage List of cultural and natural heritage sites of 'outstanding universal value'. Where the significance of a World Heritage Site is undermined or threatened, the site can be included on the List of World Heritage in Danger. The Venice Charter is the principal reference in terms of policy, and has promoted a debate on the application of conservation principles in different cultural regions. The Convention has become perhaps the most powerful international tool for the promotion of conservation of both cultural and natural heritage.

1972 Recommendation on Cultural and Natural Heritage, UNESCO

The Recommendation concerning the protection, at national level, of the cultural and natural heritage was adopted by the same General Conference of UNESCO as the World Heritage Convention, in 1972. The recommendation is conceived as complementary to the Convention, and intends to induce States to safeguard all components of their cultural and natural heritage, considering that it should 'no longer be regarded as a check on national development but as a determining factor in such development', and that such heritage be considered as one of the essential aspects of regional development plans and planning in general. Attention is given to: the organization of services and cooperation among the various bodies; definition of appropriate protective measures, whether scientific, technical, administrative, legal or financial; educational and cultural action; and international cooperation.

1975 Declaration of Amsterdam

'The Declaration of Amsterdam' was the concluding document of the European Architectural Heritage Year 1975, and was adopted by 25 European States at the final conference in Amsterdam. The recommendations were addressed to governments as well as local and regional authorities bringing to their attention the richness of Europe's urban and rural heritage, and the requirement for strengthening legal and administrative measures for its protection. In particular the Declaration emphasizes the importance of integrating conservation requirements into all aspects of the urban and regional planning process, and engaging local authorities and citizens ('integrated conservation'). Its recommendations require the adoption of appropriate legal and administrative measures, identification of financial resources, as well as the development and promotion of methods and techniques by competent professionals.

1975 European Charter, Council of Europe

As a result of the European Architectural Heritage Year, the Council of Europe prepared a 'European Charter of the Architectural Heritage', adopted in October 1975. Referring to previous European agreements and The Hague Convention, (1954), the Charter brings to the attention of the Ministers of the Member States ten considerations that broaden the concept of architectural heritage to include 'groups of lesser buildings in our old towns and characteristic villages in their natural or man-made settings'. These are defined as being of irreplaceable spiritual, cultural, social and economic value, and the Charter declares that this heritage is in danger, and that integrated conservation, education and cooperation of all parties can avert these dangers, and conserve this heritage as a common property of the continent.

1976 Recommendation on Historic Areas, UNESCO

The 'Recommendation concerning the Safeguarding and Contemporary Role of Historic Areas' was adopted by the General Conference of UNESCO in Nairobi in November 1976. The document results from the growing concern of dangers caused by uniformity and depersonalized housing in modern town planning and the impact on old town centres or districts and on traditional villages. Particular attention is given to the definition of historic areas as a coherent whole with human activities, buildings and surroundings. Recommendations are presented on the establishment of national, regional and local policies regarding safeguarding with special attention on technical, economic and social measures. The roles of research, education at all levels and of information management are emphasized in the process with reference to strengthened collaboration with ICCROM, ICOMOS and ICOM.

1976 Cultural Tourism, ICOMOS

'The Charter of Cultural Tourism' was adopted by ICOMOS in November 1976 and signed by several international organizations related to travel, tourism, hotels, architecture and heritage. Tourism is considered an irreversible, human, economic and cultural fact, and cultural tourism particularly has direct influence on the sphere of monuments and sites. An appeal is launched to the states to adopt all appropriate measures to facilitate the information and training of travellers who visit cultural heritage sites in order to promote protection of this heritage which is 'the very basis of international tourism'. Equally important is the training of specialists to conceive and implement beneficial touristic use of heritage sites through correct planning.

1978 Standards and guidelines for rehabilitation, USA

The United States Secretary of the Interior's 'Standards for Rehabilitation' and 'Guidelines for Rehabilitating Historic Buildings' are two separate but related documents first published in 1978 and revised in 1983 and in 1995. The Standards give a list of ten principles to be attended to in the rehabilitation of historic buildings, following the principles of the Venice Charter. The Guidelines expand on the Standards, specifying with the help of practical examples what is recommended and what is not recommended.

1981 Burra Charter, ICOMOS Australia

'The Charter for the Conservation of Places of Cultural Significance' ('The Burra Charter') was adopted by ICOMOS Australia in Burra, in 1981. While based on the principles of the Venice Charter, the aim of the Burra Charter is to provide a guideline of twenty-nine articles adapted to the requirements of the particular type of heritage in the region concerned. The definitions, instead of speaking of monuments and sites, adopt the concept of 'place' meaning 'site, area, building or other work, group of buildings or other works together with pertinent contents and surroundings'. The Charter has been accepted as a valid reference, and has been well received in many countries, particularly outside Europe.

1982 Deschambault Declaration, ICOMOS Canada

'The Deschambault Declaration, Charter for the Preservation of Quebec's Heritage', was prepared by the French-speaking ICOMOS Committee of Canada and adopted in April 1982. This document, like the Burra Charter, was conceived as a regional adaptation of the principles of the Venice Charter to the requirements of Canada and Quebec, in particular. In the definition of heritage, emphasis is given to the combined works of man and nature that constitute the framework of human society in time and space. The ten articles of the document underline the responsibility of citizens and their participation in decisions concerning the protection of heritage, the importance of education and understanding of heritage and its values, the fact that heritage be given priority in society, and provided appropriate maintenance and care to keep such heritage alive and dynamic.

1982 Florence Charter, ICOMOS

'The Historic Gardens Florence Charter' was prepared on the initiative of the ICOMOS–IFLA, International Committee for Historic Gardens, and adopted by ICOMOS in December 1982 as an addendum to the Venice Charter covering the specific field concerned. The Charter outlines how protection and conservation principles related to maintenance, conservation, restoration and reconstruction of cultural heritage can be applied to a historic garden that should be understood as a 'live monument', consisting of material that is constantly changing in its aspect.

1983 Appleton Charter, ICOMOS Canada

'The Appleton Charter for the Protection and Enhancement of the Built Environment' was prepared by ICOMOS Canada under the auspices of the English-speaking Committee, and was published in August 1983. As another regional application of the principles of the Venice Charter, the Appleton Charter focuses on the definition of the framework of heritage conservation as related to levels and scales of intervention, the type of activity, and the relevance to the Canadian context.

1985 European Convention, Council of Europe

'The Convention for the Protection of the Architectural Heritage of Europe' was adopted by the Council of Europe in Granada in October 1985. The Convention establishes procedures for the identification of monuments, groups of buildings and sites that should be protected in European countries, and lists statutory protection measures and sanctions. These are followed by agreements concerning national conservation policies, participation of the public, information and training, as well as European coordination of conservation policies.

1987 Carta de Petropolis, ICOMOS Brazil

The First Brazilian Seminar about the Preservation and Revitalization of Historic Centres was organized by the ICOMOS Brazilian Committee in Itaipava in July 1987. The resulting recommendations, called 'Carta de Petropolis', list ten principles to be attended to in relation to the planning and management of living historic towns in countries that are undergoing rapid social and economic development.

1987 Charter on Historic Towns, ICOMOS

'The Charter for the Conservation of Historic Towns and Urban Areas' was prepared by the ICOMOS International Committee on Historic Towns and Villages (CIVVIH), and adopted by ICOMOS in Washington in October 1987. The purpose of this Charter is to expand on the Venice Charter especially concerning the conservation planning and management of historic towns and urban areas. The aim is to guarantee a coherent development of such areas taking into account both the requirements of the existing historic fabric and of modern development.

1990 Charter on Archaeological Heritage Management, ICOMOS

'The International Charter for Archaeological Heritage Management' was adopted by ICOMOS in 1990. The Charter recognizes that protection of such heritage requires a wide basis of professional and scientific knowledge and skills, and elaborates how management and conservation should be applied in this context; it takes into account the legal and economic context, survey and documentation, and the application of various types of care, as well as professional qualifications and international cooperation.

1993 Training Guidelines, ICOMOS

'The Guidelines on Education and Training in the Conservation of Monuments, Ensembles and Sites' was prepared by the International Training Committee (CIF), and adopted by ICOMOS in Colombo 1993. The aim of the document is to promote the establishment of standards and guidelines for education and training in the conservation of monuments, groups of buildings ('ensembles') and sites. It provides a list of skills that are generally considered necessary for conservationists, as well as giving guidelines for the organization of training programmes.

1994 'Nara Document on Authenticity'

As a result of the broadening of the base for the conservation of cultural heritage worldwide, and the need to establish clear guidelines for the assessment of the significance and the authenticity of World Heritage Sites, an international expert meeting was organized in Nara in November 1994. The meeting was organized by the government of Japan, in collaboration with Norway, Canada, ICOMOS, ICCROM and UNESCO. Referring to the Venice Charter, the 'Nara Document on Authenticity' recognizes that

> conservation of cultural heritage in all its forms and historical periods is rooted in the values attributed to the heritage. Our ability to understand these values depends, in part, on the degree to which information sources about these values may be understood as credible or truthful. Knowledge and understanding of these sources of information, in relation to original and subsequent characteristics of the cultural heritage, and their meaning, is a requisite basis for assessing all aspects of authenticity.

5

PROTECTING THE WORLD'S CULTURAL HERITAGE

Henry Cleere

ANTECEDENTS: HERITAGE PROTECTION AND THE LEAGUE OF NATIONS

The work of UNESCO, since its foundation in 1946, tends to obscure the significant actions taken by the League of Nations in the field of heritage protection (touched on briefly by Jokilehto, this volume, page 57). It is worth recapitulating these briefly as a prelude to the post-Second World War developments (Titchen 1995, 12–35).

One of the objects of the League's International Committee on Intellectual Cooperation (ICIC), founded in 1922, was 'to build up international relations and contacts between teachers, artists, scientists, authors, and members of other intellectual professions'. One of its earliest projects concerned the protection of archaeological finds and works of art – 'movable cultural property' in contemporary jargon – which led to the creation of the International Museums Office (IMO) in 1927.

The IMO extended its field of concern to the 'immovable' heritage in 1931 by organizing in Athens a seminal International Conference for the Study of Problems relating to the Protection and Preservation of Artistic and Historical Monuments. The agenda of the conference embraced legislative protection, general principles of restoration, and technical aspects of conservation. The general conclusions that emerged included the following resounding statement (Office International des Musées 1933, 448–52):

> The Conference, convinced that the question of the conservation of the artistic and archaeological property of mankind is one that interests the community of the States, which are wardens of civilization:
>
> Hopes that the States, acting in the spirit of the Covenant of the League of Nations, will collaborate with each other on an ever-increasing scale and in a

> more concrete manner with a view to furthering the preservation of artistic and historic monuments;
> Considers it highly desirable that qualified institutions and associations should, without in any manner whatsoever prejudicing international public law, be given an opportunity of manifesting their interest in the protection of works of art in which civilization has been expressed to the highest degree and which would seem to be threatened with destruction.

These comments were strongly endorsed by ICIC in a resolution that was adopted by the Assembly of the League of Nations in October 1932. This was to be interpreted by the IMO as laying down the principle that 'the testimonies of art and history constitute a common heritage for which the possessor State is accountable to the international community' (Office Internationale des Musées 1939, 7). The more immediate practical outcome of the Athens Conference was the creation in 1935 within the IMO of an International Commission on Historical Monuments, which began work on several initiatives, among them a draft International Convention for the Protection of National Collections of Art and History. Agreement on the text of this instrument was slow to achieve and in the event the Second World War overtook the conference planned in 1939 to prepare the final text. A similar fate awaited the draft 'Declaration concerning the Protection of Historic Buildings and Works of Art in Times of War', also produced by the IMO: the Government of the Netherlands had been commissioned by the League to consult other states and was on the point of submitting the text for signature when the Wehrmacht invaded in May 1940.

In September 1939 the world was plunged into six years of total war, when considerations of heritage protection were thrust aside and the threats attained an intensity never previously experienced. The isolated efforts of scholar-soldiers, such as Mortimer Wheeler trying to protect the great Roman sites in Libya (Hawkes 1982, 216–19), failed to ensure the survival of such masterpieces as Mantegna's frescoes in the Cappella degli Eremitani at Padua.

Figure 5.1

The Taj Mahal, Agra, India. (*Photo: Henry Cleere.*)

(The photographs illustrating this chapter are of World Heritage monuments at which technical advice was given by Sir Bernard Feilden.)

THE BIRTHPANGS OF THE 1972 WORLD HERITAGE CONVENTION

Article 1 of the United Nations Educational, Scientific and Cultural Organization (UNESCO) identifies one of its purposes and functions as maintaining, increasing, and diffusing knowledge, 'by assuring the conservation and protection of the world's heritage of books, works of art and monuments of history and science, and recommending to the nations concerned the necessary international conventions'. UNESCO assumed the mantle of the League of Nations in this field and picked up a number of the ICIC initiatives. The draft Declaration developed by the Netherlands, for instance, became the 1954 Hague Convention, whilst ICIC efforts to control the looting of archaeological sites and historic monuments came to fruition with the 1970 Convention on the Means of Prohibiting and Preventing the Illicit Import, Export and Transfer of Ownership of Cultural Property.

Progress towards a convention on the protection of the immovable cultural heritage was somewhat slower. Initially UNESCO concentrated its efforts on the establishment of an international fund for the 'preservation and restoration of monuments', for which a proposal was made as early as 1948. An international Committee of Experts that met in 1949 recommended the establishment of a permanent international committee charged with monitoring work on sites and monuments of art and history and archaeological excavations, administering an international fund for the preservation of monuments, organizing expert missions, recovering objects of cultural interest, and protecting cultural property in the event of armed conflict. When this proposal was put before the General Conference of UNESCO in 1950, Mexico submitted a 'Project for an International Convention for the Protection of Historic Monuments and Art Treasures'.

The years that followed saw a number of abortive efforts to take these proposals further forward by UNESCO, by Member States, and by the International Council on Monuments and Sites (ICOMOS), with special emphasis being laid on the international fund. Greater success was achieved with the UNESCO campaigns designed to secure funding for specific threatened sites, beginning in 1960 with the Nubian campaign (Säve-Söderbergh 1987). UNESCO acknowledged that these campaigns established the principle that the preservation of a given site or monument, wherever located, is a responsibility to be shared among Member States in the interests of a heritage which belongs to all mankind, picking up the first enunciation of this principle in the Preamble to the 1954 Hague Convention, which states that 'cultural property belonging to any people whatsoever... [is] the cultural heritage of all mankind'.

The emphasis shifted from the international fund to the wider concept of an international policy on protection and conservation in the second half of the 1960s. An expert meeting was convened in 1968 'to co-ordinate, with a view to their international adoption, principles and scientific, technical and legal criteria applicable to the protection of cultural property, monuments and sites'. In its report the meeting asserted that:

> The logical inference is that States should make their own, effective, protective arrangements, while an international agency (which could be UNESCO) would be

> empowered, in specified conditions, to take action to protect sites and monuments of international importance.... International protection implies the existence of something more than the sum of all that is nationally important, i.e. the existence of certain items of universal importance.

Following this crucially important meeting, a series of further meetings took place in the later 1960s. These advanced to a point where the Director General of UNESCO was authorized by the General Conference in 1970 to prepare a 'Preliminary Draft Convention concerning the Protection of Monuments, Groups of Buildings and Sites of Universal Value'. This document was circulated to Member States for comment in 1972.

Meanwhile, the concept of a World Heritage Trust, mooted in the immediate post-war period, was revived on the other side of the Atlantic. President Richard M. Nixon, in his 1971 message to the US Congress, proposed that:

> It would be fitting by 1972 for the nations of the world to agree to the principle that there are certain areas of such unique worldwide value that they should be treated as part of the heritage of all mankind and accorded special recognition as part of the heritage of all mankind as part of a World Heritage Trust.... We must act together now to save for future generations the most outstanding natural areas as well as places of unique historical, archeological, architectural, and cultural values to mankind.

It should be noted that this statement makes reference to 'natural' areas as well as cultural. In fact, much of the initiative around this time comes from those preoccupied with the protection and conservation of the natural heritage. Russell E. Train, then Chairman of the US Council of Environmental Quality, proposed in 1965 the establishment of 'a trust for the world heritage that would be responsible to the world community for the stimulation of international co-operation efforts to identify, establish, develop and manage the world's important natural and scenic areas and historic sites for the present and future benefit of the international citizenry' (Slatyer 1984, 5). He made this proposal to the Committee on Natural Resources of the White House Conference on International Co-operation. The International Union for the Conservation of Nature (IUCN) picked up this notion and in its turn moved to the drafting of an international convention for the implementation of the proposed trust, completed in early 1971.

With the availability of two draft conventions, plus a draft 'World Heritage Trust Convention concerning the Preservation and Protection of Natural Areas and Cultural Sites of Universal Value' prepared by the US Government, the debate speeded up. A Special Committee of Government Experts met in April 1972 to prepare the definitive text of a convention for submission to the UNESCO General Conference later that year. The resulting text was first considered by the UN Conference on the Human Environment, held in Stockholm in June 1972, and by the IUCN Second World Congress on National Parks, held in the USA in September 1972, both of which endorsed the proposed Convention. It was finally approved by the UNESCO General Conference on 16 November 1972, though not without a struggle to achieve the necessary two-thirds majority. The World

Heritage Convention came into effect in 1975, when it had been ratified by twenty countries, led by the USA closely followed by Egypt, Iraq, and Bulgaria.

THE WORLD HERITAGE CONVENTION SINCE 1972

Objects of the Convention

The Convention is posited on the awareness that there are certain parts of the heritage whose value to the world as a whole is so outstanding that their protection, conservation, and transmission to future generations are matters not merely for individual countries in which they occur but also for the international community as a whole. It also acknowledges that many of the countries that are most richly endowed from the artistic, archaeological, architectural, palaeontological, biological, geological, and ecological points of view lack adequate resources to protect this heritage.

The carefully structured preamble to the Convention begins by noting that the heritage is 'increasingly threatened with destruction not only by the traditional causes of decay, but also by changing social and economic conditions which aggravate the situation with even more formidable phenomena of damage and destruction'. It is interesting to note that this echoes the concern expressed by the Athens Conference in 1931 about the threats to which the heritage is exposed: the 'deterioration or disappearance of any item of the... heritage constitutes a harmful impoverishment of the heritage of all the nations of the world.'

The preamble goes on to consider that 'protection of this heritage at the national level often remains incomplete because of the scale of the resources which it requires and of the insufficient economic, scientific and technical resources of the country where the property to be protected is situated'.

The concept of universality is underlined in the next paragraph: 'the existing conventions, recommendations and resolutions concerning cultural and natural property demonstrate the importance, for all the peoples of the world, of safeguarding this unique and irreplaceable property, to whatever people it may belong.' The preamble then reiterates the mutual dependence of the countries of the world in protecting the heritage: 'in view of the magnitude and gravity of the new dangers threatening them, it is incumbent upon the international community as a whole to participate in the protection of the heritage of outstanding universal value, by the granting of collective assistance which, although not taking the place of action by the State concerned, will serve as an effective complement thereto.' This paragraph makes two important points. First, it introduces the fundamental criterion of 'outstanding universal value' and, secondly, it stresses that the primary responsibility for the protection of its heritage remains with the state concerned, a factor that remains constant throughout discussions on heritage protection and conservation from the earliest days of the discussions within the League of Nations.

The Convention reflects the concerns of the period when it was written in its emphasis on the increasing level of threats to the heritage. This is equally

Figure 5.2

Ed Deir (The Monastery), Petra, Jordan. (*Photo: Department of Antiquities, Jordan.*)

recognized in the individual protection legislation concerning the cultural and natural heritage in most of the countries of the world. It also tacitly acknowledges the fact that this legislation is very variable in its effectiveness and application.

Definitions

Article 1 of the Convention defines three categories of cultural heritage:

> *Monuments* : architectural works, works of monumental sculpture and painting, elements and structures of an archaeological nature, inscriptions, cave dwellings and combinations of features, which are of outstanding universal value from the point of view of history, art or science;
>
> *Groups of buildings* : groups of separate or connected buildings which, because of their architecture, their homogeneity or their place in the landscape, are of outstanding universal value from the point of view of history, art or science;
>
> *Sites* : works of man or the combined works of nature and of man, and areas including archaeological sites which are of outstanding universal value from the historical, aesthetic, ethnological or anthropological points of view.

These definitions are broadly drawn, with great skill, so as to be comprehensive and descriptive rather than specific. There is, for example, no reference to towns or villages or to individual classes of building such as temples, churches, castles, etc., which often figure in the exhaustive lists in national legislations.

Obligations of States Parties

Under the provisions of Article 4, each State Party to the Convention must take 'effective measures for the protection, conservation, presentation, and transmission to future generations' of its own heritage. To this end it must undertake:

(a) to adopt a general policy which aims to give the... heritage a function in the life of the community and to integrate the protection of that heritage into comprehensive planning programmes;
(b) to set up within its territories, where such services do not exist, one or more services for the protection, conservation and presentation of the... heritage with an appropriate staff and possessing the means to discharge their functions;
(c) to develop scientific and technical studies and research and to work out such operating methods as will make the State capable of counteracting the dangers that threaten the... heritage;
(d) to take the appropriate legal, scientific, technical, administrative and financial measures necessary for the identification, protection, conservation, presentation and rehabilitation of this heritage;
(e) to foster the establishment or development of national or regional centres for training in the protection, conservation and presentation of the... heritage and to encourage scientific research in this field.

Administration of the Convention

At a triennial General Conference, the States Parties to the Convention elect an Intergovernmental Committee for the Protection of the Cultural and Natural Heritage of Outstanding Universal Value (usually known as the World Heritage Committee), which is the sovereign body for the Convention. It is semi-autonomous within the UNESCO system, which has made it possible for non-member countries of UNESCO such as the United Kingdom and the United States of America to retain their association with the Convention.

The Secretariat for the Committee is provided by the World Heritage Centre of UNESCO. Two non-governmental organizations (NGOs) are identified within the Convention as advisory bodies: ICOMOS in respect of cultural heritage and IUCN for the natural heritage. The third named advisory body is the International Centre for the Study of the Preservation and Restoration of Cultural Property (ICCROM), which is an intergovernmental organization (IGO). They are responsible for advising the World Heritage Committee on the suitability of properties nominated by

States Parties for inclusion on the World Heritage List, on monitoring activities, and on technical assistance requests from the World Heritage Fund.

List of World Heritage in Danger

The Convention also provides for the creation of this special list of properties 'for the conservation of which major operations are necessary and for which assistance has been requested'. The dangers foreseen as being relevant to this list are specified in Article 11.4 as 'the threat of disappearance caused by accelerated deterioration, large-scale public or private projects or rapid urban or tourist development projects; destruction caused by changes in the use or ownership of the land; major alterations due to unknown causes; abandonment for any reason whatsoever; the outbreak or the threat of armed conflict; calamities and cataclysms; serious fires, earthquakes, landslides, volcanic eruptions, changes in water level, floods, and tidal waves'.

The World Heritage Fund

The Convention also establishes a Fund for the Protection of the World Cultural and Natural Heritage of Outstanding Universal Value (known as the World Heritage Fund). This is made up of biennial contributions made by States Parties (proportional to their subscriptions to UNESCO), plus other contributions from individual States Parties, UNESCO, private bodies, etc., many of which are earmarked for specific projects.

Grants made by the Committee from the Fund (Art. 22) may take the following forms:

(a) studies concerning the artistic, scientific and technical problems raised by the protection, conservation, presentation and rehabilitation of the. . . heritage;
(b) provision of experts, technicians and skilled labour to ensure that the approved work is correctly carried out;
(c) training of staff and specialists at all levels in the field of identification, protection, conservation, presentation and rehabilitation of the. . . heritage;
(d) supply of equipment which the State concerned does not possess or is not in a position to acquire;
(e) low-interest or interest-free loans which might be repayable on a long-term basis;
(f) the granting, in exceptional cases and for special reasons, of non-repayable subsidies.

Implementation of the Convention

The Operational Guidelines

The early work of the World Heritage Committee concentrated on the drafting of guidelines for its work. The current edition of the *Operational Guidelines for the*

Implementation of the World Heritage Convention dates from January 1996 (UNESCO 1996).

The Committee's functions are defined as fourfold: the identification of properties for inclusion on the World Heritage List; the monitoring of the state of conservation of properties already on the List; the inscription of properties on the List of World Heritage in Danger; and the allocation of resources from the World Heritage Fund.

Next comes a section on general principles, which begins by reiterating the major objectives of the Convention. It then goes on to set out a number of principles which expand and illuminate the Convention. For example, it impresses the need for efforts to be made 'to maintain a reasonable balance between the numbers of the cultural heritage and the natural heritage properties entered on the List'.

No formal limit is imposed either on the total numbers of properties on the List or submitted by individual States Parties. However, the latter sentiment is offset to some extent by an invitation to States Parties whose heritage is already well represented on the List to slow down their rate of nominations, to help in making the List 'more universally representative'. Those countries not adequately represented are invited to seek assistance from the Committee in preparing further nominations.

Nominations are required to conform with a standard format. Important elements in this are juridical data on legislative protection and how it is implemented and comparative evaluations where nominated properties fall within categories that are already well represented. Stress is increasingly laid on the need for buffer zones, defined as areas 'surrounding the property which has restrictions placed upon its use to give an added layer of protection'. States Parties are also encouraged to consider serial nominations of groups of properties from the same historico-cultural group. Information about management plans are also required in respect of each nominated property.

Criteria for inclusion on the List

The *Operational Guidelines* set out detailed criteria for inclusion of cultural properties on the World Heritage List (UNESCO 1996, paras 23–42). These fall into two groups, the first being what may be termed the 'cultural' criteria. Each property must conform with one or more of the following criteria:

- (i) represent a masterpiece of human creative genius;
- (ii) exhibit an important interchange of human values, over a span of time or within a cultural area of the world, on developments in architecture, monumental arts or town-planning and landscape design;
- (iii) bear a unique or at least exceptional testimony to a civilization or cultural tradition which is living or which has disappeared;
- (iv) be an outstanding example of a type of building or architectural ensemble or landscape which illustrates (a) significant stage(s) in human history;

(v) be an outstanding example of a traditional human settlement or land-use which is representative of a culture (or cultures), especially when it has become vulnerable under the impact of irreversible change;
(vi) be directly or tangibly associated with events or living traditions, with ideas, or with beliefs, with artistic and literary works of outstanding universal significance (the Committee considers that this criterion should justify inclusion on the List only in exceptional circumstances or in conjunction with other criteria, cultural or natural).

The second group of criteria relate to the more practical aspects of nominations; nominated properties must:

(i) meet the test of authenticity in design, material, workmanship or setting and in the case of cultural landscapes their distinctive character and components (...reconstruction is only acceptable if it is carried out on the basis of complete and detailed documentation on the original and to no extent on conjecture);
(ii) have adequate legal and/or traditional protection and management mechanisms to ensure the conservation of the nominated cultural property or cultural landscape. The existence of protective legislation at the national, provincial or municipal level or well established traditional protection and/or adequate management mechanisms is therefore essential.... Assurances of the effective implementation of these laws and/or management mechanisms are also expected. Furthermore, in order to preserve the integrity of cultural sites, particularly those open to large numbers of visitors, the State Party concerned should be able to provide evidence of suitable administrative arrangements to cover the management of the property, its conservation and its accessibility to the public.

The *Operational Guidelines* include more detailed categorization of two types of cultural property. Groups of urban buildings (UNESCO 1996, paras 26–34) are divided into three categories:

(i) towns which are no longer inhabited but which provide unchanged archaeological evidence of the past;
(ii) historic towns which are still inhabited and which, by their very nature, have developed and will continue to develop under the influence of socio-economic and cultural change; and
(iii) new towns of the twentieth century.

The other group that is given special attention is that of cultural landscapes. These are also divided into categories (UNESCO 1996, paras 35–42):

(i) clearly defined landscapes designed and created intentionally by man;
(ii) organically evolved landscapes:
– relict landscapes in which an evolutionary process came to an end at some time in the past, either abruptly or over a period;
– continuing landscapes, which retain an active social role in contemporary society, and in which the evolutionary process is still in progress;

(iii) associative cultural landscapes, whose inclusion is justifiable by virtue of the powerful religious, artistic or cultural associations of the natural element rather than material cultural evidence, which may be insignificant or even absent.

THE FUTURE OF THE CONVENTION

The first nominations were received in October 1977, and twelve of these, from seven countries, were inscribed on the World Heritage List in September 1978. Eight of these were cultural sites: the city of Quito (Ecuador), Aachen Cathedral (Germany), L'Anse aux Meadows (Canada), Mesa Verde (USA), the rock-hewn churches of Lalibela (Ethiopia), the Ile de Gorée (Senegal), the historic centre of Cracow, and the Wieliczka salt-mine (both Poland). Of the six 'cultural' criteria, all were applied with the exception of (v), which was not used until the following year, in the case of medieval Cairo.

Since that time the number of States Parties to the Convention has risen to 146 and the total number of inscribed sites is 469. Of these, 350 are cultural sites and monuments, 102 are natural sites, and 17 are so-called 'mixed' sites, inscribed on the basis of both cultural and natural criteria. They are located on the territories of 104 States Parties.

The twentieth anniversary of the Convention in 1992 was the occasion for a review of what had been achieved so far. IUCN published an important volume of essays (Thorsell 1992) and UNESCO commissioned an independent review by

Figure 5.3

The Stupa Mound and Great Bath, Mohenjodaro, Pakistan. (*Photo: Department of Archaeology and Museums, Pakistan.*)

a distinguished scholar who had been associated with the Convention from its earliest days (Pressouyre 1993).

The Committee also carried out internally a strategic review of its activities over the past two decades. This identified a number of problem areas. One source of considerable concern was the fact that, whilst 132 countries had joined the Convention at the end of 1992, only 82 of these had sites on the World Heritage List. Certain countries had relatively large numbers of sites on the List: France had 20, all save 1 cultural, India 19 (14 cultural, 5 natural), Spain 17 cultural and 1 natural, the USA also 17 (8 cultural, 9 natural), and the United Kingdom 14 (11 cultural, 3 natural). The Committee therefore resolved to take steps to correct this imbalance.

One of those steps was the introduction of the concept of the cultural landscape, which would favour those countries with non-monumental cultures, such as those of the Pacific region, and those with a relatively modest architectural and archaeological heritage but agricultural or pastoral landscapes of considerable antiquity, such as the countries of sub-Saharan Africa or northern Europe and America.

Several other fields of heritage that were inadequately recognized on the List were also identified. One of these was the industrial heritage, represented by no more than a handful of sites (Ironbridge in England, the Wieliczka salt-mine in Poland, the Rammelsberg copper mining area and the Völklingen ironworks in Germany, the Engelsberg ironworks in Sweden). The assistance of TICCIH (The International Committee for the Conservation of the Industrial Heritage) was invoked by ICOMOS in the preparation of indicative lists of the most important industrial monuments in the world, sector by sector, to act as a guide for States Parties and the Committee alike. Twentieth-century architecture is also largely absent: the only examples are three masterpieces of Gaudí in Barcelona and the city of Brasilia. Here the specialized advice of the International Group for the Documentation of the Modern Movement (DoCoMoMo) has been sought.

These two specialist studies form part of a much wider project being developed by the World Heritage Committee. It has drawn up a global strategy, to identify those aspects of the world cultural heritage that are currently under-represented on the World Heritage List, or even absent from it. A statistical analysis of the sites inscribed on the List by the end of 1993 revealed a number of significant lacunae and anomalies. For example, of the 320 sites and monuments on the List at that time, 154 were situated in Europe, 67 in Asia, 43 in Africa (including the Maghreb countries), 23 in North America, 29 in Central and South America, and only 4 in Australia and Oceania. Of the 90 religious monuments (churches, cathedrals, monasteries, temples, shrines, etc.), 65 were Christian, the remainder being spread over Buddhist, Islamic, Hindu, and other monuments. No fewer than 96 were archaeological sites and 86 historic towns or town centres, whereas there were only 4 vernacular settlements inscribed on the List.

A series of regional meetings is under way, so far in the Philippines (specifically on rice-producing landscapes), in Australia on the non-monumental heritage of Australasia and Oceania, currently represented only by the sacred mountains of Tongariro (New Zealand) and Uluṟu-Kaṯa Tjuta (Australia), and in Zimbabwe on the heritage of southern Africa. As with industrial heritage and twentieth-century architecture, expert groups will also be invited to indicate what they consider to be the most significant sites and monuments within their respective fields. These will then be made available to States Parties, who will be invited to consider whether they wish to nominate those on their territories. There will be no compulsion to do so, since the right of nomination is the prerogative of States Parties, nor will the appearance of a property on such a list be an assurance that it will be included on the List.

It is hoped in this way to make the List more truly representative of the world's cultural heritage. The Committee has rejected the proposal made on more than one occasion that there should be a maximum figure imposed upon the List, since it believes that this would militate against probable future extension of the concept of what constitutes the cultural heritage. It recognizes, however, that the number of new nominations will dwindle from the present level of around forty annually to a trickle in the next century.

The other area of concern for the Committee is the monitoring of sites and monuments that are already on the World Heritage List. The natural sites have been subject to regular review by IUCN and the World Conservation Monitoring Centre, but up to the present there has been no systematic monitoring of the cultural sites on the List (apart from those in Latin America, which were the subject of monitoring activities up to 1994 as part of a joint UNDP/UNESCO project). Proposals have been made for the systematic monitoring of the state of conservation and management of all the cultural monuments on the List on a five-year cycle. The States Parties to the Convention will have primary responsibility for carrying out this work, but they can call on assistance from the UNESCO World Heritage Centre and from the NGOs. ICOMOS carried out a detailed monitoring study of the three Cultural Triangle sites in Sri Lanka (Anuradhapura, Polonnaruwa, and Sigirya) in November 1994, and has developed a methodology that will be available for all States Parties. Major monitoring exercises have been carried out by ICOMOS National Committees in Norway and the United Kingdom. Credit should also be given to those countries such as Australia and Mexico which have been carrying out regular monitoring of their World Heritage sites for a number of years.

CONCLUSION

The World Heritage Convention is a visionary yet effective instrument that has already done much to preserve and protect the cultural and natural heritage of 'outstanding universal significance' and to raise the profile of the heritage in the consciousness of most of the countries of the world. The perception of the heritage has expanded considerably since 1972, with the result that the World Heritage List now embraces a much wider range of properties. However, much remains to be done, notably in giving form to that expanded perception through the creation by the end of the century of a World Heritage List that is truly representative of the broad spectrum of human achievement and belief.

BIBLIOGRAPHY

Hawkes, J. (1982). *Mortimer Wheeler: Adventurer in Archaeology*. Weidenfeld & Nicolson, London.

Office International des Musées (1933). *La Conservation des Monuments d'Art et d'Histoire* (The Conservation of Artistic and Historical Monuments). Paris.

Office International des Musées (1939). *Mouseion*, Paris.

Pressouyre, L. (1993). *La Convention du Patrimoine mondial, vingt ans après*. Éditions UNESCO, Paris.

Säve-Söderbergh, T. (1987). *Temples and Tombs of Ancient Nubia: The International Rescue Campaign at Abu Simbel, Philae and Other Sites*. UNESCO/Thames & Hudson, London.

Slatyer, R. O. (1984). 'The origin and development of the World Heritage Convention', *Monumentum*, Special Issue; 3–16.

Thorsell, J. (ed.) (1992). *World Heritage Twenty Years Later*. IUCN, Gland (Switzerland) and Cambridge (UK).

Titchen, S. M. (1995). *On the construction of outstanding universal value: UNESCO's World Heritage Convention. . . and the identification and assessment of cultural places for inclusion in the World Heritage List*, unpublished PhD thesis, Australian National University, Canberra.

UNESCO (1996). *Operational Guidelines for the Implementation of the World Heritage Convention* (WHC/2/Revised January 1996). UNESCO World Heritage Centre, Paris.

6

THE EDUCATION OF A CONSERVATION ARCHITECT: PAST, PRESENT AND FUTURE

Derek Linstrum

The utmost care should be given to the selection of an Architect. One who is fit to be intrusted with the reparation of an ancient edifice should in the first place be a man thoroughly conversant with the architecture of which it is an example. He should be both an artist and an antiquary; and if it is a church he has to deal with, he should have a thorough understanding and respect for its ritual uses. More than this, however, he should have a full appreciation of the value of an ancient monument, and a wholesome dread of infringing upon its authenticity, though fully able if additions are necessary to carry them out in the old spirit. He should be a man of sufficient decision of character to enable him to cope with any attempts which may be made to lead him into a departure from a true conservation course.

'Conservation of Ancient Monuments and Remains. General Advice to Promoters of the Restoration of Ancient Buildings'.
Sessional Papers of the Royal Institute of British Architects 1864–65, London 1865.

THE BACKGROUND AND THE DEVELOPMENT OF EDUCATIONAL CONCEPTS

'Houses are built to live in, not to look on', wrote Francis Bacon at the opening of his essay 'On Building' (first published in 1597), and that is a good maxim for any conservation architect who believes that one of the prime objects of conservation is to enable a building to continue in use. How very appropriate then that one of the present occupants of the romantic, partly ruined, brick and flint Stiffkey Old Hall in Norfolk, which was built by Bacon's son, Nathaniel, in the 1570s with constant parental advice, should be the dedicatee of this *Festschrift*.

'If you get a dead building you must be brave and give it life', Bernard Feilden once said when we were sitting talking just after the conclusion of one of the private, one-man directed international work camps he used to organize each year for a small group of Scandinavian and North American architectural students who lived with his family, learnt to do practical work on buildings by day and attended informal lectures and discussions in the evening, while enjoying cultural tours to East Anglian towns and buildings at the weekend. These camps represented a continuation of a tradition of learning from practical experience of an older master, a theme he has often reverted to, as when writing of ICCROM missions that 'if young practitioners are sent with the eminent... experts, they can learn from the discussions. One of them should be given the task of seeing the conservation project through to completion' (Feilden 1978, 60ff). Implicit, too, is the requirement that an architect who wants to work on old buildings must be master completely of the way in which they were conceived and constructed, understanding their materials and behaviour; and there is the conviction that in the international world of conservation we all have much to learn from one another. These were all basic in his direction of ICCROM from 1977 to 1981, and equally so in any discussion about the training of a conservation architect. But how necessary is specialized training? And when should it be undertaken? Is it not sufficient to have trained as an architect? A. R. Powys, who was Secretary of the Society for the Protection of Ancient Buildings from 1911 to 1936, had no doubts:

> Perhaps no branch of architecture calls for more highly specialised qualifications in all concerned than does that which has to do with the treatment of ancient buildings. The questions which have to be dealt with are of a most intricate nature, always involving the consideration of the twin needs of structural identity and of conservation, and sometimes also the making of alterations. In the course of work on an ancient building points difficult to decide and needing instant decision constantly arise. What needs renewal? How much may be retained? What technical method may be employed? And like questions confront one at every turn. (Powys 1929, 4)

The questions have not changed, even though the expression 'ancient buildings' might sound too archaic and limiting today when an architect might be dealing with Centre Point in London or Birmingham New Street Signal Box, since such unlikely monuments may now be admissible to the canon. For many years Powys's SPAB, through its Lethaby Scholarships, literally provided the only means by which a young architect could undertake any organized training in conservation. They were named after the architect who wrote of 'a whole new art of building conservation', and who believed that 'right understanding of the ancient would make us modern and produce a form of building art proper for to-day' (Lethaby 1979; 157). The latter opinion would receive some support today, but it was not a belief that would have found adherents among up-to-date architects in the early 1950s when Dr William Singleton, the first Director of what was to become the Institute of Advanced Architectural Studies in York, was writing; 'Never before has there been so great a need for architects to study old buildings and to handle them with sympathy and knowledge.' The backlog of the wasted war years was catching up, and a new generation of architects and planners trained with different beliefs and models was engaged in bringing about a brave new world. 'Obsolescence' was a nasty word that was being bandied about with a licence to destroy or maim, and a town centre, however historic, was regarded as a potential *tabula rasa*. Already developers were setting out, with the blessing of planners and planning committees and with the connivance of architects, on a systematic demolition and renewal. There was insufficient recognition of the value of old buildings in themselves or in their context, and too few planners had read the wise words of the artist John Piper, written in 1947, that;

> The social function of architecture as delight... is to express and satisfy the imaginative life of an enormous variety of people in different stages of emotional education, and this means the requisitioning of all kinds of architecture and every available architectural resource as a means of stimulating the imaginative life of people living at totally different imaginative levels.
>
> (Piper 1948, 102)

Its successful implementation also meant, as Piper wrote, that 'architects and town-planners have another task to add to their present ones; to re-educate themselves visually'. They were also having to work within gradually increasing legislation which afforded protection, first for buildings of architectural or historic interest, and then for areas with an identifiable character. By the 1960s there had been many changes, as well as a popular reaction against much of the new architecture, all of which tended to place old buildings in a new light which showed up the lack of experienced people who had the knowledge to take responsibility for them.

> The last architects to have had a traditional training in building construction and in the history of architecture are now verging upon the age of retirement [and] circumstances have been highly unfavourable to any large intake of junior men into this field of work. The fact must be faced that this is less a career than a vocation. In a highly competitive world seeking adequate material rewards there is little opportunity for the expenditure of time or the painstaking care that are needed, both in training for this work and in putting this education into practical effect later. (Harvey 1972, 82)

That was how Dr John Harvey summed up the situation, stating the lack of financial incentive as opposed to lucrative commercial work in a very restrained manner. But the problem was not simply an English one, or a European, or a North American. It was becoming recognized that throughout the world there was a growing threat to historic towns and buildings, not least because of what Harvey called 'a sharp cleavage between the architect, as now trained in modern design and methods, and the conservator, acquainted with traditional usage and materials, and with the directing part played by professional architects as they were until recently' (Harvey 1972; 83).

Nevertheless, the place of the architect was generally upheld. Bernard Feilden's widely publicized work on York Minster had introduced structural and organizational innovations which complemented the traditional craft methods as well as offering an outstanding example of multidisciplinary cooperation, but he had no doubt that the architect was the coordinator.

> He is in a central position from which he can see a historic building as a whole and become its 'curator'. He is able to understand how it was conceived, how it was built, how it has responded to its environment, and how it has been used or abused... ! (Feilden 1978, 60ff)

> He has the most experience in making value judgments.... He must maintain the architectural unity of the building, calling on persons who are experts in various parts of the task but who may not understand the project as a whole. (Feilden 1976a, 259)

But the fact remained that in the 1960s the number of experienced conservation architects was dwindling, even though it was already apparent that existing buildings, many of them listed as being of architectural or historic importance, were increasingly becoming the responsibility of a wider section of the architectural profession in public and private practice, often for the first time. It was not only architects who were needed. In the 1970s Bernard Feilden was writing to recommend the setting up of the Foundation known as 'Pro Venetia Viva', in which he referred to:

> a shortage of building craft skills of the types and depth necessary to conserve the European Architectural Heritage, much of which is at risk from this reason, and from economic causes. This shortage has been aggravated by the lack of money to pay for the labour intensive craft skills of the past. Skilled craftsmen do exist, but as they were trained in the 1930s they are now nearing the end of their careers and it is urgent that their skills should be passed on and not lost for ever. (Feilden 1976b, 1)

INTERNATIONAL CONCERN

In 1966 a UNESCO resolution convened a meeting in the small Italian city of Pistoia to discuss 'problems relating to the training of architects and technicians responsible for the preservation of monuments and sites' (*Monumentum* 1969, 9ff) Their shortage was considered 'one of the key problems of the present [especially] in the case of countries undergoing rapid change'. The outlook was gloomy: 'historical quarters and buildings which are architecturally or historically interesting are disappearing with alarming rapidity. Classified or scheduled buildings may survive, but only as lonely archaisms for the buildings which surrounded them have given way to curtain-walled structures'. In varying degrees, it was a universal situation in the 1960s. One fact UNESCO recognized was that it was not solely a problem of architectural education, but that multidisciplinary action was needed. Questions were asked about whether architectural schools gave courses on the principles of conservation and restoration or the use of archaeological techniques and evidence, and whether architectural conservators (assuming they existed) were involved in town-planning programmes.

Predictably, the answers were not likely to be encouraging although the Chairman, Pietro Gazzola, was able to make reference to some courses that had been set up in different countries; but he focused on the facilities that had been developing under the direction of Professor Guglielmo Angelis d'Ossat, mainly on the latter's initiative, in the Faculty of Architecture of Rome University (Gazzola 1969, 15ff). The first international course in architectural conservation was organized in 1961, but following the recommendation of the 1964 meeting of experts in Venice, at which the Venice Charter was produced, the International Centre for the Study of the Preservation and the Restoration of Cultural Property (later to be known as ICCROM) was invited to collaborate in the organization of these courses and to provide them with an international framework. ICCROM was founded in 1956 by UNESCO and placed in Rome as an autonomous, scientific, inter-governmental organization, at first under the direction of Dr Harold Plenderleith. Among its varied functions, which have developed in several directions since its foundation, the most relevant to the present subject is that it should 'promote, develop and provide training relating to the conservation and restoration of cultural property and raise the standards and practice of conservation work'. Since 1977, ICCROM has organized its architectural conservation course separately from the two-year course of Rome University; it extends over eighteen weeks.

Albert Chauvel spoke at Pistoia about the quite different and unique method of training in France, which stretches back from today to those almost legendary figures, Ludovic Vitet, Prosper Mérimée and Eugène Viollet-le-Duc, although the present organization of *inspecteur-général*, *architectes-en-chef* and *architectes des bâtiments de France* dates only(!) from 1897. The training, undertaken by graduates of the course who provide continuity, provides the sole entrée to one of these senior appointments and in order to qualify to undertake it the requirements were, and are (despite some changes in the 1980s), formidable. A candidate must not only be an architect, but also an artist, a technician, an excellent draughtsman and a photographer; he must have studied social geography, archaeology, town planning and legislation; he must have published original work and be in touch with artists, scientists and research-workers; he must possess a broad general culture and show himself able in the highest degree to compose, carry out and create while displaying sensitiveness, imagination, a sense of his responsibilities, and skill in analysis and synthesis (Chauvel 1969, 57ff). Remembering how Philibert de l'Orme in the sixteenth century had discussed at length the multifaceted duties and attributes of an architect, who was represented in an engraving by a man with two pairs of hands, winged feet and three eyes, one might conclude that *le plus ça change, le plus c'est la même chose.*

There was no equivalent programme in England, although a Select Committee set up in 1841 had discussed the possibility as it was recognized that there was much 'misjudged *reconstruction* under the name of *restoration and repair*' and that 'in France great attention has lately been paid to this subject by the Government' (Crook and Port 1973, 641). Nevertheless, there was no attempt to regularize or institutionalize a professional responsibility for the treatment of old buildings. Maybe an English suspicion of excessive control survived in Donald

Insall's contribution at Pistoia, which questioned if the conservation of old buildings should become a preserve of specialists, trained exclusively in traditional methods and incapable of modern building. He could not agree with such a separation, affirming that an architect is, above all, qualified to analyse; to identify the problems of a site or a building, and to solve them in terms of the best available constructional methods and materials: 'To make the best of an old building and to evoke its special qualities requires the same approach as making the best of a vacant building site.... The training of the conservator is largely a question of developing a basic attitude of mind – and this should be a part of the equipment of every architect'. He warned of the dangers of over-specialization, especially in adapting and revitalizing old structures when there might be a temptation to recreate an original balance which time has destroyed. 'But history cannot be rebuilt, nor the clock turned back. The architect who can re-evaluate the assets of a situation or a structure and help it to meet new circumstances can sometimes achieve more for it, by bringing it back to life and use, than too historically correct a restoration' (Insall 1969, 25ff). This meeting in Pistoia was revealing how different national characteristics and traditions were reflected in attitudes towards conservation practice and education, even if the general goals were the same.

Transatlantic comments were made by Charles Peterson, one of the Directors of the Columbia University programme that had been initiated in 1964. His opinion was that far from being 'specialized', the knowledge required of a competent conservator (or 'restorationist' or 'preservationist' as he is known in the United States) must be of the widest range. It should include,

> old materials and how they were shaped by the tools of their period in order to know what to save, what to discard and how to reproduce missing parts; historic decoration, since for authenticity it is generally required to design missing or deteriorated parts that are harmonious in design on the whole; structural problems, including the effect of one hundred thousand tourists per year passing through a frame house originally designed for the use of a private family; how to select and adapt modern equipment . . . to an old structure being readied for modern uses; research techniques so as to be able to instruct researchers on the information required.

He advocated 'a licence for restorationists which would guarantee the competence of the practitioner in advance of awarding commissions', but at the same time he drew attention to the introduction of a training scheme for selected artisans who could be employed in a sub-professional capacity as 'Buildings Restoration Specialists' (Peterson 1969, 71ff). The wide range of elements which Professor Peterson included under a conservation umbrella, including all the traditional and man-made building materials, the history of service installations and structural techniques, shaped the programme of the important conference he organized in 1974 to commemorate the 250th anniversary of the founding of the Carpenters' Company of the City and County of Philadelphia. Bernard Feilden (1976a, 258ff) spoke about his work on the three cathedrals of Norwich, York and St Paul's, and I was invited to discuss (for the first time at such a large gathering) training for conservation (Linstrum 1976a, 340ff).

Returning to 1966, the final recommendations of the Pistoia meeting which had introduced a number of crucial questions, were in three parts. The first referred to the training of architects and town planners in general, which should include teaching of history and the humanities, and 'a thorough education in the preservation of historic sites and monuments, the history of art, the history of architecture and architectural techniques, and the history of town planning and of the development of landscapes and gardens'. The second part accepted the idea of specialized architects to whom preservation, restoration and presentation of sites and monuments should be exclusively entrusted; it advocated closer collaboration between architects and specialists in other related university disciplines; and postgraduate courses were regarded as essential at national, regional and international levels. The third part referred to the training of site foremen and craftsmen.

ENGLISH-SPEAKING COURSES

In the United States, the first courses were established within an architectural history programme, at the University of Virginia in 1959, by Professor Frederick D. Nichols, at Cornell University four years later by Professors Stephen W. Jacobs and Barclay G. Jones, and at Columbia University, in 1964, by Professor James Marston Fitch. Columbia was the first to establish a degree in historic preservation, consisting of two parallel curricula, one for graduate architects and one for social and art historians; then, in 1973, it was authorized to offer a new degree, a Master of Science in Historic Preservation (Tomlan 1994, 187ff; De Long 1993, 63ff).

Progress was slower in England where it was the influential conservation groups that contributed the first decisive moves towards identifying the need for training. The formation of the Georgian Group in 1937 and the Victorian Society in 1958, both as offshoots of the venerable Society for the Protection of Ancient Buildings which celebrated its centenary in 1977; the reorganization of the Ancient Monuments Society in 1952, and the founding in 1957 of the Civic Trust, to which were affiliated local protest and amenity societies nation-wide; all these introduced a formidable, uniquely English element in conservation. In 1959 a number of these organizations, as well as the Society of Antiquaries, the National Trust, representatives of the ecclesiastical architectural associations and of the building professions, came together to found the Conference on Training Architects in Conservation because 'the constituent bodies, all of whom have some practical responsibility or special concern for historic buildings, were deeply concerned at the increasing need to make better provision for professional training in conservation'. In 1961 a postgraduate course was set up at the Institute of Archaeology in London under William Eden who had pioneered such work with a one-year Certificate course at the Bartlett School of Architecture in the 1950s. The new course, which led to a Diploma in the Conservation of Historical Monuments, was a two-year, part-time course that made use of evenings and some weekends and included practical training in an approved office. The course lasted until 1969.

In 1967 the University of Manchester established an M.A. course which had a special emphasis on vernacular architecture, urban history and traditional building crafts, all subjects which represented the particular interests and skills of some staff members and the departmental head, Professor Richard Cordingley, who believed in the understanding of history as basic to architectural design. There were full-time and part-time opportunities, and the course continued until 1990. Edinburgh College of Art, in association with Heriot-Watt University, initiated a conservation specialization in the final year of studies during the 1960s under Peter Whiston, and this was developed in the 1970s and 1980s principally under an architectural historian, Colin McWilliam, who was a leading activist in conservation concern in Edinburgh. A one-year long course, followed by supplementary work on a dissertation, led to the award of an M.Sc. from 1969 (France 1993; 48ff).

THE YORK CENTRE FOR CONSERVATION STUDIES

The first students arrived at York in 1972 to launch the postgraduate course of which I was Director until 1992, but that was not the genesis of conservation studies in the city. In 1949 there had been the first of what was called the York Summer School of Architectural Study, an annual event organized by the Academic Development Committee of the York Civic Trust. This was to enable architectural students to study and make measured drawings of historic buildings, as well as offering lectures on related subjects; but the organizers were also encouraged to set up courses on the protection and repair of historic buildings. In 1953 the York Institute of Architectural Study was founded, with the additional description of the 'National Reference Centre for the Protection and Repair of Historic Buildings'. Gradually the Institute acquired a reputation as a centre where intensive short courses with a topical relevance to architectural theory and practice were offered. For instance in 1955 the Church of England Inspection of Churches Measure had introduced a new professional responsibility for architects, and the Institute put on the first of many courses dealing with church inspection and repairs. Sir Bernard was one of the course members; 'I was one of your first students', he has often said. In 1960 the University of York was founded, and the Institute became a part of it when its name was changed to the Institute of Advanced Architectural Studies, although some cynics asked why, if they were advanced, did they seem to concentrate on the past? Perhaps Lethaby had already supplied the answer when he said that a 'right understanding of the ancient would make us modern and produce a form of building art proper for to-day'.

The location of a conservation course is a matter of some importance. There was an apparent inevitability about the establishment of York as a centre for conservation studies in the historic King's Manor, itself an example of what Henry James called 'that phenomenon so delightfully frequent in England – an ancient property or institution lovingly readopted and consecrated to some modern amenity' (James 1905, 37f). In this case the readoption had been executed by Feilden and Mawson. At the beginning of the 1970s the city could be regarded as a microcosm of the conservation world. Bernard Feilden's work on the Minster had demonstrated that a large-scale project involving an historic building, even a world-famous cathedral, could benefit from innovative techniques and lessons in site organization, as well as emphasizing the necessity for a multiprofessional team. The newly founded York Archaeological Trust's excavations of the Roman and Viking city had added another dimension to urban history and conservation; the pressure group, York 2000, was planning a classic and ultimately successful campaign against a proposed, potentially destructive inner ring-road (Cummin 1973), while the government-sponsored conservation report on the historic city, prepared by Lord Esher (Brett 1968), had underlined the need for policies that covered many more aspects of legislation, design and economics than the hitherto relatively straightforward repair of old buildings.

In 1968 the Institute hosted the important Historic Towns and Cities Conference, which appropriately followed the passing of the Civic Amenities Act that brought into being the concept of a Conservation Area. The purpose of the Conference was to clarify 'the concept of conservation as the reconciliation of change and preservation in a historic area, within a planning context, and on an economically viable basis, and to demonstrate the possibilities of environmental improvements' (Ward 1968, ix). It was self-evident, and confirmed by two questionnaires, that such subjects were not being included in training in schools of architecture and planning, and at the instigation and with the support of the Radcliffe Trust, a full-time postgraduate course was planned which would complement a strengthened programme of mid-career courses of the type that had helped to build up the Institute's reputation. An entry requirement, at that time unique, was that applicants should have had at least four years' experience after qualifying or graduating. This eliminates the perpetual student who cannot face up to the real world, and it means that participants can build on their own experience. A course is of far greater value to them at that stage of a career, and they are often able to offer valuable contributions of their own as well as taking part in discussions based on reality. They are able to make fuller use of the experience and advice obtainable from the visiting lecturers, and they are able to collaborate in working out what are, in effect, individual courses of study and research built around the programmed lectures, seminars and visits. Fifteen was regarded as the ideal number of participants, but up to thirty have been accepted on occasions. A written dissertation is obligatory in order to qualify for an academic award – a Diploma for the first ten years of the course's life, and from 1982 a Master's degree.

NATIONAL STANDARDS AND DIVERSIFICATION

The number of institutions in the United States which offered courses increased during the 1970s, and at the end of the decade the National Council for Preservation Education was set up to establish training standards for all graduate programmes leading to a degree (Tomlan 1994, 191). The nucleus of a similar organization in the United Kingdom was the Conference on Training Architects in Conservation (COTAC), which as already recounted, had been in existence since 1959. In 1973 it collaborated with a lukewarm RIBA in a joint group, at first chaired by Bernard Feilden, to encourage part-time education. Its objectives were set out as being:

1 to provide a deeper understanding of architectural history in its aesthetic, cultural, structural and socio-economic aspects;
2 to develop the skills of analysing, assessing and recording buildings;
3 to provide a grounding in old and modern craft techniques and standards;
4 to enable the student to diagnose and remedy structural and material defects in old buildings;
5 to develop the student's ability to preserve, restore and modify old buildings with accuracy and sensitivity;
6 to familiarize the student with the laws, regulations and contractual procedures peculiar to conservation work;
7 to develop the student's understanding of the social and planning implications of building conservation work;
8 to enable the student to give authoritative advice on the maintenance and management of old buildings
9 to equip the student with a full understanding of the principles of conservative repair.

Twenty years later this still seems a sound basis for a training course although today there would probably be a greater emphasis on the relationship of conservation and design, as well as on economics and laboratory techniques.

The result of this move by the RIBA/COTAC group was the setting up of three part-time courses; one was in London at the Architectural Association, under the direction of Reginald Wood (France 1993, 53ff); a second was in Leicester at the Polytechnic (now De Montfort University) where Peter Blakesley planned a course with a practical and traditional character (France 1993, 56ff); and the third was in Liverpool at the Polytechnic where Cecil Wright's initiative was not supported by private or local government staff despite the apparent potential of a large conurbation, and after three years it was closed. The establishment by Bernard Lane in 1980 of a day-release certificate course within the Department for Extra-Mural Studies at the University of Bristol offered another pattern (France 1993, 61ff). Each of the courses in existence by the early 1980s had its own character, partly because of its location in the university or polytechnic departmental system and the availability of lecturers in related or integral subjects; partly because of its entry requirements, its ratio between taught and practical work; partly because of the necessary differences between full-time and part-time programmes; partly because of the type of conservators for whom the course was designed; but also very largely because of the discipline and experience of its founding director.

It can be argued that these differences in character were strengths of the courses, and preferable to a standard pattern or curriculum, although there was some confusion about the relative values of the different academic awards. The more recently founded courses, of which there is an ever-growing number of realizations or proposals, have maintained this diversity. Bournemouth University, for example, offers a one-year postgraduate study in architectural stonework conservation, an M.Sc. in building conservation and a B.Sc. in heritage tourist management. The well-established M.A. course in Architectural Building Conservation at De Montfort University, Leicester, is now being complemented by an M.Sc.

programme in Conservation Science designed with full-time or part-time alternative routes for candidates with either scientific or non-scientific backgrounds. The provision of a conservation laboratory is becoming regarded as essential in many centres of training. Plymouth School of Architecture (Polytechnic of the South-West) has a part-time three years' course which includes builders and engineers in its membership, while Oxford Brookes University in association with the University of Oxford's Department for Extra-Mural Studies has an optional one year full-time and two years' part-time modular programme that is attractive to Conservation Officers although not specifically for them. It also has a programme of short courses (France 1996, 54). The Royal Institution of Chartered Surveyors has a course based at the College of Estate Management in Reading which is designed not only for all the categories of surveyors but also for architects, town planners and engineers; this is by-and-large a distance learning programme with study weekend, but it includes group site visits. Other institutions are considering setting up courses, including one which it is intended will concentrate on the conservation of modern buildings. English Heritage's decision, first to list buildings of the 1930s, and then in 1995 to propose a selection of forty post-Second World War commercial and residential buildings, means that architects of my generation (or more precisely and regrettably their work rather than themselves) are brought within the ever-expanding conservation umbrella, and this increases the importance of including the recording of oral architectural history in any programme while first-hand memories survive.

But how many more postgraduate courses are needed? The country is reasonably well covered with the exception, principally, of North-East England and South Wales. How many more can be expected to attract participants or find appropriate lecturers? There are questions about the nature of training courses which need to be addressed in a changing situation.

THE NATURE OF ARCHITECTURAL CONSERVATION AND EDUCATION

Although there are common elements, both in philosophy and practice, between the work of conservators of painting and sculpture and those of architecture, there are obvious but significant differences. The west front of Wells Cathedral may be sculpture, but it is also a structure exposed to the elements; painted interiors of churches and palaces may be related to easel paintings, but they are on walls subject to rising damp, water penetration and dry rot. The architectural conservator's role is more complicated than those of his fine art colleagues, simply because he is dealing with a structure that is used (unless it is a ruin) and has to meet many utilitarian requirements while preserving, as far as possible, its integrity and authenticity. It is a responsibility which, apart from the practical considerations of fabric repair, maintenance and environmental conditions, incorporates two major elements that are virtually unique in the conservation world; one is an understanding of architectural history, and the other is creative design.

Viollet-le-Duc (1869, viii, 32) wrote that when dealing with an historic building, 'the best plan is to suppose one's self in the position of the original architect, and to imagine what he would do if he came back to this world and had the programme with which we have to deal laid before him'. But would the medieval architect in that case be restricted by medieval thought and practice in this later age? Or would he be exposed to current ideas and possibilities? That raises the sorts of question that are familiar in arguments about authenticity in other arts. Would Bach and Handel have preferred performances of their works on our present-day instruments if they could have heard them? And so on. But Viollet-le-Duc's recommendation does presuppose a deep understanding of the often complex history of a building and the original intention, in its planning, its construction, its decoration and furnishing, its original use, and the changes that have occurred to it over the years.

It has to be understood in its time, which means having a knowledge of comparable contemporary buildings and the sources of their design and construction, including the invaluable building manuals, ranging from the eighteenth-century Batty Langley and Peter Nicholson to the more recent Mitchell, and the Heavenly Twins who guided my generation of architectural students, Jaggard and Drury. A building is an open book to those who know the language, but it can be more easily comprehensible by reference to volumes that were probably used by the original architect. It has sometimes been suggested that an intimate knowledge of a building's history and its condition will more or less automatically suggest potential solutions to defects that have been identified; and up to a point this may be true. It is like the relationship between a man and his personal physician, who has his patient's history plainly in front of him. But it would be misleading to think that, as on a computer screen, a specification will be instantaneously revealed by pressing the button marked 'knowledge'. But we might imagine another button marked 'history', which acts as a caution. Once destroyed, historic fabric has gone forever, and so has all the evidence contained within it.

There is a long tradition of the architect as historian. Sir Christopher Wren was curious enough about the origins of Gothic architecture to attempt to elucidate it even though he was not especially sympathetic to it; and until relatively recently all architectural history was written by architects, who can bring a special knowledge to the investigation of a building, as well as an understanding of how their predecessors thought about their design problems, taking into account current limitations, and how they organized the preparation of drawings and the control of the building work (Linstrum 1993, 229ff).

In such an investigation, today's conservation architect may have the benefit of the experience of other experts, of art, architectural and structural historians and of archaeologists; but there is one responsibility that is the architect's alone, since he is the one member of the team who has been trained to think as a designer of a whole building. Good design and good taste are as necessary in conservation practice as in new commissions, whether we are considering the presentation of an ancient monument and its site, the repair or rehabilitation of an historic building, an addition to an existing building or group, or the very common conversion to a new use. But it must be recognized that by the very nature of an architect's vocation and training he is a creator rather than a conservator. So are craftsmen, who would rather re-carve than consolidate and repair decaying stonework, rather replace old timber than treat it for beetle or rot, rather regild than accept worn gold leaf. One of the reasons why the growth in popularity of conservation thinking was not immediately welcomed by the architectural profession was the fear that creativity would be restricted; but while self-discipline and respect for the building under treatment are required, there is ample evidence that designing within an existing architectural, urban or rural context can be a stimulus. In his ARCUK study of conservation in schools of architecture Roger France concluded that:

> Both staff and students are sure of the importance of conservation in terms of context. This is, however, seen as a matter of relating whole new buildings to their urban setting... It is a perceptual rather than a conceptual phenomenon; in other words a matter of visual compatibility rather than of principle... There was universal ignorance with respect to the existence of the Venice Charter.
>
> (France 1994b, 36)

Just how much architectural design in general has been influenced by the expansive spreading of the conservation umbrella is arguable, but outside the scope of this chapter. Its advent happens to have coincided with Post-Modernism and New Classicism, and all three probably have common roots, especially a dissatisfaction with the commercialized versions of the Modern Movement which became more or less standard practice in the post-war years; but it is worth quoting Sir John Summerson's thought that if the Modern Movement is dead (1980) it is an interesting idea – perhaps the first really inspiring new idea since the Movement was born:

> It is, anyway, liberating. It means that there may be, once again, some point in discussing architectural language, in trying to define the nature and value of ornament and entering into the whole question of architecture as a vehicle of social meaning. (Summerson 1980, 114)

All of which is entirely relevant in an analysis of old buildings and the application of conservation ethics. But conservation policies have been directly responsible for one frequently used and often abused design expedient in an attempt to maintain or enhance the character of a conservation area by preserving a building shell or a street frontage and inserting a new structure behind the façades. That too, it might be argued, is related to the emphasis on façades which is a characteristic of Post-Modernism. It has become so common in practice that it might symbolize a popular conception of conservation in the 1970s and 1980s; organizational methods and structural techniques have been evolved to meet the needs of the integration of old façades and new structures, but while it offers opportunities for creative designing, it also brings real dangers of trivialization and the lessening of concentration on the proper respect for and care of buildings of architectural and historic importance which must remain at the centre of any conservation training course.

Nevertheless, that central concern has acquired a modern image. As the conclusions of the 1989 ICCROM meeting in Ferrara summarized the situation:

> Scientific rigour is making new demands; specific research on materials and structures, the application of new products and of new methods of protection and repair, increasingly sophisticated instruments; so many new scientific contributions making the profession more and more complex and demanding to practice, therefore making indisciplinarity all the more necessary. (ICCROM 1991, 52)

The report continues this theme and provides objectives for training courses:

> On the one hand, experienced and very competent experts are needed to deal with each specific problem; treatment of humidity in walls, preservation of stone against deterioration, structural reinforcement, protection of woodwork and panelling, etc. The knowledge acquired in laboratory research and long, practical experience are essential to a specialist's supplementary training.
>
> On the other hand, the primary need remains; that is, to instruct professionals capable of formulating correct diagnoses as a result of thorough training in which both the theory and the practice of conservation are included; to have experts open to dialogue with all those who share in revitalizing the heritage; and, even more, to have people talented enough to leave positive traces of their intervention among the signs of history.

The reference to 'all those who share' reinforces the multidisciplinary collaboration in conservation projects and the responsibilities of the professionals involved. A COTAC document, (COTAC 1994) based on a first draft by Bernard Feilden and made available in 1994, identifies sixteen different professions and then details the special skills each needs for conservation. These are: Administrator or owner, Archaeologist, Architect, Art/architectural historian, Builder or contractor, Conservation or historic buildings officer, Conservator, Engineer (civil or structural), Environmental engineers, Landscape architect or historic gardens conservators, Master craftworker, Materials scientist, Building economist (quantity surveyor), Surveyor, Town planner, Curator.

THE GROWTH OF TRAINING COURSES AND EDUCATION

If we look at the ICCROM Training Index we find there are courses all over the world, but predominantly in Europe, where 57 per cent of the total are located, and in North America which accounts for 25 per cent (ICCROM 1995; 7). The 1994 edition lists 453 training programmes, but some obvious questions have to be asked before we conclude that even in these apparently well-served parts of the world, conservation education is being taken seriously. What are the objectives of these courses? At what level are they being taught? For whom are they designed? What are the qualifications and experience of the teachers? The crucial questions are – for whom? by whom? and with what objectives? Here it might be useful to make an analogy with the medical profession and ask if the courses are intended to produce general practitioners or specialists. As in the medical world, so in the building professions both are needed, but probably the general practitioners are more necessary today than the specialists, both in practice and in education. It should also be recognized, as it was at the 1989 Ferrara meeting, that a specialist is only likely to become one through long experience allied to education, and not through the latter alone. Even the best courses can only be sound foundations on which to build. But the general practitioner with a knowledge of conservation principles, values, repair techniques and a sensitive approach (the man COTAC thought should be in every practice) is too often absent, although at least 50 per cent of the workload in many offices comes within a general definition of conservation. He should be involved in the general planning and development projects, and with the detailed design.

Another analogy that might be made with the medical profession is the requirement that doctors must attend up-dating seminars, usually regionally based. This is compulsory, and the seminars make use of the latest information and up-to-date research which will be of value to general practitioners. In England, Continuing Professional Development (CPD), which offers a means of ensuring that a new or greater responsibility for conservation is properly recognized and catered for, is now a condition of RIBA corporate membership, theoretically occupying 35 to 70 hours per year. This is intended to maintain existing skills and knowledge, to enhance and broaden skills bases and areas of expertise, while searching for new directions in which to develop knowledge. A survey carried out for the RIBA and published in 1995 shows that conservation is well placed in a ranking assessment of various optional subjects. 34 per cent thought it very important, 46 per cent quite important, and only 20 per cent not important (RIBA 1995, 10ff). But what was understood by 'conservation'? Architectural history, the foundation of conservation, was at the bottom of the poll – 57 per cent thinking it not important, and only 10 per cent very. But what was understood by 'architectural history'? In connection with CPD requirements there are now one- or two-day seminars on offer in various centres, on conservation-related subjects, including those organized by English Heritage under the title 'Conservation in Depth', many of which continue the pattern of York's pioneering short courses, but there is a question about how seriously CPD is taken, and how effective it can be in 'developing knowledge and skills'.

It is understandable that courses initiated since the 1960s have been at a postgraduate level and based on the assumption that participants are already trained and qualified in one of the relevant building professions as well as having had some site experience. This was a natural response to the situation when conservation was still a minority element, although increasing, in most offices. But since then the balance between new work and the re-use of old buildings has changed in the light of a recognition that apart from other qualities they represent valuable resources of energy and materials that are irreplaceable. This situation suggests that all general practitioners should have been introduced to conservation principles and technology during their undergraduate and graduate education. But it still seems important, I would say obligatory, that a conservation architect should first be one who has undertaken a recognized course of training; without having had that comprehensive education he or she seems unlikely to be prepared to deal with a building as a whole and make the best use of it while respecting its identifiable character. Within that programme there are opportunities for internships in appropriate offices where a student can benefit from working with an experienced architect in a limited form of the old master/pupil relationship which had a great deal to recommend it.

Twenty years ago I concluded a paper on training (Linstrum 1976b) with the thought that one day we would perhaps be able not to have to make a special plea for conservation because it would have been accepted as an integral element in architectural practice and design. In some ways, in some parts of the world, we might think we are a little nearer that goal today. A welcome change in RIBA policy towards this viewpoint is implicit in the holding of a National Conservation Conference in 1995 and the setting up of regional working groups which would consider, *inter alia*, training techniques and theory, accreditation, and the raising of the conservation architects' profile. This was followed by a declaration by the President-elect in favour of 'significant changes in the schools of architecture' which would produce a 'wide variety of different types of architects . . . not only in the traditional skills but to be architectural project managers, architect planners, *architect conservationists*, architect contractors, architect administrators' (my emphasis) (*Building Design* 1995, 1).

It is to be hoped that there will be significant changes, principally in the compulsory integration of conservation principles and practice in Part II programmes, and in the appointment of experienced architectural conservators on the Institute's Visiting Panels. But it has to be a genuine, informed change if it is to be of value. Simply to fill a slot in a programme with something called 'conservation' is not enough. In 1985 ICOMOS recommended that:

> University training in conservation must be obligatory for future architects, town planners, art historians and archaeologists. Architects and town planners must be taught theory (i.e. the doctrine and philosophy of conservation, the history of architecture, art and town planning), and they must receive practical training (conservation projects, documentation and analysis of an architectural monument, technical means of conservation). (ICOMOS 1985)

A comprehensive programme which brought in all the components – construction, traditional crafts and materials, environmental services, tourist management, legislation, landscaping, and design in all its aspects – could be devised and linked with an office placement system. It would not create a specialist, but it would enable a newly qualified architect to play an informed part in a wide range of everyday conservation projects. I recall writing, twenty-two years ago:

> I believe a realistic assessment of the place of conservation work in general practice and the present tendency to examine the potential use of old buildings would emphasise the importance of revising the emphasis on new buildings and techniques in the majority of undergraduate courses... one can see a whole year of a course could profitably be occupied by such a comprehensive programme based on conservation. (Linstrum 1976a, 351)

It still could; but why confine it to one year, which seemed a daring idea in 1974? Should it not begin at the beginning before prejudices develop, to encourage a respect for old buildings, whether they are patients or neighbours in a project, and to establish that conservation and new design are two complementary halves of a whole in practice? There is no reason why the two should be regarded as mutually exclusive and irreconcilable in fact or in education. Nor is it ever too early to begin to understand the importance of knowing when to leave a building alone and to understand its significance and authenticity. Once having been established, conservation should remain as an ever-present consideration throughout undergraduate and graduate programmes along with all the other ingredients.

A similar proposition has been developed by Professor Michael Tomlan of Cornell University:

> In such a model, design courses deal not only with restoration, reconstruction, rehabilitation, new additions, and providing new contexts, but also with sympathetically retrofitting historic interiors and treating ruins. Architectural history would be rewritten in part, incorporating information gained as a result of research stimulated by preservation efforts. Examining the evolution of mechanical and electrical systems would provide the student with a better context in which to make comparative evaluations, and retrofitting structures of wood, masonry, steel and concrete would offer an almost unending spectrum of challenges.... It is an idea whose time has come... a faculty need only dare to be different.
>
> (Tomlan 1994, 93ff)

But such a programme would call for a radical adjustment in teaching methods and in the perception of many teachers.

The basic educational objectives set out above by COTAC would be recognized as obligatory by the mid-Victorian writer whose advice is quoted at the head of this chapter, and by A. R. Powys when he referred to questions 'of a most intricate nature', as well as by the group of distinguished scholars who drafted the Venice Charter in 1964. But in all these statements, and many others, the emphasis was on what were generally referred to as 'monuments', even though the Charter did enlarge the definition to embrace 'not only the single architectural work but also the urban or rural setting in which is found the evidence of a particular civilization, a significant development or an historic event'. The image of a 'monument' has evolved during the last thirty years, through the efforts of such groups as DoCoMoMo (Documentation and Conservation of Buildings, Sites, and Neighbourhoods of the Modern Movement) internationally and The Twentieth Century Society in England. But already by 1964 there were signs that the meaning of 'conservation' (or 'restoration' or preservation', depending on which country one was from) was taking on a wider interpretation.

The protection and rehabilitation of historic urban areas was the subject of a separate document at the Venice meeting where the Charter was conceived, and eleven years later the Amsterdam Charter became a documentary support in favour of integrated conservation and conservation planning in historic towns. However, the idea of a *secteur sauvegardé*, a protected area in which every proposed change is subject to strict control, had become law in France in 1962. As a result of this, certain selected central areas of towns and cities were being designated, protected and surveyed by teams of architects, planners, traffic engineers, sociologists and historians working under the direction of an architect who was appointed to assume overall control. In some cases the cost of this survey was borne entirely by the government; in others, where there was to be an extensive programme of restoration and controlled new building, the financing was partly the responsibility of a development company which was going to do the work (Linstrum 1982, 9ff). Thus, commercial interests were being introduced into the world of conservation; similarly in the Netherlands, where areas of historic towns were designated and subject to financial help, private developers were investing money in restoration, conservation and new work, while working with the authorities (Linstrum 1984, 259ff). Such innovations were introducing conservation practice into general architectural and planning practice, and there was a similar result in the United Kingdom after the 1967 Civic Amenities Act formulated the concept of the conservation area, and as the number of buildings on the statutory list of those of architectural and historic importance continued to proliferate. A direct result of both of these was the emergence of the Conservation Officer as a new member of the building professions team and an often powerful influence on new design. Just how important this responsibility was to become is apparent at the present time when there is debate about whether the very active Association of Conservation Officers (now numbering over 1 300 members) should become a recognized Institute.

THE INTERNATIONAL DIMENSION

When UNESCO initiated the training centre in Rome, which became known later as ICCROM, the avowed intention from the beginning was that it should be international in its intake, its composition reflecting the interests and concerns of its member states – originally 6 in number but now approaching 100. There was no such intention when the course in York was planned, but it quickly began to attract applicants from outside the United Kingdom, and thanks to grants and bursaries from the British Council, the European Commission, the Charles Wallace Trust, and other organizations, that too became international in composition, although with a predictably greater slant than ICCROM in favour of the Commonwealth countries. In 1977 the courses at ICCROM and York, together with that which had originally been housed at the College of Europe in Bruges and then transferred to the Katholieke Universiteit of Leuven where it was under the direction of Professor Raymond Lemaire, all achieved recognition and financial support from the European Commission. This still continues.

There can be no doubt that at one level participation in an international course can broaden a member's knowledge and understanding of what is happening in other parts of the world; but it does raise questions about pre-entry experience and understanding. To provide, for example, postgraduate training and experience for British-trained architects working in Britain within an established pattern of legislation and public participation is one thing; but to have a group from several very different cultural and climatic backgrounds is quite another. Even in Europe, where the aims in general are similar, there are different interpretations in evaluation and conservation treatment which are related to cultural traditions. Again, those of us who have been privileged to live and work in a stable society in a country, even in one part of that country, in which our ancestors have lived for generations, have put down strong roots and feel an almost anthropomorphic attachment to buildings which are part of our history. But there are relatively few countries which have enjoyed such security. Voluntary or forced transmigrations which have prevented the putting down of roots and the growth of a sense of identity with a place and its buildings are common in many parts of the world; and in all the continents which were colonized by European powers there were interruptions into a culture of continuity which have complicated the concept of conservation. As an Indian colleague expressed this:

> A lack of faith in the indigenous language, values and traditions was inflicted and the idea remained deeply rooted. As a result, even today, the architectural heritage (in the widest sense) is often considered as an obstacle to economic and social development – development is synonymous with modernization, and conservation with conservatism. It is not possible to reverse the trait or regain faith in traditional resources overnight, nor is it advisable directly to indoctrinate the idea that the built heritage is important to our present and future society. Rather has conservation to be redefined, and a re-evaluation of its objectives in this particular project has to be made. (Basu 1989, 123)

The contributions made by colonization have in their turn been integrated into the urban and rural fabric and become an element in that diverted line of continuity; but these comments by an Indian architect-planner about the complexity of conservation in her country could be echoed in other Asian, African and South American countries, and they represent an important, even crucial, consideration in education. Another writer has pointed out an important psychological and intellectual difference between the Muslim concern for the religious past and the European humanists' interest in its legacy which 'has affected more and more the process of identification, restoration, and conservation of ancient monuments or cultural legacies' (Arkoum 1990, 25). Almost a century ago Lord Curzon, the Viceroy of India who initiated practical care of the country's monuments, wrote to the secretary of the Society for the Protection of Ancient Buildings that while he was in 'general agreement with the principles which it advocates, [their application] to Indian monuments is necessarily limited by political, religious and traditional considerations and by a variety of local conditions which... render it impossible to lay down any general rule which shall be applicable to all cases (Linstrum 1995, 1ff). Similarly, at an important meeting convened in Rome in 1982 of coordinators from most of the leading courses, it was recognized that although there are common methodologies and philosophies, 'the specific history and culture of a given country, as well as the relative importance of various technical and sociological problems, determine different regional situations, and teaching should take the practical aspects of these factors into account' (ICCROM 1983, 220).

Nevertheless, there are many common denominators in conservation practice and these provide a basis in an international training programme, avoiding too great an emphasis on practice in the country where the course is located and seeking to establish what might be usefully integrated in a wide variety of consciousness and experience in an even wider social, cultural and economic range. The sophisticated techniques practised in the West are of little significance in poor countries, even in some of those European countries which are not a part of the European Union. Even if they were, they would probably be as inappropriate as imported Western architectural technology and fashions are in tropical countries. Essentially practical advice has been published by Bernard Feilden in his *Guidelines for Conservation* (1989) based on Sir John Marshall's Conservation Manual (1923) for Indian sites, and in *Management Guidelines for World Cultural Heritage Sites* (1993), of which Dr Jukka Jokilehto was joint author.

A WORLD-WIDE NETWORK

The setting up of a rational chain of regional training centres coordinated by ICCROM is not a new idea. It was being proposed in the 1970s (Feilden and Linstrum 1975, 127ff), and again at the Rome meeting in 1982 and at the Ferrara meeting in 1989. There have been several valuable moves in this direction by ICCROM in the development of regional programmes in Latin America, North Africa, the Middle East and Central Asia. A course has been set up in Tunis which could become a model for similar regional networks, and another is being planned in Teheran which will serve neighbouring Islamic countries too. The Aga Khan Foundation has also promoted seminars and courses relating to its special interests in Islamic buildings and towns; but a pattern of regular courses linked to ICCROM remains, as yet, an ideal.

There is little doubt that a national or regional training course should be directed by a native with a wide knowledge of the traditional culture and a sympathetic understanding of social needs, expectations and ambitions. Working from within a situation is preferable to imposing an external viewpoint derived in part, consciously or unconsciously, from another set of circumstances. But that does not mean ignoring the experience of conservation thinking and practice elsewhere. Can we then identify that there is a need to train teachers who will be responsible for organizing and supervising courses and for passing on to their teachers basic advice on how to teach? Is that a future role for ICCROM, York and other centres which have built up an international reputation? The urgent need for such teaching has been emphasized at several international meetings during the 1980s and early 1990s.

What form might such a training workshop take? It would be essentially one of participation and critical analysis, with led discussions on practical topics, didactic methods subjected to group criticism, potential exercises tried out with an analysis of the results, basic research and presentation methods offered for criticism, and experimental supervision of written work. All these are necessary elements in conservation education, in Europe and North America as much as in Asia and Africa. The development of sound judgment based on ascertained facts is a major requisite of any conservator, and such a workshop would concentrate on how to achieve it by education allied to experience. Just as integration has been seen as an increasingly acceptable element in all aspects of conservation in the 1980s and 1990s, so should it be in education, in recognizing the roles of the different contributors to each stage of a project, from scientific investigation of the building fabric, documentary research, recording, analysis and assessment, proposed works, interventions and additions, to the final use and presentation. Such a preparation for an educational programme in any part of the world would then allow it to develop to suit the local conditions, but with reference back to common interests in other countries.

ENVOI

Opening the 1982 Rome meeting of coordinators of training courses, Professor Raymond Lemaire said that it was doubtless because he was the last survivor of the 1966 meeting at Pistoia that he had been asked to preside. I feel I am in a similar position as a veteran attender of meetings and conferences in many parts of the world (often in Sir Bernard's company), in respect of this contribution to his *Festschrift.* By the time it is published, a quarter of a century will have passed (very enjoyably and very instructively) since I took up my appointment as Director of Conservation Studies at the University of York. My successor has kindly described that event as 'incredibly courageous... at a time when building conservation was not particularly fashionable' (Poole 1991, 16f). I cannot remember feeling particularly courageous, but certainly the many changes in attitude, especially of the building professions, to the acceptance of conservation could hardly have been foreseen. From being thought of by many practitioners as being something negative and restrictive, it has been more widely recognized as economically sound, socially beneficial, potentially positive and creative; but that only increases the need for informed knowledge about how to treat old buildings, whether or not they are listed as being of architectural or historic importance.

The world of conservation has changed out of all recognition from that inhabited by the author of the 'General Advice to Promoters of the Restoration of Ancient Buildings' in 1865, whose words are quoted at the head of this chapter, but certain elements are immutable. There is new technology, about which a conservation architect must be well informed, but that does not replace a working knowledge of traditional craft techniques and materials; and 'a wholesome dread of infringing upon [a building's] authenticity' is currently and appropriately a popular subject for international conferences. And despite all the changes, an educated and experienced conservation architect will still be 'a man of sufficient decision of character to enable him to cope with any attempts which may be made to lead him into a departure from a true conservation course' based on the traditional principles, *understanding* of a building and its continuous life, *respect* for its history, fabric and environment, and *intellectual honesty, harmony* and *taste* in its treatment.

ACKNOWLEDGEMENTS

I would like to acknowledge the help I have received in various ways from Richard Davies, Roger France, Jukka Jokilehto and Michael Tomlan in preparing this chapter.

BIBLIOGRAPHY

Arkoum, M. (1990). 'The meaning of cultural conservation in Muslim societies', *Architectural and Urban Conservation*, Aga Khan Trust for Culture, Geneva.
Basu, S. (1989). *Interventions in the Urban Core, Calcutta*, unpublished thesis, University of York.
Brett, L. (Lord Esher) (1968). *York: A Study in Conservation*. HMSO, London.
Building Design (1995). 28 April.
Chauvel, A. (1969). 'La formation, le recruitement et l'administration du personnel specialisé dans la restauration et la preservation des monuments et sites historiques en France', *Monumentum*, iii, special issue. (See also *Monuments Historiques* no.113 (Janvier–Février 1981) which is devoted to 'Profession Architecte en Chef des Monuments Historiques', and Yves Boiret's contribution to ICCROM (1983).)
COTAC (1994). *Multi-disciplinary Collaboration in Conservation Projects in the United Kingdom*.
Crook, J. M., and Port, M. H. (1973). *The History of the King's Works* (ed. H. M. Colvin) vi, HMSO, London.
Cummin, D. (ed.) (1973). *York 2000: People in Protest*. York.
De Long, D. (1983). 'Training in architectural conservation at Columbia University', in ICCROM 1983.
Feilden, B. M. (1976a). 'Restoring three cathedrals', in *Building Early America* (C. E. Peterson, ed.), Chilton Book Company, Radnor.
Feilden, B. M. (1976b). *Pro Venetia Viva*. ICCROM, Rome.
Feilden, B. M. (1978). *An Introduction to Conservation of Cultural Property*. UNESCO, Paris.
Feilden, B. M. (1989). *Guidelines for Conservation: A Technical Memorandum*. Indian National Trust for Art and Cultural Heritage, New Delhi.
Feilden, B. M., and Jokilehto, J. (1993). *Management Guidelines for World Cultural Heritage Sites*. ICCROM, Rome.
Feilden, B. M., and Linstrum, D. (1975). 'Training for conservation: A European view', in *Architectural Conservation in Europe* (S. Cantacuzino, ed.), Architectural Press, London.
France, R. (1993). *Postgraduate Courses in Architectural Conservation*, unpublished M.Sc. dissertation. Department of Educational Studies, University of Oxford.
France, R. (1994a). 'Historic preservation courses. Innovation alongside architecture in the United Kingdom', *ASCHB Transactions*, xix, 44–56.
France, R. (1994b). *Architectural Conservation in Schools of Architecture*, research report, Architects Registration Council of the United Kingdom.
Gazzola, P. (1969). 'The training of architect-restorers', *Monumentum*, iii.
Harvey, J. (1972). *Conservation of Buildings*. John Barker, London.
ICCROM (1983). *International Meeting of Coordinators of Training in Architectural Conservation, Rome 1982*.
ICCROM (1991). *Development of Postgraduate Training in Architectural and Urban Conservation*. Ferrara (1989).

ICCROM (1995). *Training Strategy in the Conservation of Cultural Heritage Sites.*
ICOMOS (1985). *Report on the Conclusions adopted by the Specialists in Training.* 20–1 November 1985, Paris.
Insall, D. (1969). 'Training for architects and planners in conservation methods and techniques', *Monumentum,* iii.
James, H. (1905). *English Hours.* Heinemann (1960 edition).
Lethaby, W. R. (1979). *Philip Webb and his Work.* Raven Oak Press, London (first published 1935).
Linstrum, D. (1976a). 'Training architects for conservation in Britain', in *Building Early America* (C. E. Peterson, ed.), Chilton Book Company, Radnor.
Linstrum, D. (1976b). 'Education for conservation', unpublished paper delivered at Ernest Cook Seminar, Leicester Polytechnic.
Linstrum, D. (1982). 'The world of conservation: Yves Boiret', *Monumentum* (March 1982) xxv, 9.
Linstrum, D. (1984). 'An interview with Jan Hengeveld', *Monumentum* (December 1984), xxvii, 4.
Linstrum, D. (1993). 'The uses of architectural history today', in *Companion to Architectural Thought* (B. Farmer and H. Louw (eds), Routledge, London. (A shortened version of the Inaugural David Saunders Memorial Lecture delivered at the University of Adelaide in 1987 and published in the *Society of Architectural Historians Australia and New Zealand: Papers* (1987).)
Linstrum, D. (1995). 'The sacred past', *South Asian Studies,* xi (discussion of Lord Curzon's contribution to Indian monument conservation).
Monumentum (1969). 'Working paper of the meeting in Pistoia', *Monumentum* iii, special issue.
Peterson, C. E. (1969). 'Comments on the Pistoia working paper', *Monumentum* iii, special issue.
Piper, J. (1948). 'Pleasing decay', *Buildings and Prospects.* Architectural Press, London.
Poole, M. (1991). 'Interview with Peter Burman', *Aspect: Views from the King's Manor,* i.
Powys, A. R. (1929). *Repair of Ancient Buildings,* J. M. Dent, London (reprinted SPAB, 1981, 1995).
RIBA (1995). *RIBA Journal* (April).
Summerson, J. (1980). *The Classical Language of Architecture.* Thames and Hudson, London.
Tomlan, M. (1994). 'Historic preservation education: alongside architecture in academia', *Journal of Architectural Education* (May).
Viollet-le-Duc, E.-E. (1869). *Dictionnaire Raisonné de l'Architecture.* Morel, Paris.
Ward, P. (ed.) (1968). *Conservation and Development in Historic Towns and Cities.* Oriel Press, Newcastle.

7

ASSESSMENT AND RECORDING: A PRACTITIONER'S VIEW

Jo Cox

INTRODUCTION

> No man is a stranger to habitation, or structures; but every one is well acquainted with some sort or other (Roger North, *c.* 1695–6)

The assessment and recording of buildings, linked to conservation projects and undertaken as a full-time occupation, is a relatively new professional specialism. There is no obvious or single training route; no specific professional body and few practitioners who do this kind of work exclusively. It has been and still is undertaken by antiquarians, architects, surveyors, engineers, historians of many different complexions, or archaeologists, each inclined to single out his or her own interest and employ his or her specialist techniques on a site. It is also still carried out, without payment, by enthusiasts, whether individual or in groups, whose interest in buildings fills spare time and who may be more knowledgeable than any professional about a national building type or buildings in their locality. Our practice, begun seven years ago, is one of an increasing number working exclusively on assessment and recording and earning a modest living from an activity that many individuals still undertake for free, or carry out as a background to their own work on site.

Written and drawn accounts of buildings have a long and honourable history stretching back far beyond any modern discipline of architectural history or archaeology. They originate in the common human impetus to understand and transmit an understanding of buildings and structures, whether for intellectual curiosity and pleasure, as suitable subject matter for paintings and drawings, to record and justify the complicated and expensive business of construction or, more urgently and more recently, as some kind of memorial to (although never a substitute for) structures that are decaying or on the point of change or destruction.

The Roger North quotation above reminds us that buildings are familiar to all people and it is difficult to find, or indeed imagine, anyone who has no interest in the various forms of shelter we inhabit and shift between. Specialists have fine-tuned, systematized and established some of the conventions of the art of finding meaning in buildings. This, however, has only been a development of a common interest for non-specialists, whose aspirations and observations reiterate that buildings tell important tales about ourselves and other people. Two accounts of buildings divided by seven centuries illustrate this point. The first is an extract from Abbot Suger's account, written in about 1144–7, of the rebuilding of the Abbey Church of St Denis in France:

> Whenever the columns were hauled from the bottom of the slope with knotted ropes, both our own people and the pious neighbours, nobles and common folk alike, would tie their arms, chests, and shoulders to the ropes and, acting as draft animals, drew the columns up; and on the declivity in the middle of the town the diverse tradesmen laid aside the tools of their trade and came out to meet them, offering their own strength against the difficulty of the road, doing homage as much as they could to God and the Holy Martyrs. There occurred a wonderful miracle worthy of telling which we, having heard it ourselves from those present, have decided to set down in pen and ink for the praise of the almighty and His Saints
>
> (*De Consecratione*, translated by Erwin Panofsky)

In a medieval Western Christian setting, the act of rebuilding the church is perceived as a miracle of shared humility where physical labour unites all levels of society in homage to God. The second account is the analysis of materials of artisan house construction in Manchester in 1844 by Frederick Engels:

> The outer walls, those of the cellar, which bear the weight of the ground floor and the roof, are one whole brick thick at most, the bricks lying with their long sides touching; but I have seen many a cottage of the same height, some in process of building, whose outer walls were but one-half brick thick, the bricks lying not sideways but lengthwise, their narrow ends touching. The object of this is to spare the material, but there is also another reason for it; namely the fact that the contractors never own the land but lease it, according to the English custom for twenty, thirty, forty, fifty or ninety-nine years, at the expiration of which time it falls, with everything upon it, back into the possession of the original holder who pays nothing in return for improvements upon it. The improvements are therefore calculated by the lessee as to be worth as little as possible at the expiration of the stipulated term.
>
> (*The Condition of the Working-Class in England in 1844*, English translation of 1892)

Here Engels understands the economic relationships that make the cottages as they are in the modern capitalist world of 1844. Both accounts use buildings to confirm a view of each writer's world at a particular time, although each author had other considerations uppermost. Suger, as it has been suggested, was eager to defend himself from any charge of personal vanity or extravagance in the campaign to rebuild St Denis and Engels was identifying, in brick bonds, the injustices to which he saw communism as a solution. There is nothing static or objective about finding meaning in buildings and being engaged in it full-time does not transform it into a science, nor does it fix its boundaries.

Each culture will have its own list of heroes and standard-setters and they will be drawn from diverse backgrounds. Here and now, that list would include William Stukeley, Wenceslas Hollar and John Britton for drawings at the 'accurate' end of the scale, as well as Augustus Pugin for correcting too much faith in 'accuracy' and showing that interpretation can be propaganda, even in drawings. Sir Banister Fletcher would have a place for a methodology that gives credit to more than one culture and the Batsford publishing house for nourishing a popular interest in the buildings of one nation. Sir Nikolaus Pevsner would be included for mapping out the buildings of a whole country in the *Buildings of England* series and being there to answer the earnest question, 'what does Pevsner say?', repeated by countless individuals in every parish church in the land, alongside Osbert Lancaster whose wit in *A Cartoon History of Architecture* (Figure 7.1) makes as much fun of our attitudes to historic structures and buildings as it does of the buildings themselves. T. Hudson Turner and James Parker would be included for the revelation of illustrative source material in *Some Account of Domestic Architecture in England* (1859–82) and L. F. Salzman for bringing written records to light in *Building in England down to 1540* (1952). Some pioneers have simply opened up whole new fields of interest, such as C. F. Innocent's extraordinarily early and brilliant account in England of what he calls 'humble' buildings, *English Building*

Figure 7.1

Very Early English, from Osbert Lancaster's *A Cartoon History of Architecture*. With perfect economy Lancaster reveals the comedy in the relationship between old buildings and living people. (*Copyright: John Murray.*)

Construction (1916), which is still an important work of reference for the historians of vernacular buildings today. Others have translated a specialism into terms that non-specialists can comprehend, for example Rowland Mainstone's *Developments in Structural Form* (1975) which kindly takes the mathematics out of engineering history. The Royal Commission on Historical Monuments both in England and Wales have raised, by example, the standards for the analysis of buildings and have educated specialists and laymen alike, as anyone reading the three recent volumes on the medieval houses of Kent will testify.

The elements of recording and assessment, then, are neither fixed nor new. They are to be found in antiquarianism, all sorts of history including architectural, industrial and engineering history, drawing and painting, and archaeology as well as in sheer curiosity about the eloquence of buildings, whether it is about the social life and ambitions of our predecessors or changes in materials, craftsmanship and methods of construction. Good assessment and recording is a hybrid, uniting different methods of inquiry: drawn and photographic record (both making use of whatever technology, budgets and circumstances permit), documentary history and good presentation, all in the context of interpretation. The balance between the elements and techniques will differ from country to country depending on diverse cultural pressures. What is novel in the 1990s is the way these elements and techniques can be combined to service conservation ends.

The increasing relevance of assessment and recording when conservation is to the fore is one manifestation of the intense interest in old buildings today and, in some cultures, the profits to be made out of the conservation business. The sheer number of specialists in the arena has increased dramatically. Conservation architects and archaeological conservators rub shoulders with lime specialists and roofing consultants. All are divided by the myopia of expertise and rank, and each speaks his or her own specialist language. However well organized and however small the job in hand, this tends towards the condition of the Tower of Babel. We must add to this the fact that most building conservation jobs involve a level of bureaucratic procedure, whether it is listed building or ancient monument consent, planning permission or public funding. This draws in any number of national and local conservation officers and inspectors who must report back to, or represent the views and conform to the policies of, political masters and grant-awarding bodies as well as negotiating terms with the owner and listening to ginger groups or irate members of the public, who also have a voice in a democratic system.

In a conservation context, the recorder or assessor is there, it seems to us, to make some sense of old buildings in terms that all these individuals can make use of and that is directed towards preservation. If an assessment is work that does not involve drawings, measured drawings always involve assessment of some kind. The fundamental questions are the same: what is it? What does it mean? How important is it? The answers are limited by the long-term aim of conservation, which is finding the best way of maintaining what it is and what it means, based on how important it is. The impetus has to be not just the record, nor even the fullest understanding, but the survival of the structure or group of buildings in good conservation terms. The fullest understanding is likely to be achieved when a building is slowly dismantled and examined, piece by piece. The best conservation project is likely to leave as much as possible of the same building intact. It is more important to sacrifice understanding than the building. There is no point in this kind of work unless you are passionate about, or at least have some kind of commitment to, a philosophy of conservation and believe that the survival of the fabric and meaning of a structure is worthwhile. This means understanding some, at least, of the conservation system within which you are working, some kind of engagement in the debate about what 'good conservation terms' are, and an appreciation of the decisions and activities that will follow or complement your work.

Assessment and recording must have their own conservation principles whilst being elastic enough to satisfy different conservation ends. These can range from extensive work to assess a whole class of buildings across a widespread area in order to assist in the creation of conservation policy, to the minutely intensive recording of an individual structure or feature to provide information that may determine the level and nature of public funding, and aid the process of designing repair or additions by revealing the meanings that can be found in any building. Rarer, unfortunately, than either of these, is work carried out to record actual conservation projects in progress and evaluate them, providing information to assist future management of the site, as well as making the most, in general educational terms, of projects where particular results can be used to judge the effectiveness of both philosophy and practice.

AIMS

The high theoretical ground, as with most endeavours, is only occupied by practitioners from time to time, but it is worth standing on it occasionally. From this viewpoint, it seems that our particular practice sets out in search of the 'quiddity' or 'essence' of an individual building or group of buildings, although neither are words we would ever include in a report. It does seem to us, however, that this will be precisely what a good conservation architect will also be seeking: to preserve quiddity that is the result of historical change and, more than likely, to add to it. Quiddity is found by two completely different ways of approaching a building, sorting for sameness and sorting for difference. We believe that these two approaches have to be part of any worthwhile project, whether the business on hand is extensive or intensive.

Sorting for sameness is to discover the broader contexts of the building. This involves using existing knowledge and existing conventions, which may be drawn from any number of available 'ologies' and 'isms'. They may be the typologies, parallels or comparisons that are familiar from the universe of architectural history, or what is nowadays called 'building history'. The latter seems to be a more neutral term and able to encompass a pigsty or pantile factory as well as a Palladian house. Existing knowledge from building history will help to raise and answer questions such as: 'how does this Cistercian abbey relate to what is known and understood about similar types of building?' 'how does a particular Gothic

Revival programme of restoration fit into a concept of the Gothic Revival?' 'how does the work of a named architect relate to the rest of his *oeuvre* in date or quality?' 'how does this vernacular building show similarities of plan and construction detail with buildings in the region, or across national boundaries, or simply a group of two or three in the neighbourhood?' or 'how early is this use of cast-iron construction or how complete is the equipment that this building was erected to house?'. This is the search for local, national and international value, and it is value judgements that determine what is conserved. The conventions of building history may need to be augmented by any other sources of information relevant to the building in hand. It may be difficult to explain the vast numbers of late nineteenth-century cattlesheds in East Kent and determine a conservation policy for them without some knowledge of the workings of the London meat trade or developments in transporting cattle at the time. A cooperage will not make much sense unless you know the processes involved in making a cask, and the form and details of a medieval castle tower will require some grasp of the developments of medieval weaponry.

In sorting for sameness the best you can hope for is to be on a permanent learning adventure, combining study of criticism and examples with a willingness never to turn down an invitation to look at a building with anyone who can educate you in any aspect of its design and development. It is good fortune if you have twenty years' experience of the type of building or group that is in hand. If you do not, you must forge and maintain good contacts with those that do. This does not have to be a matter of commercial self-interest, but can be the welcome exchange of knowledge amongst enthusiasts, and there is nothing to compare with it. The critical element in this use of a network of expertise, which will go far beyond what any individual can gather in a lifetime, is proper acknowledgement of others in anything you produce.

The second approach in the search for quiddity is sorting for difference in an intense concentration on the site in hand. What are the elements that make this structure, or building, or group of buildings unique? Careful observation, measurement, drawings, photography and documentation may each, or all, establish that the original design, materials of construction, craftsmanship and use are far from being a textbook exemplar and may even debunk received wisdom about the site in hand, or about a whole category of buildings. If measured drawings are required, observation will be all the richer and the act of measuring and drawing a building is probably the best single way to interpret it.

The impact of local circumstance and environment – geology, remoteness, climate – will all play their part. However exemplary a building or site may be, large- and small-scale changes will have taken place over time as a result of the physical, cultural or personal histories on the site. Physical change may be the result of anything from the racking of a roof, to the impact of a new transport system that brought a new material to the building. Cultural change may have sent the servants, or the livestock, off to a different location or involved the removal or addition of features associated with redundant religious or social rituals. Personal change may affect the building as it changes hands, or as a single owner had a contracting or expanding purse, or more children, or invested in a new industrial process or piece of technology that needed to be accommodated. It is the overlay of these that makes a site particular at this time and gives it quiddity, which may include disfigurement and structural damage as well as beauties and structural soundness. If the first approach to quiddity is a learning adventure so, too, is the second, more intense approach in search of the *genius loci.*

The balance between these two avenues of approach will be different, not only from culture to culture, but also from job to job. In one culture the recreation of buildings in replica may be considered to involve no loss of value whilst another may come close to sanctifying the physical evidence of the past, seeking to preserve everything from signs of physical failure to paint or wallpaper fragments from a long-lost scheme of decoration. Particular buildings may lend themselves more to the first approach. For instance, late nineteenth-century terraced housing or a hutted encampment on a military site will require a more intense consideration on sameness than an ancient house or church, rebuilt or amended time after time in response to peculiar local needs and fashion.

Finding quiddity is simply the act of defining, as clearly as possible, what the building or site actually is, what it means and therefore what there is to conserve. Once that has been established the arguments and compromises can commence.

PRINCIPLES

As with architectural solutions to a particular problem, there is no single 'correct' way of undertaking assessment or recording. Buildings and their circumstances are just too various to yield to a single approach. Working to a formula has many advantages, especially if it has developed out of experience of diverse buildings, but it is worth bearing in mind that the next project might be better served by abandoning the formula or re-inventing it to suit the work in hand. There are general principles, however, and very general they are, which are likely to underpin successful assessment and recording.

Non-intervention

As a rule, the initial stages of assessment and recording should be wholly noninterventionist and the work should preferably be non-interventionist throughout. This is a complete departure from the premise of excavation archaeology, that understanding travels hand in hand with destruction and decontextualization.

Non-intervention is a frustrating principle, especially when you know that structures on the site are likely to be interfered with and opened up by others for the purposes of repair or alteration. It makes for awkward working since it does mean that even driving a nail in to set up a level must be a considered decision. It involves distinguishing between dirt and decoration and judging whether an act of physical intervention is acceptable because it can be reversed without any damage, lifting a floorboard for instance, or unscrewing the boxing round a chimney-piece.

Quite apart from these decisions on site it does mean that you must be satisfied with an imperfect account, relative to the information that would be available if you were dismantling the building piece by piece. It is humiliating to discover that a problem area with which you have wrestled for weeks, exercising every care not to damage rotting plaster, is made crystal clear with one sharp blow of a contractor's hammer the day after your report is handed in. Nevertheless, however tempting it may be to investigate destructively, it must be avoided. The search for knowledge is not a good enough reason for causing damage and it should be possible to make a useful initial assessment without any damage whatsoever. Destruction, ironically, is what you leave to the conservation architect, the engineer and the contractor, hoping that your analysis will keep this to a minimum by indicating particularly sensitive areas or the chance of finding hidden features.

There are, of course, exceptions to the rule of non-intervention, apart from the question of reversibility, but if it is taken as the norm it means that every act of interference with the building has to be mentally justified and preferably discussed with a colleague. There are still far too many examples of wall-paintings known only from a few fragments of coloured plaster in the skip, old floorboards smashed beyond re-use, and decently-designed chimney-pieces destroyed in search of information that might, or might not, be there. On the best projects unnecessary destruction may be kept to a minimum at every stage by the presence of a good conservation architect or a persuasive bureaucrat with good systems of supervision and communication. It is far more common however, for unthinking damage to occur at every point. The stripping away of old surfaces and textures that make up so much of the beauty and character of old buildings is considered vandalism by many conservation architects, as well as being valued by owners and caretakers, and it is not the job of the assessor or recorder to vandalize a building.

Once the preliminary decisions about the conservation scheme on the site have been made, you may have an opportunity to observe and record as a result of other people's intervention. You may specifically be asked to pursue the search for understanding by dismantling parts of the building that have to be sacrificed because they cannot be repaired or because a new use simply cannot accommodate them. There may be cultural reasons, for instance the presentation of the building, for following up a destructive avenue of understanding. In these cases stripping or an archaeological dig may be justified. There may also be projects where an element of destruction is properly understood and anticipated by all concerned, as with fire-damaged buildings where the engineer or architect may need to inspect parts or all of the structure or where it is known in advance that parts of the building are going to be demolished. When such opportunities arise, they should be taken, of course, but their limits should be clearly defined. In principle intervention is the job of the architect, the engineer and the contractor and 'unscrupulous archaeological curiosity', as Feilden has described it, should be avoided.

There are practical advantages to non-intervention additional to philosophical propriety. If it is understood from the outset that your initial assessment, and the production of scale drawings, will leave the building in the condition you found it, you can be trusted to get on with your work at the time when it is likely to prove most useful. This is before the architect has developed his or her ideas and budgets have been finalized and when options are still open. If you insist on meddling with the fabric of the building, or digging holes in the ground, the proposals for your work will have to proceed through the relevant paperwork and permissions, which may delay your arrival on site to the point where your work will be only of use to historians and of little assistance to the building.

Independence

As a concentrated body of information and educated opinion about a site, an assessment should identify what there is to conserve, perhaps by finding original drawings, clarifying rarity, disentangling the changes that have occurred or indicating the likelihood of hidden features. At its best, assessment metaphorically draws out the boundaries of the arena within which other individuals can debate, make compromises and generally operate most effectively. A useful assessment should begin at a discreet distance from the vested interests, professional and commercial, that inevitably occur in conservation.

Any worthwhile report (or drawing) may contain information, as well as judgements, that are inconvenient or even a positive nuisance to its readers, including the client. If fixed ideas about the nature and future of the site have already emerged before you are contracted, your work may present a challenge to the status quo and is likely to involve, at the very least, additional thinking time and possibly far more expensive additional costs to the project. It is important not to give way to pressure to make snap judgements before you have become properly acquainted with the site. It is also important to check that you are not using borrowed spectacles or tailoring information or judgement to please. These are all pitfalls. If the building is allowed to tell its own story you have the best chance of making sense of it and you will be able to do your job in allowing the conservation programme to be as well-informed as possible. The pitfalls are less likely to occur if initial recording and assessment is properly programmed into the project and its results delivered on time.

The importance of independence might be a statement of the obvious, but assessors and recorders are not only needed by building conservationists. A client may employ you hoping that your assessment or record will support an argument in favour of destruction. Here all the questions of what 'good conservation terms' are, arise. If the ideal is the preservation of as much of the fabric and meaning of the building as is consistent with budgets and practicalities, there will always be arguments about the weighting and nature of budgets and practicalities. Nobody sets out to damage a building without some kind of reason (usually connected with status or profit) for doing so. Real independence and a reputation for it may allow you to carry out a level of work for a client who is hoping for a defence of damage or demolition, in precisely the same way as you would for a client with a different end in sight, although there will be many projects that are better turned down. You may be able to provide an educated and well-informed evaluation and/or record of the site, that establishes, say, that the building is not the earliest cruck-framed barn in Gloucestershire, or that an apparently seventeenth-century timber-framed house was mostly rebuilt in the 1920s, but be unable to support an opinion – either privately, or to the client, or at a public inquiry – that this means that it should be mutilated or knocked down. This is a difficult line to tread. If the client is informed in advance of the fact that your conclusions may not necessarily suit his ends, and is quite clear of the limits of what you are prepared to do, it may be possible to undertake work for him or her on the basis that good information should not be the exclusive province of those committed to conservation. It may even be possible to persuade the client that a conservation option might be acceptable.

A healthy distance from vested interests in the site should not mean remoteness from others working there or from the large and small decisions that have to be made by other individuals. Watching briefs are simply impossible to carry out successfully unless good personal relationships are developed, along with due respect for the pressures under which developers and contractors, particularly, have to operate. Contractors are unlikely to be helpful unless they understand the nature and purpose of the work of recording, or of watching briefs, but are often extremely helpful if they are properly acquainted with what is being done and why.

Open-mindedness

Open-mindedness is essential. It is the attitude that allows the abandonment of a favoured approach if the site and brief require it. It encourages a healthy scepticism about the mythologies that surround many individual buildings and which can lead down well-trodden but misguided paths. It ensures a readiness to pick up on some new aspect of a building that you had previously neglected or failed to observe, whether it is the way water is brought to and carried away from a medieval house or the evidence of how a nineteenth-century plasterer turned the corner of an elaborate ceiling cornice. Open-mindedness does mean accepting that if your eyes are sharper today, they must have been less attentive yesterday. In extensive survey work, where understanding is a process of slow accumulation, it may mean accepting that early efforts were relative failures and you may have to revise work undertaken at the beginning of a project. There is so much to be seen in and understood about old buildings that you never achieve a perfect understanding and it is rare to look back on any job without wishing it had been done with the additional knowledge that has been acquired since.

Good communication

It must be plain to all that there is no point, in conservation terms, in achieving the most perfect understanding of a site within the limitations of non-intervention, if you are unable to pass on that information and influence the decision-makers, the designers, and the menders and makers. The principle of good communication should penetrate recording work, report writing and formal and informal contacts with others concerned with the building. The practical aspects of this are dealt with below. What can be forgotten, is the value of communicating well, not only at all levels of the hierarchies that develop within a conservation system, but with individuals who may appear to have nothing to do with the conservation system but who do know the site well.

It is a fact of life that a maintenance man who has worked on a site for twenty years will know some aspects of it a great deal better than you do, and also better than any of the professionals who have made a small number of preliminary visits. A long-term user of the building, especially one who has seen parts of it that are now concealed, but have been exposed, is a valuable source of information. What a maintenance man knows may prove to be of interest to a structural engineer or an architect, and on a large and administratively complex site you are likely to be one of few who can carry that information across the social and physical divides. An ability and willingness to talk to anyone is a great advantage.

In the south-west of England architects (conservation or otherwise) are known privately by many contractors and craftsmen as 'a pair of shiny shoes', and plans are known as 'comics': thus 'a pair of shiny shoes came on site today and showed us the comic'. On the large conservation sites with a high profile there is more likely to be an atmosphere of mutual respect, but certainly in the English system, issues of social class are still an obstacle to good communication. In spite of the recognized worth of the opinion of the hands-on individuals on site, there can be a real reluctance by contractors to communicate constructive ideas or anticipated problems to those decision-makers who work principally with paper and who are not on site every day. To date, recorders and assessors are not well-enough established or high-ranking enough on site to be dubbed anything in particular and, especially during the slow and often dirty process of measuring for drawings, may have better opportunities than anyone else for collecting and sensibly sifting a wide span of useful knowledge and acting as a go-between at a critical time. All sources of information should be clearly acknowledged in any discussion or report.

PRACTICALITIES

On site

This is not the place to give an account of site safety or proper insurance. However, a client commissioning recording or assessment work might do well to remember that these issues may need to be checked if anyone undertaking the work (a local historian, for instance) is likely to be unfamiliar with the relevant requirements.

Time on site is a precious commodity and it is worth trying to design as much of it into the timetable as possible, whether the objective is a brief assessment or a minutely detailed account. The attempt to understand all the phases of a complex building history in a matter of weeks or even days requires a mixture of vision and doggedness. At its best it has a touch of genius to which few can lay claim. Whether it is inspired or just conscientious, it does need enough time to feel really familiar with the site. Journals and books can always be consulted a second time but the building is nearly always on the brink of change and most must be made of time on site.

A quick assessment may take no more than a day, but in our experience a second visit is always rewarding enough to be worthwhile, since it allows a useful intervening period of thinking time that nearly always raises issues that can be resolved on a second occasion, when the basics of the building are better understood. If it is only a part of a building that is the subject of the brief, it is still important to try and see all of it at an early stage. Detailed recording work is most comfortably done in good weather and long days but, when the length of the project allows, there are sound arguments for choosing to be on site in variable weather, including wind and rain. Changes of light may simply allow more to be seen, or help to explain the ways in which rooms have been lit and perhaps used, bad weather may clarify drainage problems or help to explain structural failures, or even the design of chimneys or planting used for shelter.

A systematic approach makes sense, although it will not be the same for every individual nor for every job. Some may prefer to arrive for the first time armed with some documentation or at least historic Ordnance Survey (OS) maps, others may prefer to make the first visit unencumbered by any documentary props, or even a camera, lest these should influence a close look at what is actually there. If the nature of the repair or redevelopment is uncertain at this stage, it is worth trying to put yourself in the shoes of a conservation architect and consider what is most likely to be a problem area for him or her and how the place might be redeveloped or re-serviced. An idea of which parts of the building are most at risk or most likely to be worked over will show where the most attention from you is likely to be important.

At an early stage, and preferably before writing notes, making sketches, or whatever system is preferred for recording first observations, it is important to get a general feel of the place. This means a good walk round the building and its general environment. This is rarely wasted time. Although the brief may express an interest only in the building, the nature of that building will be determined, to a greater or lesser extent, by its immediate environment. A large-scale OS map can be useful, in case features relevant to the site might otherwise be missed. It is important to get a grasp of the landscape in which rural buildings are set and to find as many different viewpoints onto the building as possible. For vernacular buildings this may identify the source of the materials of construction, the topographical reasons for a peculiar design, or the archaeological evidence of associated structures. For an industrial building it may establish, for instance, the system of transport that brought raw materials to the site and took goods from it. For an urban building it provides what is often a valuable understanding of the character and date of the area around the site in question. If photography is involved in the project, a good walk round pinpoints distant high places from which photography may be effective and some judgement of the times of the day when it will be possible to photograph the building in good light. In towns parts of a building may be relatively inaccessible to the view except from neighbouring properties and it may be worth mentally listing buildings or gardens from which an informative look or a photograph might be taken with the permission of the owners.

A close look at the building should determine the appropriate scale of drawings, if these are required, and what available techniques are likely to be most useful, whether it is the use of electronic surveying equipment or how much lumber may have to be moved in order to reach everything of interest. If the brief is already written and requires drawings at an inappropriate scale, a reasoned request to change it is usually, in our experience, accepted without renegotiating costs. The notion of a 'template' that will be adequate in every case, including the scale of drawings, should be resisted. If the building presents unexpected historic treasures on first acquaintance, or it seems that preliminary clearing or building work might put anything of interest at risk, the client should be informed immediately to make sure that nothing is lost unnecessarily.

Assessment and recording work is sometimes undertaken by gluing together the activities of different parties, with a presumption that existing plans drawn by an architect can be used, or that existing research can form the basis or take the place of a new documentary investigation. An initial appraisal of the building can help to establish how practicable this is likely to be. The conventions of architects' drawings often render them of little use in producing phased plans; they do not need the level of measured accuracy or the same kind of information that a building historian requires, especially for a complicated building. Existing documentary research may have masses of useful information, but if it is unreferenced it is probably unusable. Site observation should clarify the likely sources of primary documentation and help to judge how useful existing research may be.

Drawing and photography

There are plenty of published guidelines on the conventions of measured drawings of buildings, of different levels of detail. It is worth checking the expected final form of the drawings at the outset. Drawings produced for publication cannot have the same level of detail as large-scale drawings produced for the record. It is not usually possible to convert an A0 size drawing, done by hand rather than using a CAD system or plotter, into an A4 size for publication without losing detail, as should be plain from the Keystone drawings illustrating this text. If it is clear that publication is a possibility, hand-drawing technique can be amended for reduction.

Given the same brief and the same scale at which to draw, no two recorders will produce identical drawings, however similar their methods may be. Drawings are selective and reflect the personal idiom of the draughtsman and the way in which the building has been understood. Drawing can be an invaluable aid to understanding but the better the building is understood to start with, the better the drawings are likely to be. In conservation it is more economic to have a small number of good drawings than £10 000 worth of masonry detail or timber-framing that is not understood at all and therefore drawn badly.

Any skilled recorder is likely to pay particular attention to parts of the building that are difficult to understand or particularly complex in the hope, often fulfilled, that the problem will be solved on the drawing board, or after a period of thinking time. This is very different from setting out to produce measured drawings of a building with absolutely no experience of construction history or without that peculiar imagination which seeks to 'see' the building as it was being erected, or in the course of alterations that may have occurred centuries ago, and mentally reconstructs it with its long-dead series of occupants going about their business, passing through long-blocked doorways or counting out the money to pay the glazier for the new set of windows of which only a couple of jambs survive. This may sound fanciful but it is part of the acquisition of 'X-ray eyes', as they are known in the trade, that distinguish the best practitioners and the best quality drawings.

It can be difficult for a recorder with a background in excavation archaeology to understand that the first building they record is not a large 'small find' destined for a quiet life in a museum collection. Dutifully transferring all the irregularities and surface marks of a building on to paper or screen with the expectation that someone, sometime, somewhere, will interpret and categorize it correctly, is of very little use to a building conservation project. Unlike excavation archaeology it is not a question of recording something that is either going to disappear or be removed. When buildings are drawn in conjunction with conservation, they are recorded at only one point in their evolution. The significance of what can be seen has to be judged, in order to know how much attention it should receive on the drawing, and the building has to constantly re-imagined, in use at different times, in order to focus on what is likely to be concealed. A subtle change in the plane or thickness of a wall may be far more telling than an obvious alteration that, with a little thought, can confidently be attributed to a 1970s change in the central-heating system. A set of clean pegs in the timber joints of a medieval smoke-blackened roof, may be the only physical evidence of a massive post-medieval reconstruction that resolves a whole host of interpretation problems on the site and may influence the nature of the conservation scheme.

Good scale drawings have numerous different uses. They are a portable representation of the building, and multiple copies can take the building, and its conservation problems, into the far-flung offices and committee rooms where decisions are usually made, saving the expense of site meetings and allowing for detailed discussion that would be difficult based only on visual memory. However, scale drawings are extremely expensive to produce, whether by hand or using electronic methods. The need for them depends on the job in hand and the question of why they are needed must always be asked. Rough site drawings will be adequate for some purposes (Figure 7.2), especially extensive surveys, where there may be little time or need for the immaculate presentation that technology has taught us all to expect. The Kent Farmstead Survey, commissioned by Kent County Council and some of the Kent district councils and supervised by Jane

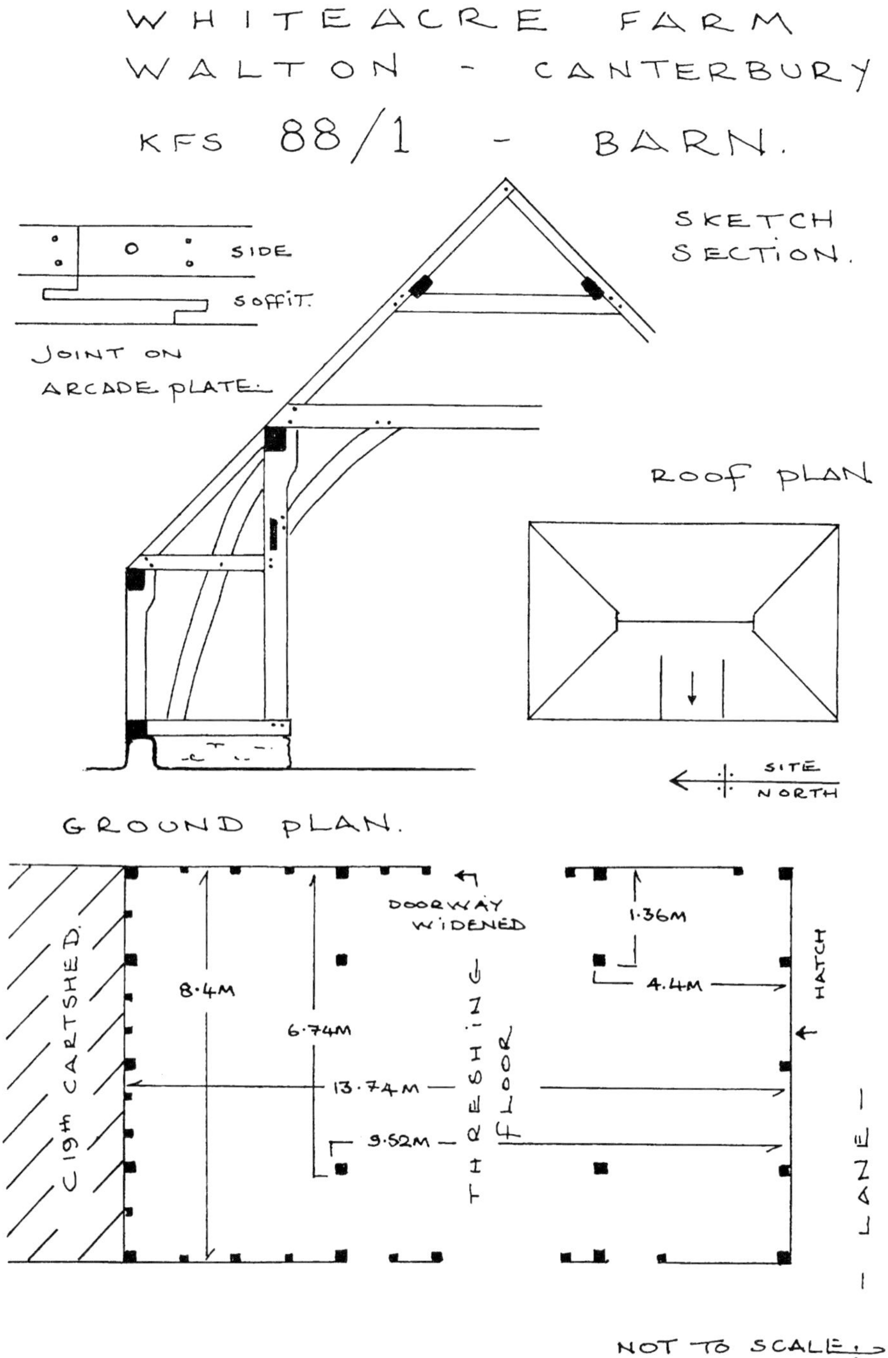

Figure 7.2

Low-level record drawing of the type produced for an extensive survey, the Kent Farmstead Survey, for Kent County Council, directed by Jane Wade. (*Copyright: Keystone Historic Buildings Consultants.*)

Wade, was designed to record as many traditional farmsteads as possible. With limited funds and the historic evidence vanishing in conversion or dereliction, coverage was felt to be the best policy. Two farmsteads per day were found, photographed and recorded at a low level with floor plans only and schematic indication of features and interesting detail. Careful detailed drawings of parts of a building only may be perfectly adequate for the job in hand. At Cadhay, a manor house near Ottery St Mary, Devon, external sculptures of 1617 of the Tudor monarchs (of whatever religious complexion) on the courtyard walls needed conservation and drawings commissioned by the owner helped both to muster grant-aid and were used by the architect to show precisely what needed repair (Figure 7.3).

There is usually a good case for floor plans, cross and long sections with details at an appropriately large scale if major repairs are to be carried out to the whole building and, although the reverse is not true, there is no reason why drawings at the accurate, rather than diagrammatic end of the scale, should not also serve the architect or surveyor who should be able to use them as the basic drawings for his or her part of the work.

At Cullacott, a neglected medieval house in Werrington parish, Cornwall, English Heritage commissioned scale drawings (Figure 7.4) with an accompanying report and these were produced before any decisions about budgets or the appointment of an architect or contractors. It was clear from the outset that the whole structure of this Grade I house was fragile and that it would need partial rebuilding as part of the repair and, consequently, a percentage of its existing fabric would disappear in the course of the work. Every part of the building was likely to be closely considered and worked over. Quite apart from recording parts of the building that could not be saved and illustrating the interpretation of the building, the drawings were used in the discussions about grant-aid and the negotiations about the nature and form of re-use for the building. The same drawings served for locating the samples taken from the roof timbers for tree-ring dating, which may help to date other local houses with roof trusses of a comparable design. The surveyor in charge of the repair was able to re-trace the drawings and re-annotate them for his purposes of information to the contractors.

The drawings on this site were hand-drawn with the minimum of intervention, which was a matter of knocking in nails for line levels as carefully as possible, and only into substantial timbers or mortar joints, or using very light pencil marks on the limewashed plaster. In spite of an archaeological consultant who recommended, in the strongest terms, knocking off wall-plaster to establish whether a blocked doorway was primary or secondary, it was left well alone and, as it turned out, a major scheme of wallpainting that was fragmentarily visible covers a good area of the medieval hall including the lintel of the door in question. This would have been damaged by investigation. The wall-painting conservators were able to use the drawn elevations for their preliminary report to indicate the extent of suspected wall-painting hidden under layers of limewash and, as more has been carefully revealed, to show this on the drawings as guidance to the contractors.

If a part of the building is likely to be destroyed in a repair scheme, or if it seems that there is no immediate prospect for conservation, however humble, large-scale

Figure 7.3

Scale drawings of the external statuary in the Court of Sovereigns, Cadhay, near Ottery St Mary, Devon. (*Copyright: John R. L. Thorp.*)

drawings, including large-scale details, may be justified. At Beetor farm, North Bovey parish on Dartmoor, the Dartmoor National Park Authority was concerned that a traditional farmstead with a fine set of disused granite farm buildings was at risk from collapse and dereliction. The farmstead included a wide range of farm buildings, in various stages of disrepair, including a butterwell and a goose-house, the

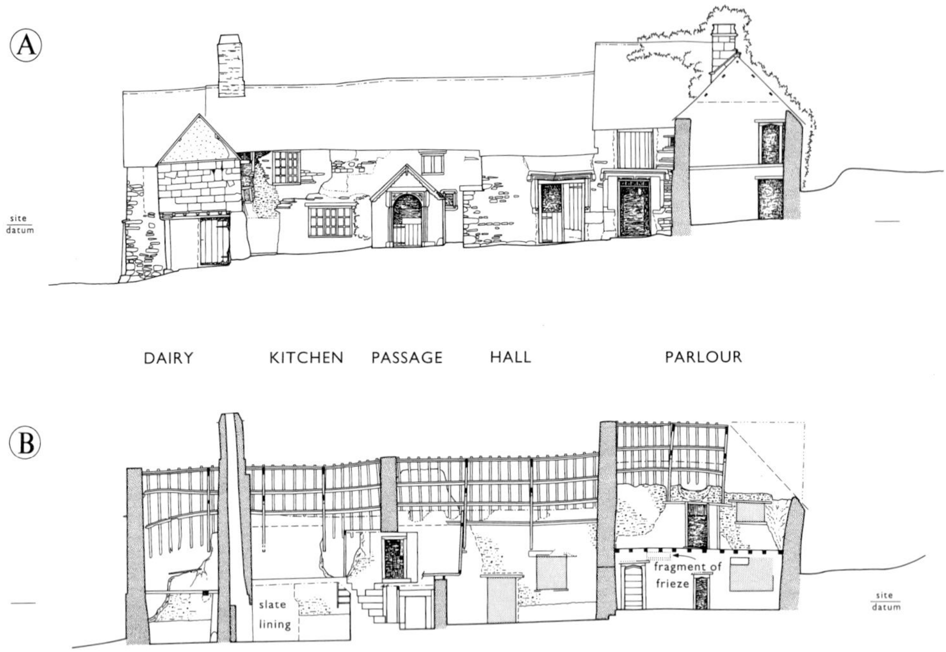

Figure 7.4

Elevation and long section of Cullacott, Werrington, Cornwall, produced as part of an assessment preceding a major repair scheme. (*Copyright: Keystone Historic Buildings Consultants.*)

latter constructed in granite and protected from the depredations of the fox by a pond in front. Although there was no immediate prospect of repair, it was decided that a detailed record (Figure 7.5) would be a good insurance policy. If the buildings can be repaired at some future date, when more of their structure has collapsed, the drawings will provide good evidence for a sensitive scheme. The double linhay, a Devon cowbyre to separate cattle of different ages, collapsed a few months after it was recorded. For this building, and any others that fall apart, the drawings as a record may help to make sense, as comparative evidence, of other traditional farmsteads, disappearing at speed, on the moor as they are all over Europe.

At Langford Court, a large listed Grade II* farmhouse near Cullompton, Devon, a thatch fire had destroyed most of the upper storey of the house, including most of the roof timbers. This was an emergency job for English Heritage. Sifting through the charred remains and knowing what type of timbers we were looking for, it was possible to retrieve and reconstruct, on the ground, the form of medieval roof trusses. The mortise at the top of one of the principal rafters showed that the truss had a saddle at the apex and we were able to search specifically for this in the heaps of timber and found it. This was one of those cases where archiving a drawn record is a good deal more convenient and practicable than storing the badly-burnt timbers for future reference. We decided that this justified a level of detail, and of naturalism that would not have been used had the timbers been *in situ* and likely to remain there (Figure 7.6).

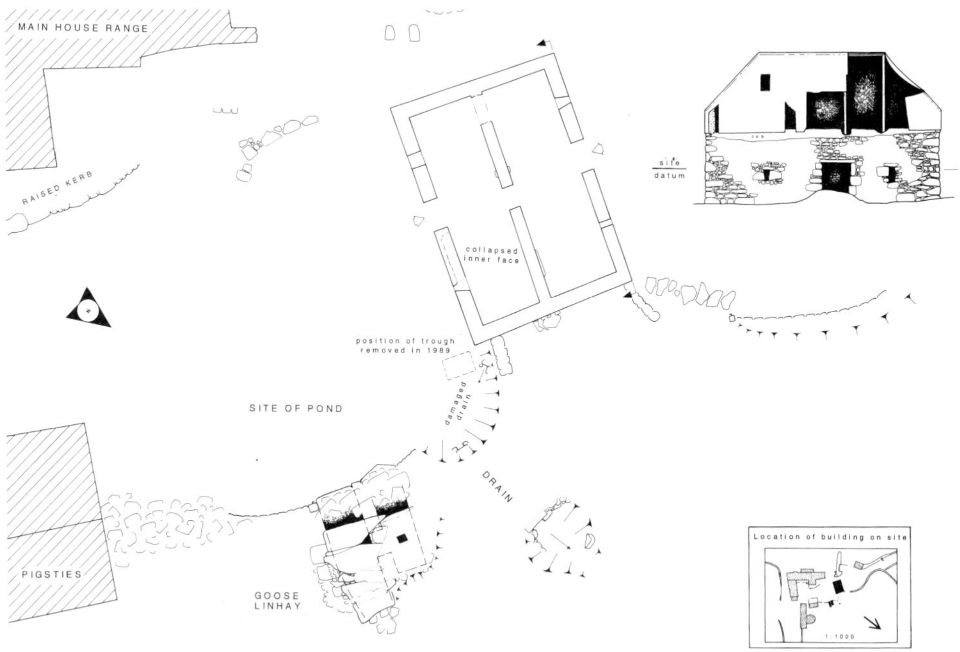

Figure 7.5

Part of a Dartmoor farmstead at Beetor Farm, North Bovey, Devon, showing a double cattle linhay and goose linhay. The roof of the cattle linhay collapsed shortly after recording. (*Copyright: Keystone Historic Buildings Consultants.*)

If there is a particularly tricky conservation passage in the building, the same impetus for detailed drawing is likely to apply. At Membury Court in East Devon, a fourteenth-century detached chapel on a farmstead that had been owned by Goldcliff Priory in Wales, had been remodelled in the sixteenth century and eventually fetched up as a cider and then a chicken house. However, amongst the high quality fourteenth-century details that survived, was a very fine east window, in a poor state of repair (Figure 7.7). This was an obvious candidate for drawing at a scale of 1:10 with sections at 1:2. The geometry involved in the drawing of the window was a good education in recovering the design methods of the masons, making use of dividers. A good recorder, with the luxury of more single-minded time on site than many others, will notice what may be less obvious small or rare items that might be forgotten in the course of a major repair scheme and deserve to be drawn in order to bring them to the attention of others (Figure 7.8).

At the current stage of technical innovation electronic means of drawn record (which are getting better all the time) are by no means always more accurate or cheaper than a hand-drawn record. It depends on the building. The advantage of measurement by hand is that it presents an opportunity (for which there is no substitute) for an intense examination of the fabric of a building, uninterrupted by the lenses, keypads, physical lumber and hierarchy of personnel that are part

Figure 7.6

A medieval roof truss reconstructed from charred timbers retrieved following a thatch fire at Langford Court, Cullompton, Devon. (*Copyright: Keystone Historic Buildings Consultants.*)

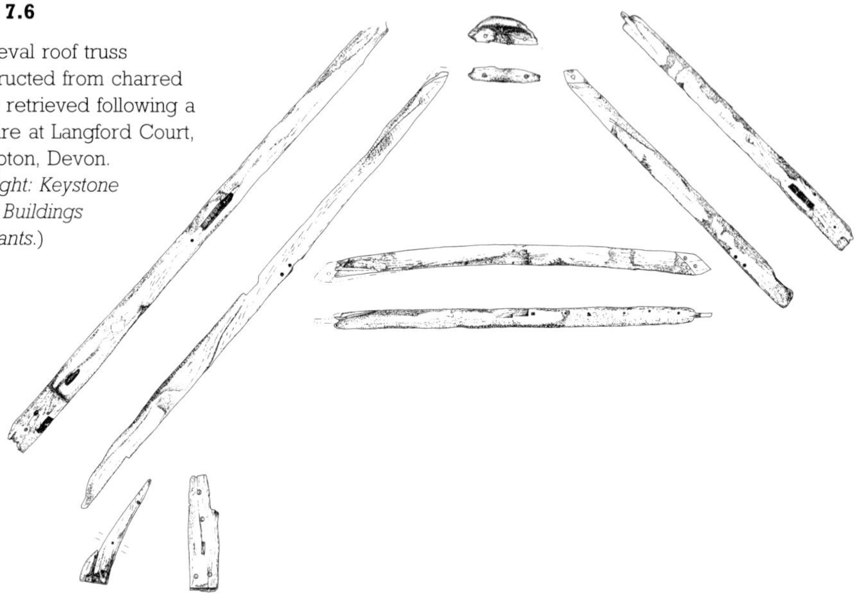

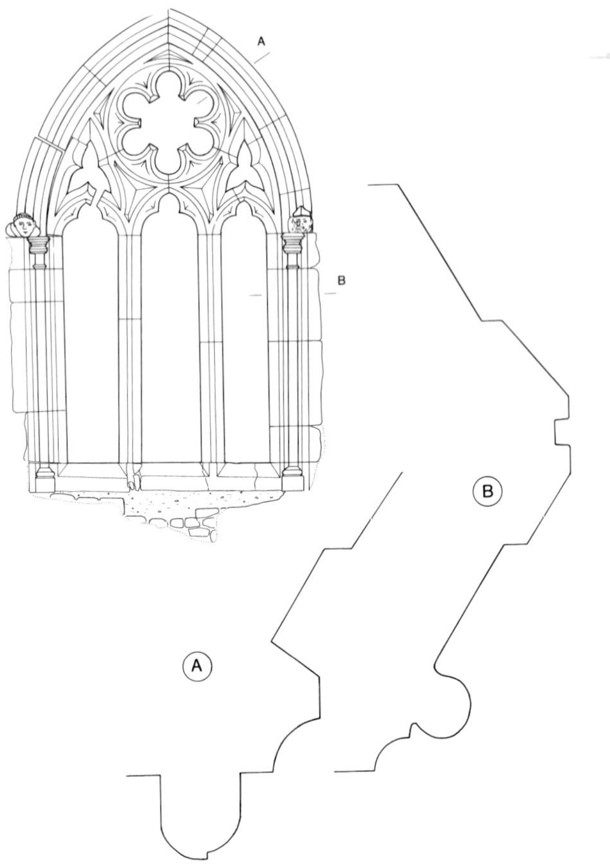

Figure 7.7

The Decorated east window of a private chapel at Membury Court, Membury, Devon, in poor condition. (*Copyright: Keystone Historic Buildings Consultants.*)

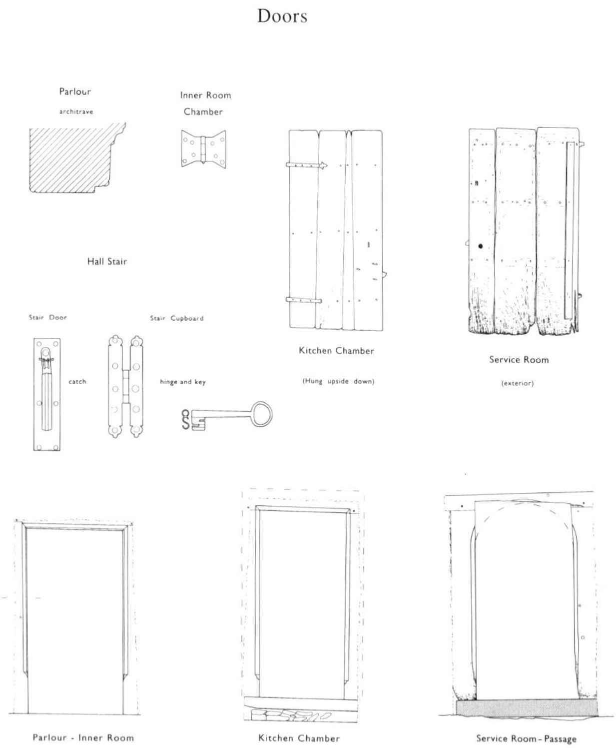

Figure 7.8

Detailed drawings of surviving features at Broomham Farmhouse in Kingsnympton, Devon. (*Copyright: Keystone Historic Buildings Consultants.*)

of the various high-technology systems. For large buildings electronic methods are often preferable and archiving and copying at different sizes, an important consideration, is certainly less cumbersome. In many cases a combination of the two systems may be the most effective and efficient.

Drawing may also include hypothetical reconstruction, if this is likely to prove useful, although the distinction between how the building appears today and what might have been its appearance in the past must be absolutely clear. At Athenaeum Street in Plymouth, smart terraced housing, built from 1829 when Plymouth was a thriving naval city, lined one of the city centre streets. The houses were occupied mostly by naval officers, gentlemen and the professional classes. Changes, consisting of repairs following bomb damage and recent (illegal) alterations to the original windows, porches, doors and iron railings left not a single house with a complete frontage. Plymouth City Council and English Heritage commissioned a project to look at the houses in detail. A reconstruction of the surviving details of the front elevations showed just how impressive the houses might look with their missing architectural detail reinstated. The eccentric glazing-bar pattern of the sash windows on the principal floors was recovered only from the scars of the original glazing bars which survived on frames of the sash windows (Figure 7.9). This helped to make the case for enforcement action to replace features that had been removed.

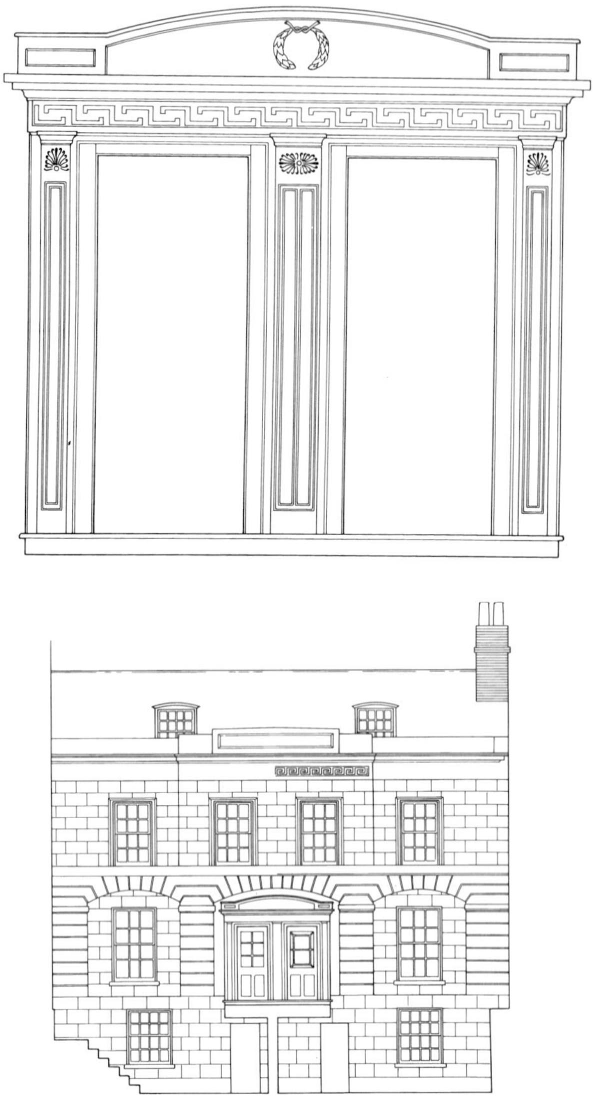

Figure 7.9

Porch and front elevation of a typical pair of houses in Athenaeum Street, Plymouth, reconstructed from surviving detail and evidence of former features from several different houses. (*Copyright: Keystone Historic Buildings Consultants.*)

Photographic recording, including the most sophisticated rectified and digitized photography, is very expensive and usually undertaken by or subcontracted to specialist organizations. It often travels hand-in-hand with assessment and drawn record and this is most effective when you are in personal contact with the photographers, rather than expecting a third party to see that the work is coordinated. Photographs can be invaluable aids to drawn record, as well as good substitutes for it in some circumstances. If a building is on the point of collapse or about to be disguised by masses of scaffolding, the speed of photography is a blessing. For the purposes of illustrating a report, or an *aide-mémoire*, a good deal can be done with quite modest equipment. It is still conventional to use black-and-white prints for record photography, for archival reasons, although colour prints may be more effective in a report and transparencies are preferred for colour publication.

Documentary research

The cost of documentary research and the skills involved are probably the least celebrated of all the effort that goes into a good assessment or record. Rather like the wicket keeper in a cricket team, the documentary historian is often the neglected member of the recording team, rarely in the limelight but crucial for a successful outcome. Research involves a vast amount of time rejecting, but having looked for, particular pieces of information as well as a continuous process of selecting what to examine and what to ignore, then choosing between the laborious business of transcription or tracing, or the swifter, but often less useful, technique of choice quotation and notes. As with the interpretation of buildings there are individuals who have a sixth sense for documentary research and this usually means an obstinate refusal to believe anything in secondary sources until the primary sources have been covered. In a perfect world, documentary research might produce drawings charting original design and subsequent changes, historic photographs and written material explicating what was done to the building, by whom and when. These are never all forthcoming.

Paper documentation has its own logic of creation and survival, and one that is often quite different from the survival of physical evidence in a particular building. Coordinating the two strands of information, without distortion of either, is not an easy task. For the buildings that, historically, have had a low status, the most that may be found is a series of historic maps and perhaps a good sequence of the names of owners and you, and indeed the client, may wonder whether the expenditure of money and time was worthwhile. However, a good researcher will almost always manage to come up with something more than this and there are exceptions to the rule that the grander the building and its owners were, the more fruitful research will be. A cowhouse may be costed and dated in a book of estate accounts while a large part of a castle, a parish church, or a seminal example of technological innovation may have left little, if any, surviving paper records of relevance to the building historian.

There is a level of documentary research into buildings that can be carried out by anybody with an interest and some grasp of the basic primary and secondary sources that a local record office, local studies library or county sites and monuments archive will provide. The tithe maps and different editions of the OS maps and plans can provide terminal dates for phasing the building. It is usually possible to find something of the sequence of owners, making use of Census Returns, parish records and various taxation surveys. Very rapidly, however, one moves into highly specialized areas, particularly for early buildings. The Public Record Office is not a place to be visited lightly. It is not easy for the untrained to acquire the skills of reading medieval Latin and only the very patient and intrepid are likely to crack some of the Byzantine indexing systems that stand between the researcher and the piece of vital information. An enormous amount of information is there, however, and also exists in less well-known libraries and archives. There is also information in private hands, and a documentary researcher should combine the skills of persuasive letter-writing to gain access with an ability to work quickly in unfamiliar surroundings, often after a gruelling journey and under the eagle eye of a solicitor or simultaneously carrying on a conversation with an enthusiastic keeper of the records.

If you do not carry out your own documentary research it is not a luxury for the researcher to spend time on site with you, discussing the building. This means that time in archives, which is always less than one would like to have, can be focused on the apparent periods of construction and change. It also makes real, for the researcher, the business of trying to solve a problem of judging status or highlighting a problem area where perhaps it seems that the building has been truncated or expanded.

At the Royal William Yard, Stonehouse, Plymouth, documentation was crucial in unravelling the development of a very large and complex site, created between 1824 and 1832, to supply food, drink and clothing to the navy and army when abroad. The Plymouth Development Corporation and English Heritage commissioned an assessment to clarify the importance and various phases of re-use of the yard and its buildings, destined for redevelopment. The complex included a bakery, brewery, cooperage, slaughterhouse and officers' housing and most of the buildings were scheduled ancient monuments. Eleven public and semi-public archives from Edinburgh to Cornwall were discovered to contain relevant documentation, and were visited for this project. Original drawings (some uncatalogued) by the engineer, John Rennie, came to light and, in conjunction with Rennie's own papers, held at the Institution of Civil Engineers, proved that the elegant cast- and wrought-iron roofs (Figure 7.10) to the buildings were original although, given their technical sophistication, it had previously been suspected that they might have been secondary. Knowing enough to recognize the importance of these roofs and enough to be aware of our own shortcomings in this specialized area of engineering history, we decided to subcontract the evaluation of the roofs to an engineering firm, Alan Baxter and Associates, whose in-house historian, Robert Thorne, has a special interest in early iron roofs.

Documentation also revealed that the massive brewhouse on site, which Rennie had designed to supply (when at full stretch in wartime) 120 tuns of beer daily, had never been used for brewing at all. In January 1831, more or less at the same time as the first equipment was being installed in the brewery, the Victualling Board decided to discontinue the beer ration of one gallon per sailor per day. Contemporary secondary sources in the shape of various visitors' guides to Plymouth repeat (what they had no doubt been told by the navy) the production capacity of the brewery and only one has a hint of the fact that the building was never used as such. This is a perfect illustration of the necessity for returning to primary sources for information.

The fact that the building was not, after all, one of the largest brewhouses of its day but had been partly mothballed, partly used for the storage of water tanks and, later, used as a slaughterhouse and rum store, was a discovery of some impact for

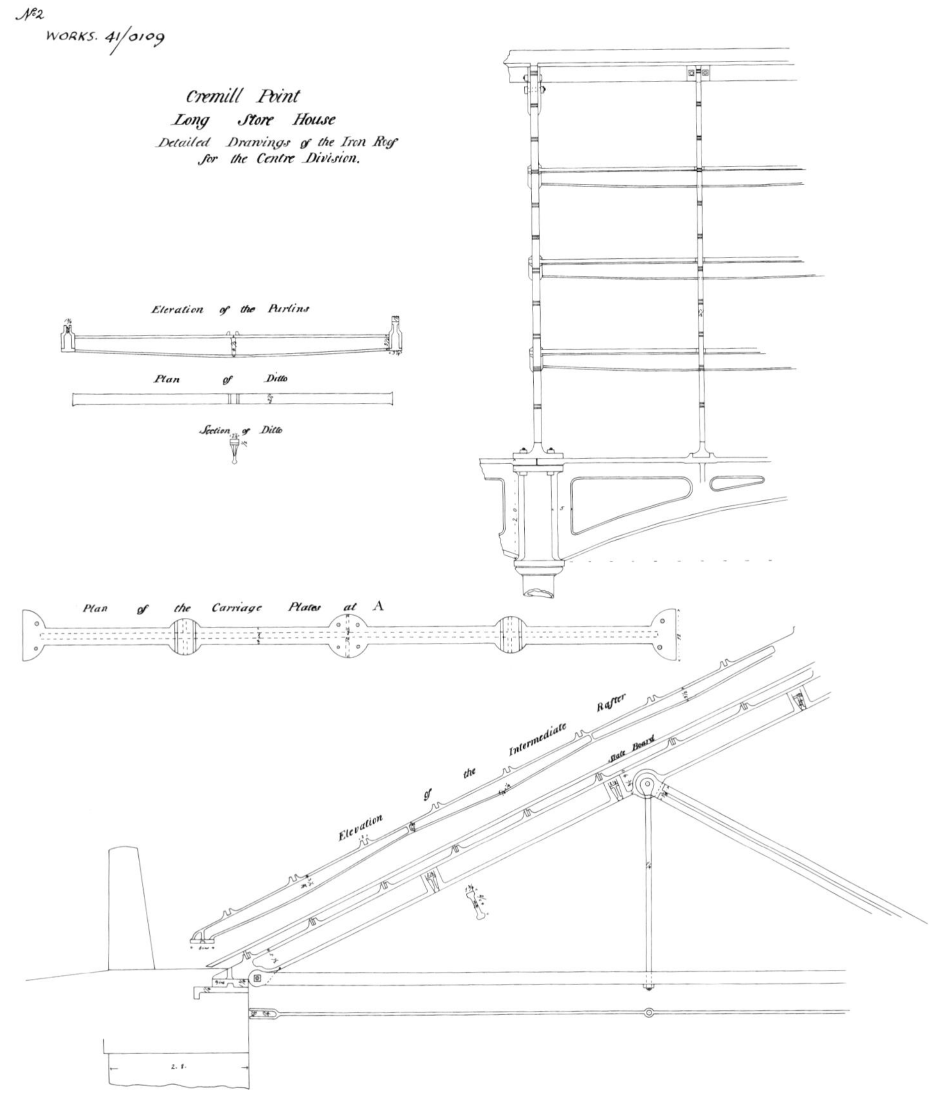

Figure 7.10

Tracing of part of one of John Rennie's drawings of the cast- and wrought-iron roof, Clarence store, the Royal William Yard, Stonehouse, Devon. The original is fragile and the tracing has been made from a photograph. (*Original held at the English Heritage Plans room. WORKS 41/109.*)

conservation purposes. It meant that instead of anticipating physical evidence of a once-active industrial process (with the implication of needing to record that evidence prior to re-use), the interior proved less important than had been expected. Other documentary evidence relating to this building has also influenced judgement of the level of acceptable change to the building. In Rennie's original elevation

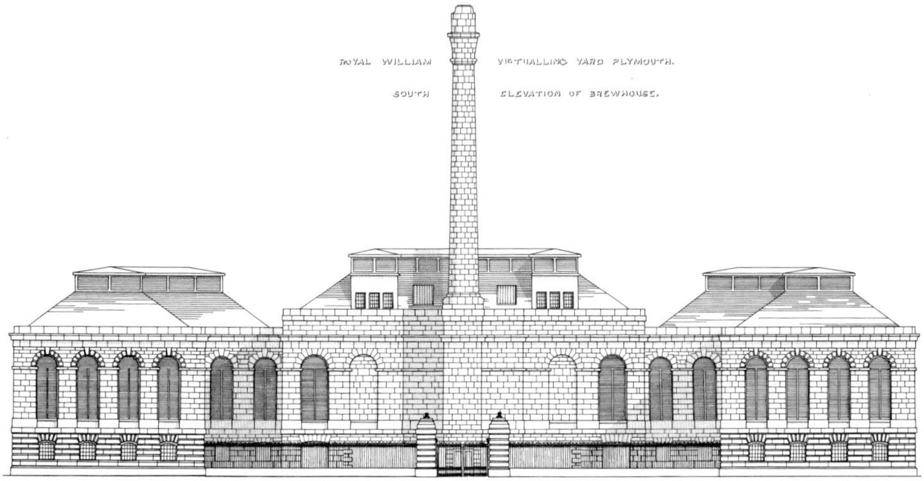

Figure 7.11

Tracing from John Rennie's elevation of the south front of the brewhouse at the Royal William Yard, Stonehouse, Devon. The original is damaged and the tracing has been made from a photograph. (*Original held by the English Heritage plans Room but archived at the Public Record Office. WORKS 41/48.*)

(Figure 7.11), the brewery is shown as a U-plan building, with a courtyard between projecting wings, the yard enclosed with iron railings. That yard is now infilled with a triple-gabled block. It had been assumed that this infilling dated from 1936, when drawings of the existing infill roofs were supplied and a cartridge case factory began operating here. In fact the origin of the infill is far earlier, and is shown on an 1872 engraving of the yard, part of a wonderful panorama of Plymouth published in the *Illustrated London News* (Figure 7.12). The infill may well have been added far earlier and was possibly the first part of the brewhouse to be used. This evidence provides a strong argument for keeping the infill in the redevelopment scheme.

At The Cobb, Lyme Regis, Dorset, West Dorset District Council, charged with the maintaining an historic breakwater on which the physical survival of the town depends, were keen to have a documented history and assessment of the structure. Initially it was hoped that this would turn up some information on the historic paved and cobbled surface of The Cobb, which was in need of repair. Some useful evidence of the paving did come to light but, as we worked through several centuries of documentation, it became clear to the Local Authority engineer liaising with us that we were also uncovering historical evidence for the long-term cost of maintaining The Cobb and the frequency of major breaches from storm damage, recorded in the old borough records of Lyme Regis.

The existence of The Cobb can be traced back to the thirteenth century, with seven centuries of costly mending, maintenance and major rebuilding. The origins of its present curious and beautiful shape, little changed since the OS map of 1903 (Figure 7.13), can be seen by 1539 at the latest, the date of its first-known illustration in a map among the Cottonian Manuscripts in the British Museum. As a breakwater, The Cobb is subject to daily erosion from an ever-changing combination of tidal and wind patterns, including the storms that, from time to time, have broken down parts of the structure altogether. Documentary research was revealing about the gradual evolution of The Cobb from its beginnings as a loose stone structure through which the water could flow easily, held in place by a cribwork of oak posts. Over a long period of time it was faced up with a mixture of undressed local stone and rather smarter Portland stone blocks, and transformed into the more solid structure of today, incorporating an active fish quay and housing a range of buildings, as well as what is now a treasured, if somewhat risky, walk for tourists along the snaking length of one of the arms. A series of reports and drawings by engineers from the late eighteenth century onwards came to light, including one of 1805 by William Jessop. The shift from an open to a more solid structure (Figure 7.14) has altered the way in which The Cobb functions as a breakwater, affected the rock on which it is founded, and influenced the silting up or scouring out of its harbour. Local fishermen,

Figure 7.12

Photographic enlargement of part of a panorama of Plymouth, published as a frontispiece to the bound annual of the *Illustrated London News* (1872).

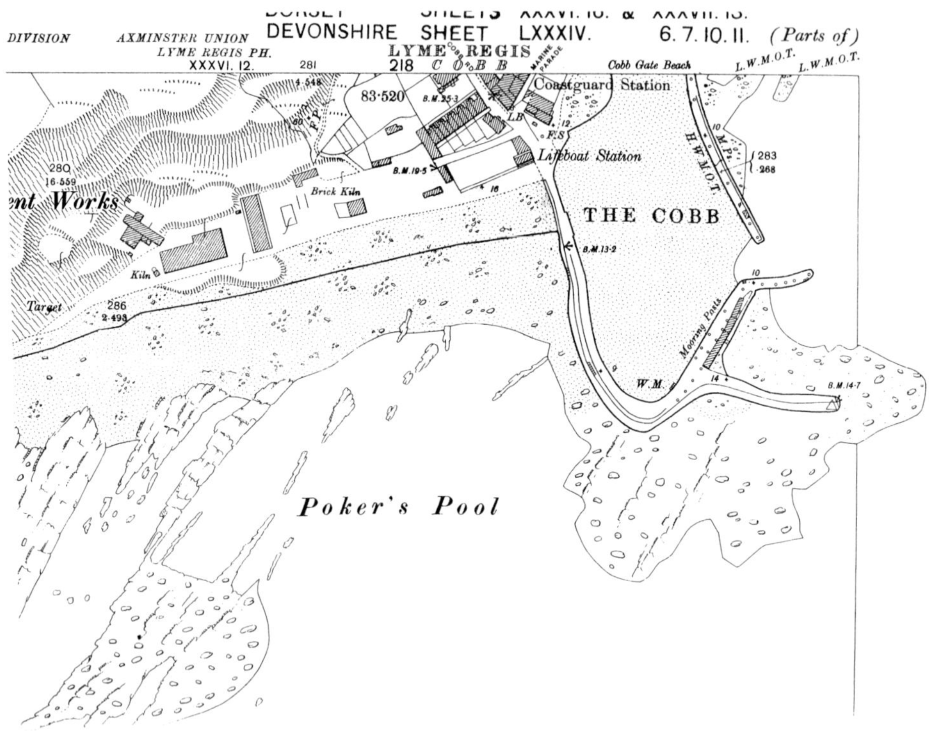

Figure 7.13

The Cobb, Lyme Regis, Dorset, shown on the 1903 OS map.

whose livelihood and safety depends on The Cobb, were intensely interested in the research and extremely helpful in directing us to documentary material and oral history that we would never have uncovered without their help.

Some of the material generated by research informed the case for grant-aid from diverse bodies, including the Ministry of Agriculture, Fisheries and Food, who were able to see how crucial the good condition of The Cobb is to the economy and physical survival of Lyme Regis town, and digest some hard facts, rather than speculation, about just how expensive it has been to maintain in the past. This was a project where, thanks to the interest shown by the client, it was possible to put the documentation to conservation and fund-raising uses that had not been contemplated at the outset of the project.

At a building of the status of Windsor Castle the extent and richness of documentation is almost overwhelming. Quite apart from the medieval records of construction, some of which have been translated and published in W. H. St John Hope in his *Windsor Castle: An Architectural History* (1913) and the various volumes of *The History of the King's Works* (1963–82), there is an abundance of surviving nineteenth-century architectural drawings, illustrating the nuts and bolts of prestigious work carried out by a number of different architects. These primary sources, some of which were seen in the course of an exercise to revise the

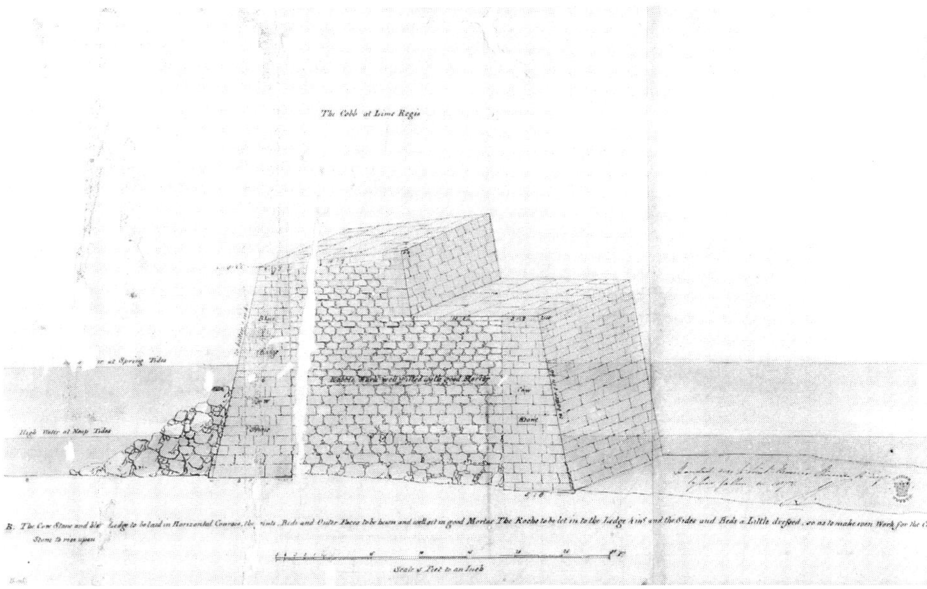

Figure 7.14

A cross section of The Cobb at Lyme Regis, Dorset, showing hearting material labelled 'Rubble Work well-filled with good Mortar' and the stone casing as 'Blue Laize [lias] & Cow Stone'. An annotation includes a date of 1872 but the original probably dates from a repair scheme of the southern arm by the Board of Ordnance following storm damage in 1792. Treasury outlay on the scheme was £9 802 5s. (*MS ADD 33231.A2. By permission of The British Library.*)

assessment of Windsor Castle in the list descriptions, suggest that there is more to add to existing judgements about castle buildings and their architects. Sir Jeffry Wyatville has been accused of transforming Windsor Castle, on which he worked in the years 1823–39, into a Hollywood version of Gothic. Certainly, his imitation of stone mouldings and bosses using armatures of wire and string to take cast plaster smacks of a theatrical set. However, like John Rennie, he also used at least one daring example of cast- and wrought-iron roofing at the castle in the late 1820s, to span the courtyard into which the Waterloo Chamber range was fitted.

The castle archives also reveal the value of old photographs. Queen Victoria and Prince Albert had a personal interest in early photography and had a photographic studio installed at Windsor. Photographs of the castle buildings show parts of the Lower Ward before work carried out by Anthony Salvin in the 1860s, as well as a providing a photographic record of the 1859 demolition of medieval buildings on the site of Henry III's great hall in the Lower Ward. Work on the North Terrace in 1883 was also recorded photographically (Figure 7.15), when the terrace was excavated to provide rooms for generators in association with what seems to have been an expansion of the electricity supply to the castle. The excavation photographs show something of the development of the north terrace,

Figure 7.15

An 1883 photograph from the Royal Archives of the North Terrace, Windsor Castle during excavation to provide rooms for electricity generators. The buttressed masonry wall of the Elizabethan phase of the terrace can be seen (to the left) and some of the brick walling of the vaults below the late seventeenth-century extension of the terrace for Charles II. (*Reproduced by gracious permission of Her Majesty the Queen.*)

regarded as a national wonder in the sixteenth century. It had been begun for Henry VIII as a timber structure for recreational and sporting use, but was rebuilt in stone for Queen Elizabeth I. The buttressed stone wall of this phase is revealed in the photograph, south of the bowed extension for Charles II, which was associated with his Star Building overlooking Eton across the Thames.

The understanding of the buildings at Windsor is advanced by the existence of the drawings and paintings of the Sandby brothers in the Royal Library. These show the castle, mostly in the 1760s and 1770s. Comparing a Sandby painting of the Horseshoe Cloister of about 1765 (Figure 7.16) with a recent photograph (Figure 7.17), the character of Sir Gilbert Scott's work here is given real force. The Sandby water-colour shows the cloister, built in 1477–81 to accommodate the priest vicars of St George's Chapel, in a state of picturesque neglect with ramshackle additions and access to some of the first-floor rooms across the leads. This was, incidentally, a common method of getting about at Windsor Castle where even Queen Victoria made her away across the leaded roof of the Dean's Cloister to gain access to the royal pew in the chapel. Scott, so often maligned

Figure 7.16

A photograph of a painting of about 1765 by Paul Sandby of the Horseshoe Cloister, the Lower Ward, Windsor Castle, built 1477–81. The masons working stone in the foreground provide some nice detail of contemporary building technology. (*RL 14555, reproduced by gracious permission of Her Majesty the Queen.*)

Figure 7.17

A modern photograph, showing the same view of the Horseshoe Cloister after restoration by Gilbert Scott (with some 1951 alterations). Scott advised restoration rather than demolition and rebuilding and left a good deal of the early timber-framing and the roof timbers intact.

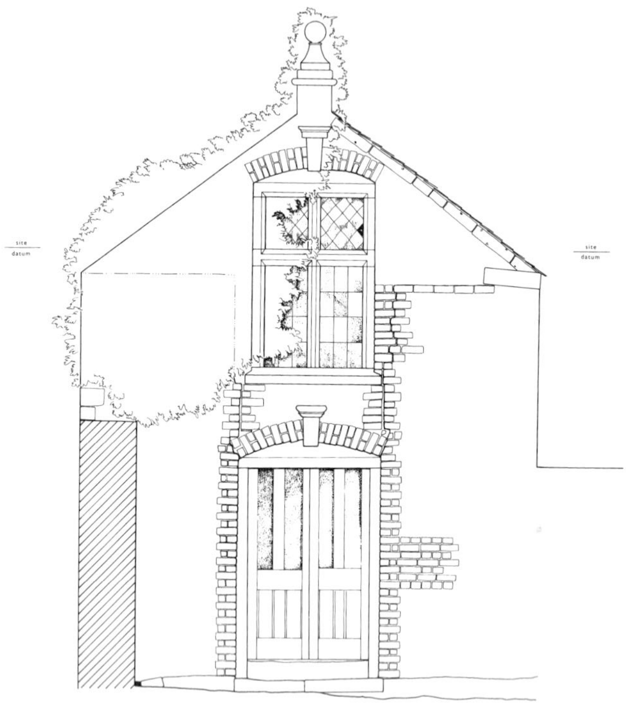

East Gazebo - North Elevation

Figure 7.18

One of a pair of gazebos in the former fountain garden at Langdon Court, Wembury, Devon. (*Copyright: Keystone Historic Buildings Consultants.*)

for acts of destruction that were, in fact, ordered by the client, recreated the cloister as a consistent piece of architecture, in conjunction with his recasting of the great west door of St George's Chapel. Although much of the romance of the cloister was lost, close observation of the building today shows that Scott preserved rather more of the fifteenth-century original timber-framed structure than the dramatic exterior alterations might suggest, and certainly more than would have survived if the Dean and Chapter had been able to afford a previous scheme which involved total demolition.

If the Sandby painting helps to identify the nature and scale of Scott's work, historic paintings can also help to put a building in context. At Langdon Court, Wembury, Devon, a pair of gazebos survive as part of the remnants of an early eighteenth-century garden. A brief to record these (Figure 7.18) for the Devon Gardens Trust led to some remarkable evidence for the design of the garden at the time the gazebos were erected. A fine provincial painting of about 1705–7 shows not only the house at Langdon Court but also its whole landscape setting (Figure 7.19). A lengthy contemporary provincial poem in praise of the house and gardens also came to light, describing some of the architectural features of the fountain garden and its statuary (long-since removed) associated with the gazebos:

Figure 7.19

Langdon Court and its garden and landscape setting shown in an anonymous painting of about 1705–7. (*Reproduced by permission of the Plymouth Museums and Art Gallery.*)

Look through the Window you'll a Garden see,
And in it read a Tragick-History,
The Hero Perseus in Effigie lives,
Andromeda with endless sorrow grieves,
Chear up, fair Nymph, nor of hard Fate complain,
To your Relief appears a valiant Swain;
But oh! She's done! She's wept herself adry,
More Tears, there's left no Matter to supply.
This Scene of Woe wou'd ev'n Medusa move,
Nay sensibly affects the God of Love;
See in the Fountain there, the am'rous Blade,
With frequent bathing cools his leach'rous Head,
Ne'er thinks of Love again.

(*The Country Seat or A Description of Langdon one of the Seats of Shilston Calmady Esq, situate near Plymouth in the County of Devon.* London, 1715. Price 6*d*.)

A 1780 lease to a gardener lists some of the planting outside the formal garden at that date (this could underpin any restoration work on this garden):

> First: 'Fruit trees planted in Langdon Gardens November 1780 whare the are to remain.
> 80 Dwaf Peach Nectarins plumbs Apricotts pears & Cherrys; NB these are against walls
> 21 standard Duke Cherrys
> 40 Dwarf Trees ye Best Appels For table
> 100 Goosberrys & Currants ye Best sorts
>
> NB these are dispersed over the Garden and not be remov'd.'

To be realistic, documentary treasures of these kinds are rare but we are surprised how often wonderful material does turn up. If sometimes a series of owners seems only ancillary to an understanding of the building, a successful documentary enterprise can give a sudden glimpse into the social context of the building. Working on a small medieval farmhouse in North Devon, Broomham, in Kingsnympton, research turned up a rare late sixteenth-century family history. This established that the erection of the house was an intrinsic part of the family's rise from humble farmers to administrators. It was built by John and Agnes Furse:

> Note here that John Furse and Annes his wyfe were too of the beste laborers that came ynto our harveste for we enjoyede moste of our londs from them. And by there menes therefore we have good case to hae them in memorye for in truthe theye were the ffyrst ffondasyon of all our credytt prayse God for hit.
> (Devon Record Office, D2507)

In fact, it seems to have been Agnes, rather than her husband, who had a real hand in erecting the house, as she is described as 'a great builder' going out with her plough team to collect the timber for the house. This information is not, of course, directly relevant to untangling the physical development of the building, but gives it a particularity which is both precious and rare for houses of this status.

Report writing

Good information and educated opinion will not count for much if it cannot be digested by a wide range of interested parties. The chances are that an assessment will have to make sense not only to a specialist who will use it as a building block in the process of their understanding, but equally to a private or corporate owner perhaps suspicious of conservation systems, unfamiliar with the jargon of architectural terminology and anxious about funding or the level of constraints that may be forthcoming. Reports are best when free of jargon and designed to interest the general reader. It should be possible for anyone who can read a newspaper to read and understand (if they wished) most of a specialist report assessing and interpreting a building. References, however, should always be properly academic, so that anyone can return to any of the sources that have been used. This is one of the few unbreakable rules of assessment and recording, as is the duty to fully acknowledge any help, whether from colleagues or the cleaner. A general reader, who will probably not bother with footnotes and the bibliography, will not be put off by their presence if the report is a good read. It can become tiresome to explain, again and again, the general understanding of the evolution of, say, the medieval house, or explain in simple terms the facemarks of medieval carpentry but some of your readers will know nothing of this and those that do know can always skip passages of familiar information.

Reports, as Feilden has argued, should be short. This is easy enough when a brief assessment is all that is required. However, when a team has spent weeks, or even months, getting to grips with a complicated and intensely interesting site, it is difficult to concentrate all the findings into a quick read. If the report is long, there is a good argument for a synopsis that is only a couple of pages (at most) at the beginning; this is probably as much as is likely to be read by members of a committee, for instance, or many of the professionals on site, although they may dip into the rest of it over a period of time. A synopsis at the beginning of each section can also be useful.

Illustrations can be a really effective shorthand for communicating information. A single page from a proposed English Heritage publication to explain and evaluate the conservation work by the Properties-in-Care section at Bowhill House, Exeter, combines text, drawings and photographs (Figure 7.20). This not only documents the work, useful for future managers of the building, but is also a teaching aid for anyone else embarking on a similar repair. Contractors, architects and historians all have immediate access to the techniques and methods of repair used, which so quickly become invisible as a conservation scheme proceeds.

There is a tendency to present documentation as appendices consisting of a list of raw references and quotations, and make little attempt to integrate it into the structural interpretation of the building. Appendices are very useful for long transcripts or lists, but it is worth making every effort to integrate the documentary evidence with the evolution of the building.

Archiving

Archiving today presents the practitioner with a real headache. Proper archiving in a busy office is often a low priority and the requirements for archivally acceptable conditions for transparencies, photographs and videos as well as drawings and reports are expensive, whether they are electronic or low-technology.

In earlier times, when information was not always considered to be a saleable commodity, public bodies (and so-called amateurs) were likely to deposit material in public archives, where it was accessible for research purposes and it was possible to avoid the all-too-common experience of the same work being carried out twice. Today, when projects are carried out for government bodies, it is still usual for the client's working copy, or an extra copy of any work, to be archived in a

1.

2.

3.

In due course, a third lift was added, beating by mallet and stick, as before. This third course was in turn stopped, quite close up under the cornice-purlin inside. This was in order that the final lift would undergo only minimal downward shrinkage when eventually placed and gone off. As can be seen from the illustrations, this strategy worked well. When downward shrinkage had apparently ceased the upper section of beamfill was added, finishing between each rafter in a sharp arris of cob.

The key lesson learnt here was that with careful planning and workmanship cob can be beaten into spaces below obstructions - in this case eaves purlins - to a level where shrinkage is minimised.

4.

7.

6.

5. Photo-record. 1, Two lifts, unpared; 2,3,4 Paring-back; 5, Final shrinkage under cornice-plate; 6, 7, Completion

Reinstatement No 2: rebuilding cob within a redundant flue, Inner Chamber

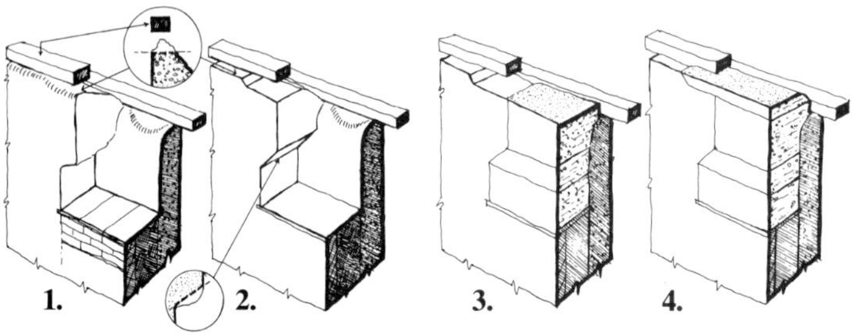

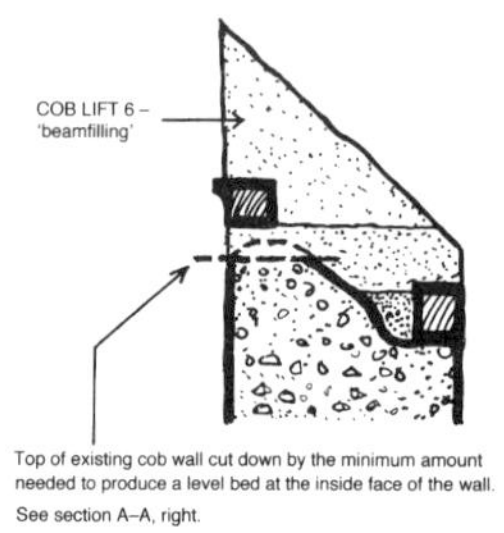

In the corner of the first floor "inner chamber" for the south range a chimney flue had been inserted in the cob wall thickness in the 18th or 19th Century. This void was to be filled. Building on the successful approach to the first reinstatement, the "requisition" for this work included sketches, reproduced here, showing the nature and extent of the operation. The flue, set within the thickness of the wall, was to be capped low down and filled in with cob above in a series of small lifts.

Illustration sequence. 1, flue capped; 2 hole edges regularised; 3, three lifts; 4, four lifts, top thin lift for minimal shrinkage; Above, cross-section of wall-head. Final acute-arris beam filling above, between rafters.

publicly accessible place, once the job has been completed. A copy in the National Monuments Record or the local County Sites and Monuments Register means that it can be consulted by any member of the public. In our experience, the systems for ensuring this are informal and do not always work. Even when they do, there is often a long time-lag before the material is deposited, since a conservation project may take several years, with drawings and reports, increasingly dog-eared, used over a long period of time.

Work undertaken for private clients is quite different. They may not want information about their property to find its way into a public archive and, even when they do, are unlikely to be willing to pay for this when a well-presented report may cost over £100 to copy. There is no easy answer to the problem of archiving. Given competitive tendering, it can be difficult to fulfil the scholarly responsibility of ensuring that work finds its way into the public domain and that something is given in return for the information that was deposited in public archives in a kinder and less competitive past. This is exacerbated by rumours of, at worst, the imminent privatization of archives or at best, the increasing commercialization of archives, which are forced to be more commercially-minded, or starved of the funds for cataloguing and making new accessions.

Practitioners should make their best efforts to see that work is deposited somewhere where it can be made use of by everybody, not only the well-off or the experts. When the client is a public body it makes sense to ask at the beginning whether they will carry the cost of an extra, clean copy of drawings or written reports to be archived in the appropriate place and perhaps to remind them of such an agreement at the end of the project. If a private client is reluctant to allow material to be deposited, or is willing, but not to pay for the costs, the practitioner may find that the only solution that can be afforded is to arrange to deposit the office archive with a County Record Office or the National Monument Record on the demise of the business or the directors.

OBSTACLES

There are two main obstacles in the path of the recorder or assessor. The first is the question of timing. Good timing is crucial to a successful project and in our experience it is rare for anybody to get this quite right. The timing for a conservation building project is an odd business. From an owner's perspective it can seem grindingly slow, as negotiating grant-aid and all the required permissions and meetings can easily take a year or more and any number of delays from any

Figure 7.20 (left)

A page from an unpublished research report to record and assess the repair of Bowhill, Exeter, Devon, for English Heritage, Properties in Care, South-West Section. Record photographs, drawings and explanatory text illustrate precisely what was done. This page shows some of the pioneering repair of cob (mud) walling. (*Copyright: English Heritage, Properties-in-Care.*)

quarter can be expected. Suddenly, however, the contractors are on site, information is needed yesterday and if it is not available, cannot be used.

Some aspects of the timing of recording and assessment are in the hands of the client. If work is commissioned too late into a conservation project it usually means that the best people for the job are not available and the work is likely to be matter of recording in mitigation of demolition, with no likelihood of influencing the conservation scheme. If the client timetables assessment and recording perfectly into the conservation scheme, timing can still go awry, perhaps the level of work and time needed has been underestimated by the practitioner, or there has been a reluctance to bring the work to a close, being in search of an unachievable perfection. Knowing when to stop is quite an art.

Most practices, including our own, have been guilty of failing to deliver a report on time, perhaps because we have disastrously misjudged the time needed or, an occupational hazard, emergency jobs (often from the same client) which have a higher priority crop up midstream. Sometimes, delays elsewhere in the project mean that a late report is still perfectly serviceable. To be legalistic, it is, of course, the responsibility of the recorder or assessor to comply with contractual obligations but, more important, what is put at risk by a failure to deliver work on time is the building (as well as one's own reputation). If it is impossible, for whatever reason, to deliver to the deadline, it is usually possible to make sure that as much of the job as is ready is transmitted to the client, particularly drawings, and perhaps an interim report where less attention than usual is paid to presentation.

The second obstacle is often connected with the first. It is a matter of how assessment and recording is used by others in the conservation process. There are cases where this kind of work is seen as a dutiful or legally-required nod in the direction of the conservation mafia and, once received, the material is never looked at, let alone integrated into the scheme in progress. This is hardly surprising if the work is not much good, or commissioned or delivered too late to make an impact. It is an unacceptable waste if good work is not exploited to the full.

CONCLUSION

We have yet to find anybody working in this area of endeavour who can lay claim to a job with which they are perfectly satisfied. This is probably an experience shared by everyone working in the conservation of old buildings. From the perspective of assessment and recording, no building or structure is ever completely understood and each project leaves frustrating puzzles. If the physical evolution of every part of the structure appears to be perfectly clear, there is always some uncertainty as to why certain changes took place. If the documentation supports the interpretation of the building at one point, you can be quite sure that there will be another major phase for which there is no supporting documentary evidence whatsoever. Everybody, however skilful, misses available information on site, or has to revise their opinion as evidence emerges in the course of works, or from comparable buildings which are better understood.

Sometimes this information arrives too late to prove useful for this particular scheme. If good quality work is produced but the building is butchered nevertheless, it also counts as a failure. No report, and no record, can be the last word on any building. If we all make use of the work of the pioneers of the past and triumphantly point up their short-comings, this is precisely what our successors will do with our efforts.

Acknowledgements

I would like to thank John Thorp of Keystone, Ray Harrison, Francis Kelly and Peter Child for advice and help with this piece.

8

USING AND RE-USING BUILDINGS

Sherban Cantacuzino

With the exception of Byzantine architecture, the first flowering of which belongs to the first millennium of our era, nearly all European architecture is the product of the millennium which is about to end. It is this second millennium which has seen the firm establishment and diffusion of familiar building types like the church, the monastery, the castle, the palace, the town house, the country house, the theatre, the hospital, the school, the railway station, the concert hall, the office block and the factory in its many and various forms.

In his introduction to *Six Memos for the Next Millennium*, Italo Calvino sees the last thousand years as the millenium of the book, and suggests that it is a sign of our millennium's end to frequently wonder what will happen to literature and books in the so-called post industrial era of technology.[1] My confidence in the future of architecture, like Calvino's in the future of literature, 'consists in the knowledge that there are things that only literature [architecture] can give up, by means specific to it'.[2] I would rather ask what future can we hold out, not for what we are able to create, but for what we have inherited, when we no longer educate our children in the beliefs which inspired that inheritance, nor even teach them its history. By converting a church into a museum or a monastery into a conference centre, we unwittingly aid and abet a new generation's loss of interest and understanding, even if, as Giancarlo De Carlo has shown at Urbino when re-establishing an old city route by unblocking an entrance in the basement of the Teatro Sanzio, the collective memory of a city is not easily destroyed.

In conserving a building, it must be desirable, therefore, to retain and even enhance wherever possible the existing use. It is perhaps fortunate that medieval and Victorian Gothic churches, with their relatively narrow nave and aisles, and particularly theatres, with their very form determined by highly specialized and specific uses, are difficult to convert to other uses. The conversion of a theatre to another use is indeed almost inconceivable and I know of no such case. With churches the recent tendency has been to maintain the building as a church but to adapt it so that it can also be used for concerts, plays and conferences.

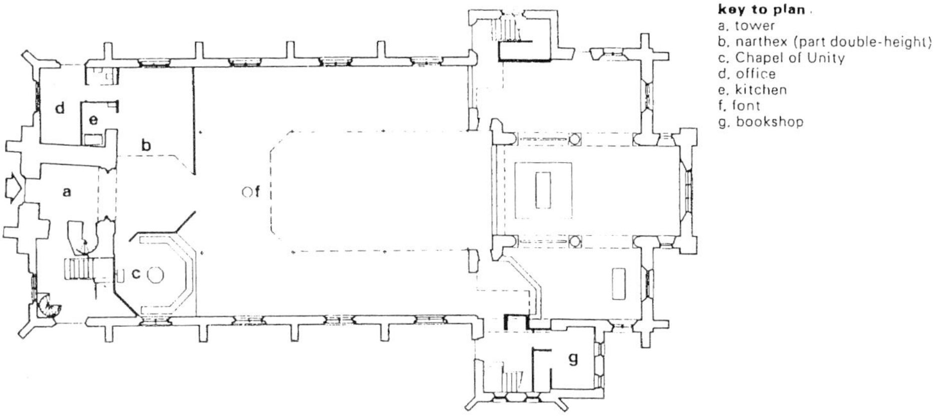

Figure 8.1

St Leonard's Church, Streatham. Plan after the restoration by *The Douglas Feast Partnership*, 1976–7.

Figure 8.2

View of the restored interior of St Leonard's Church, Streatham, with the Chapel of Unity at the back of the south aisle. (*Photo: The Douglas Feast Partnership.*)

Figure 8.3

St Mary's Church, Barnes, as rebuilt after the fire by Edward Cullinan Architects, 1983–4. The restored medieval south aisle and west tower can be seen on the left, and the vestry and choir changing room, making new use of fragments of the old walls, on the right. (*Photo: Martin Charles.*)

Examples are two fine suburban London churches, both of which were gutted by fire and extensively rebuilt, St Leonards at Streatham (Figures 8.1 and 8.2) and St Mary's at Barnes (Figures 8.3 and 8.4). The work to St Leonard's demonstrates a more conservative approach than the radical reconstruction of St Mary's. After the fire only the tower and spire, the external walls, stone arches and cast-iron columns were left standing. The window tracery was faithfully renewed in Bath stone and a new timber-trussed roof erected with its underside left exposed over the nave as a substitute for the more costly stone vaulting lost in the fire. Only in the octagonal Chapel of Unity at the end of the south aisle, with its floor-to-ceiling plate glass window, is it possible to identify an addition which is unquestionably of the twentieth century, both in concept and in the way it is detailed. At St Mary's, on the other hand, only the medieval south aisle was retained and restored to provide a narthex to a new and re-orientated nave also roofed with a timber truss, but of a highly imaginative and wholly contemporary design.

Figure 8.4

The new nave of St Mary's Church, Barnes, showing the exposed timber roof truss. The two piers in the foreground belong to the medieval south aisle which has been retained and restored. (*Photo: Martin Charles.*)

CONSERVATION VERSUS PRESERVATION

A broad definition of architectural conservation might be keeping a building alive. It may become necessary, in keeping a building alive, to make changes and to introduce new elements. There is no reason why conservation should exclude either demolition or new construction. The application of these principles is fully evident in the two aforementioned churches. The same principles can be applied to city quarters and whole cities, area conservation to distinguish it from the conservation of individual buildings. The conservation of Venice includes controlling the *aqua alta*, cutting down on the pollution of air and water, protecting the setting of the city by preventing the construction, for example, of a rapid transit system in the lagoon, repairing and maintaining existing buildings, finding new uses for redundant old buildings, (Figure 8.5) and building new buildings where necessary (Figure 8.6). In this way conservation becomes an integral part of urban planning policy and it becomes possible not only to conserve individual buildings but to

Figure 8.5

The Molino Stucky, a nineteenth-century flour mill on the Giudecca in Venice, for a long time has been considered worthy of preservation, but so far none of the new uses proposed has found favour with developers. (*Photo: Sherban Cantacuzino.*)

Figure 8.6

Recently completed housing in Venice which clearly belongs to the late twentieth century, yet in form and character continues the city's building traditions and fits in well. Facing the lagoon at the western end of the Canale di Canareggio, the housing is clearly visible to anyone arriving in Venice by train or car. (*Photo: Sherban Cantacuzino.*)

protect their setting and look after the spaces between buildings and the wider environment as a whole. Much of the good of conserving buildings will be undone if those buildings then find themselves in the flight path of a busy airport, overwhelmed by vehicular traffic or overshadowed by high buildings.

Even the relatively distant presence of high buildings can be disastrous, as is only too evident in the view of Jerusalem from the Garden of Gethsemane (Figure 8.7). For high buildings, if ill-designed and wrongly sited, can do great harm by destroying famous skylines and overshadowing fine streets or buildings, and open spaces like parks, squares or river banks. Yet there is little effort by planning authorities to study the effect at planning application stage and so prevent unpleasant surprises. The larger the open spaces and the lower the surrounding buildings are, the greater is the threat. The vast courtyards of Beijing's Forbidden

Figure 8.7

Jerusalem seen from the Garden of Gethsemane. The skyline is penetrated by tower blocks. (*Photo: Sherban Cantacuzino.*)

Figure 8.8

The Forbidden City in Beijing, with its vast courtyards and relatively low surrounding buildings, is particularly vulnerable to high buildings, even at a considerable distance. (*Photo: Sherban Cantacuzino.*)

City (Figure 8.8), one of UNESCO's World Heritage sites, will remain vulnerable to high buildings in the surrounding areas of the city as long as the Chinese economy is booming and planning regulations based on the simple geometry of lines of vision are not applied.

If area conservation is an integral part of town and country planning policy, it follows that it must come under the direction of the authority which exercises planning control. This authority, of course, must include among its senior staff properly qualified members for whom development and conservation are not incompatible and protecting the setting of a monument does not exclude the construction of new buildings in its vicinity, provided these are appropriately designed. It has been particularly difficult to persuade Third World countries of the need to bring architectural conservation into planning because conservation in these countries has been limited to isolated monuments or sites and directed by a department of antiquities staffed with archaeologists, while planning is in the hands of regional or municipal authorities. Conservation is too often equated with conservatism and so an obstacle to development. It is gratifying, therefore, for the many international experts who began advising the authorities in India more than ten years ago, to learn that the Urban Development Department of Bombay has recently sanctioned the listing of 633 buildings, the designation of 8 conservation areas including the very large Fort precinct, and the setting up of an heritage conservation committee.

The concept of conservation areas and area conservation are of particular importance because they help to make conservation part of overall planning policies. They are also important because they discourage the treatment of buildings as monuments seen in isolation and direct attention to the setting of buildings and to the spaces between buildings. Conservation policy as part of planning policy must take account of social and economic factors and address itself, not only to the problem of historic preservation, but also to new development and therefore to the problems of height and density, and of the infrastructure. For the real problem is not the conservation of the individual monument or historic building, but of the built environment as a whole where it is threatened by new infrastructural needs. Nations everywhere are attempting to reconcile the conflicting demands of maintaining their own cultural identities and yet achieving technological progress. One way of helping to resolve this conflict is by finding new uses for old buildings, thus providing the opportunity not merely of maintaining the fabric of the old building, but of introducing completely new standards and services.

THE PROBLEM OF AUTHENTICITY

The difficulty of approaching the conversion of old buildings in such a radical way is the inevitable loss of, or damage to, original fabric. It is a paradox that, in order to maintain the fabric of an old building, it is often necessary to destroy some of it in the first place. This approach is at loggerheads with another narrower definition of conservation: preserving or restoring the integrity of the interior spaces of a building, while at the same time keeping, and where necessary consolidating, the existing fabric with the addition of as little new material as possible. Raymond Lemaire has remarked that the value of a building has always depended on its use, beauty and ability to provide an historical source. It is the last, a building as the essential means of discovering and understanding the past, which has dominated twentieth-century conservation doctrine and given rise to the conflict between artistic and historical authenticity.[3] The restoration of Uppark after a fire has resulted in a building which remains artistically authentic but is only partly historically authentic because so much of the interior fabric has had to be renewed. Cesare Brandi in his *Teoria del Restauro* distinguishes works of art from industrial products, the restoration of which could make them function again.[4] In the same way it would not be unreasonable to regard all those buildings which came out of the Industrial Revolution – mills, warehouses, maltings, railway stations, etc., – as industrial products of a functional kind, to which new functions must be given when they become redundant in order to keep them alive.

Single-space structures, like churches, railway sheds or exchanges, should not be subdivided and made into cellular structures. To do so is to change their very nature. It may sometimes be necessary to preserve a church for its landmark qualities and, if no other use can be found for it, accept the subdivision of its interior into flats. It must be recognized, however, that in such an irreversible conversion the artistic authenticity of the building – the single unified interior space – would be sacrificed and, assuming no change to the outside, the correlation between exterior and interior would be lost, so that it would no longer be possible to 'read' and understand the building.

The widely accepted principle of reversibility in conservation, of being able to reinstate a building by removing any additions with the minimum of damage to the original fabric, clearly could not be applied in such a case. But the removal of later additions, however reversible, is discouraged nowadays because these additions tend to be considered as part of the history of the building, their removal resulting therefore in the loss of historical authenticity. These apparently contradictory principles can only be resolved by treating each case on its own merit and making a personal judgement. Rooms conceived as single spaces but subsequently divided with partitions to suit an inappropriate use, for example the great rooms of a castle, converted into army barracks, must obviously be restored by the removal of the offending partitions, however much these partitions may be part of the history of the building and whatever the damage to the original fabric. I would like to think that the same would apply to a church subdivided into flats, if the opportunity ever arose of turning it back into a church. In such cases artistic integrity is more important than historical authenticity.

Decisions however, are not always so straightforward. In converting All Saints at Oxford into a college library, the floor of the church was raised to give more headroom to the floor below. The danger, all the greater in classical architecture, lay in changing the proportions of the church space. In fact the classical proportions of the church interior had already been compromised by the removal of the old box pews, on the backs of which the main pilasters rested. By lifting the new floor to this level the proportions were actually improved and it is possible to conclude that the decision to raise the floor has been fully justified by the result.

Dealing with redundancy and finding new uses for old buildings clearly do not fit into my second and narrower definition of conservation which puts great emphasis on authenticity, especially authenticity of fabric, including even plaster, paint and other surface treatments that used to be considered expendable and renewable. Three celebrated examples, the Maltings at Snape in Suffolk (1967 and 1970) (Figures 8.9, 8.10 and 8.11), the Bush Warehouse at Bristol (1976) and the Cannery at San Francisco (1968), were all gutted and their roofs and interiors reconstructed, so that only the walls and their foundations remain authentic. All three are examples of industrial buildings which had become redundant and for which new cultural and commercial uses were found: a concert hall for the

Figure 8.9

Snape Maltings, Suffolk, seen across marshland and the River Alde. Some of the buildings were converted by Arup Associates in 1967, and again in 1970 after a fire, for the use of the Aldeburgh Festival of Music and the Arts. (*Photo: Arup Associates.*)

Maltings at Snape; an art gallery, theatre, restaurant and offices for the Bush Warehouse; and a market centre with shops, cafes and restaurants for the Cannery.

Converting buildings in this way accords both with the principles of finding new uses which will make a redundant industrial building function again, and with my first definition of conservation as keeping a building alive. It is a fact, moreover, that the majority of conversions to new uses are of industrial or commercial buildings which are rarely of the first rank in an artistic sense and which are usually listed the lowest grade if they are listed at all. The loss of original fabric is therefore more easily acceptable, especially if the new work has flare and imagination, and makes a distinctive contemporary contribution which might itself become worthy of listing in the future. Such might have been the case with Eva Jiricna Architects' brilliant proposal to turn the redundant Grade I listed church of St Mark's in North Audley Street into the London office of a Swiss bank (Figures 8.12 and 8.13), a proposal which, though fully reversible and approved by English Heritage, was refused planning permission by Westminster City Council. In determining applications for listed building consent, it is necessary, I believe, to make the quality of the design of the conversion work or of the new building, if it is to replace an existing listed building, a material consideration. But to do this would require those in authority to know what constitutes a good design and to have the confidence to make judgements.

Figure 8.10

The large malt house at Snape Maltings which has been converted into a concert hall. The lower building on the left has been converted into dressing rooms at ground level, and a restaurant and kitchen on the first floor. (*Photo: John Donat Photography.*)

Figure 8.11

The auditorium and stage of the concert hall at the Snape Maltings, which was created out of the large malt house. All intermediate floors and dividing walls were removed, and a new roof structure was designed. (*Photo: Arup Associates.*)

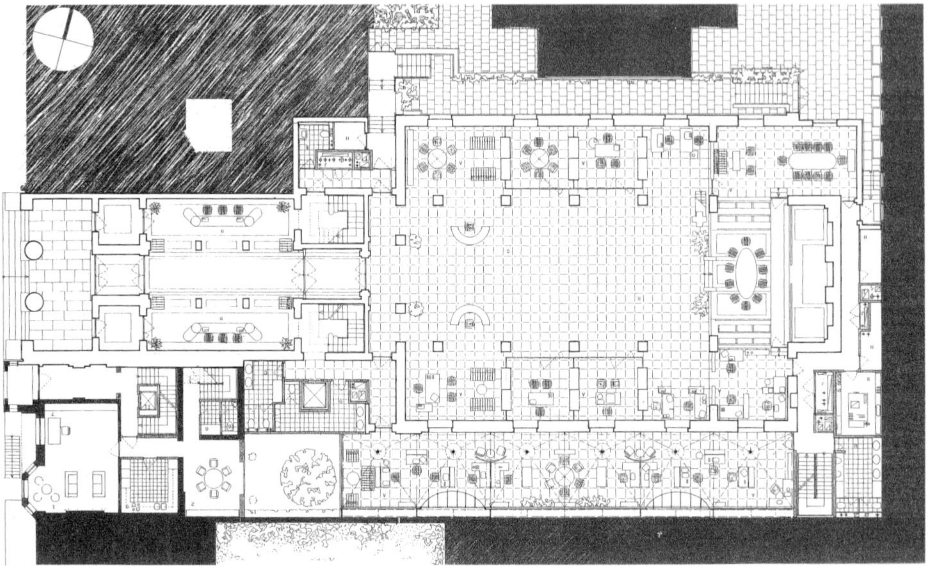

Figure 8.12

Ground-floor plan of St Mark's Church, North Audley Street, London, as converted into offices for a bank. The plan shows the nave used as a waiting area, the aisles as open but screened offices and the chancel as a conference area. An extension to the south, containing open offices on two levels, has been designed with an independent structure to have the minimum impact on the historic building and to be more readily reversible.

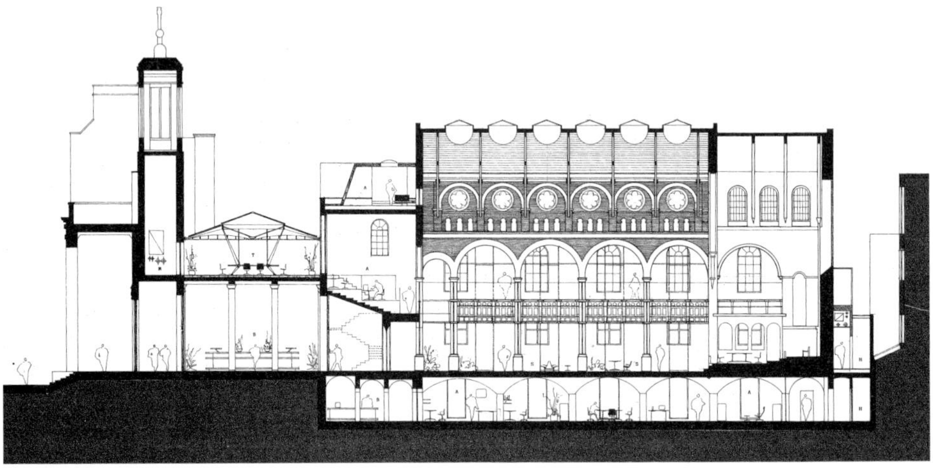

Figure 8.13

Longitudinal section of St Mark's Church, North Audley Street, as converted into offices, showing the crypt used as offices and a boardroom inserted under a new roof above the reception area.

DEVELOPMENTS SINCE 1967

To make the quality of the new work a material consideration would go some way to refute the view, which is increasingly voiced, that conservation has gone too far and is stifling creativity. Conservation has certainly come a long way since 1967, the broad review of which is of some interest. Attitudes began to change following the Civic Amenities Act of 1967. Soon after I went to the *Architectural Review* as assistant editor in that same year. The editor, J. M. Richards, who had published in the 1950s, with John Piper as photographer, a special issue and book entitled *The Functional Tradition*, illustrating a large number of mills, warehouses, maltings, etc., which were then mostly still in use, suggested that I might go back to some of these by now redundant products of the Industrial Revolution and examine the possibility of converting them to new uses. The result was a special issue entitled *New Uses for Old Buildings*, published in October 1972, and a book of the same name, published in 1975 to commemorate European Architectural Heritage Year.

In the 1930s the *Architectural Review* had been the mouthpiece of the Modern Movement but it had also, at the end of that decade, fought successfully for the preservation of Carlton House Terrace. In the 1950s and 1960s Nikolaus Pevsner's presence on the editorial board ensured the inclusion of architectural history as the subject of regular essays, while Gordon Cullen's 'townscape' articles broadened our perception and understanding of urban spaces, and Ian Nairn's 'outrage' column taught us how to defend our environment. A special issue on housing, edited by Nicholas Taylor, appeared in November 1967 and with its passionate attack on high-rise and equally passionate support of low-rise, especially in the form of the traditional terrace, was influential in promoting a more humane attitude to the form and design of public housing. In the early 1970s Covent Garden, bereft of its market and threatened by a destructive GLC road plan, was the subject of a special issue which directly caused a large number of buildings to be listed and the road plan to be stopped. Similarly, a special issue on Bath was influential in preventing the destruction of the city's minor Georgian architecture which even so eminent an historian as John Summerson had considered unimportant. The emphasis in all this was on group and area value, on spaces between buildings, on the setting of monuments rather than on the monuments themselves. The Civic Amenities Act 1967 had indeed identified the need for conservation areas and had required local authorities to designate these.

The oil crisis of 1973 brought about the notion of sustainable development. It was realized that old buildings often do their job better than new ones, that they tend to be more flexible, more amenable to change, more durable and especially more efficient in energy conservation. The oil crisis made it clear that we could no longer afford to demolish buildings which still had life in them. This was not only because old buildings were better at conserving energy, but also because the actual work of repair and rehabilitation in energy terms cost a fraction of new building. Relating to this was the fact that conversion and rehabilitation were labour-intensive. They created employment and kept both the small and the large builder busy, whereas comprehensive redevelopment and tower blocks relied on factory-made parts and mechanical assembly, which only the few very large builders could handle. When, at the beginning of the 1970s, people began seriously to re-use old buildings, the cost of repairing and converting them was considerably less than that of equivalent new buildings, especially if 'equivalent' was taken to mean the use of structure materials and finishes similar to the old buildings. But as conservation became more demanding and conservation techniques more sophisticated, so repair and conversion work became more costly until by the mid-1980s it reached the level, or even surpassed the cost, of new building. It has become necessary, therefore, to put values to the historical and artistic qualities of the old building, and to the psychological benefits derived from retaining the familiar and so maintaining continuity with the past in order to argue that preserving old buildings by maintaining them in their old use or converting them to new uses is good value for money. It is hardly necessary to add that in that period there has been an enormous growth in cultural tourism, so that looking after a country's architectural heritage and encouraging tourism, which has become a significant source of income and wealth, adds weight to the economic argument.

Within this broad review of the progress of architectural conservation in this period, and with particular regard to the adaptation of old buildings to new uses, it is interesting to note certain developments in the multifarious and diverse Muslim world, and in two countries, Italy and America. In 1976 the Aga Khan called for a better awareness of the aesthetic, cultural and social aspects of architecture on the part of both architect and client; and in 1978 the Aga Khan Award for Architecture was established to encourage an understanding and awareness of the strength and diversity of Muslim cultural traditions which, when combined with an enlightened use of modern technology, will result in buildings more appropriate for the Islamic world of tomorrow. Prizes totalling US $500 000 would be given for the best buildings every three years. These prizes, however, were the tip of the iceberg and had to be seen both in the immediate context of Award activities as a whole, and in the wider context of the Aga Khan's other initiatives related to the built environment. Award activities included seminars, field trips, architectural surveys and documentation; other initiatives included the Aga Khan Program for Islamic Architecture at Harvard University and the Massachusetts Institute of Technology, and the quarterly architectural periodical *Mimar*, which looked beyond the Muslim world to the whole of the Third World. All these activities constituted, in Mohammed Arkoun's memorable phase, 'a space of freedom'[5] in which ideas could be discussed without any of the restrictions on freedom of thought that political totalitarianism invariably imposed.

From the beginning the Aga Khan Award for Architecture gave restoration, conservation, re-use and rehabilitation due importance. For the first award in 1980 the jury created categories which included 'Restoration' and 'Preservation of Traditional Heritage'. In 1980 these categories included four projects, among them the conversion of Sinan's Rustem Pasha caravanserai at Edirne in Turkey into a modest hotel (Figures 8.14 and 8.15) and the extensive restoration of Safavid monuments at Isfahan in Iran, the documentation of which by the Italian Institute for the Middle and Far East (ISMEO) had a parallel in the 1983 award

Figure 8.14

Sinan's Rustem Pasha caravanserai at Edirne in Turkey has been converted into a simple hotel, receiving an Aga Khan Award for Architecture in 1980. (*Photo: The Aga Khan Award for Architecture; Mustafa Pehlivanoglu.*)

Figure 8.15

One of the courtyards of the Rustem Pasha caravanserai at Edirne after conversion into an hotel. (*Photo: The Aga Khan Award for Architecture.*)

in the work carried out by the German Archaeological Institute in Cairo to the Darb Qirmiz monuments. These numbered seven and included the eighteenth-century Fountain of Abd al-Rahman Kathuda, of which the upper-level madrassa was to have been converted into a coffee house (Figures 8.16 and 8.17). For the 1983 award, among the nominations which did not receive prizes were the Dar Lasram, a nineteenth-century house in the medina of Tunis, converted into a cultural centre; the Ksar Haddada at Tatahouine in Tunisia, formerly a communal Berber grain storehouse in the form of multiple vaulted cells, converted into communal housing (Figure 8.18); the mediaeval Khan Marjan in Baghdad, converted into a restaurant: and the large sixteenth-century Palace of Ibrahim Pasha in Istanbul, converted and extended into a museum and conference centre (Figure 8.19). More ambitiously, in Jidda a number of old houses had been restored by the municipality and were awaiting new uses, while in the UNESCO pilot study for Fez the possibility had been examined of using traditional houses for primary schools, dispensaries, and social centres.

Figure 8.16

The Fountain of Abd al-Rahman Kathuda in Cairo, part of a programme of restoration undertaken by the German Archaeological Institute in Cairo which received an Aga Khan Award for Architecture in 1983. (*Photo: The Aga Khan Award for Architecture.*)

THE ITALIAN CONTRIBUTION

In Italy, a country with a long and unique tradition of creative design, it was demonstrated early on that new elements in a conversion, which are designed in an unmistakably contemporary manner, add value and enhance the historic interior where new elements imitating the historic interior confuse and detract from its quality. This bold and experimental approach first manifested itself in exhibition design, both temporary and permanent, all of it setting a trend which influenced the approach to converting historic buildings when this gathered momentum in the 1970s. Toni del Renzio writing in the *Architectural Review*[6] singles out G. Pagano's Mostra Leonardesca in Milan (1939); the exhibition of old masters drawings at the Brera in Milan (1942); Silvio Radiconcini's permanent exhibition room at the Istituto Centrale del Restauro in Rome (1942); and, in 1952, the restoration and rearrangement of the art gallery in the sixteenth-century Palazzo Bianco in Genoa by Franco Albini. Then, in the middle of the same

Figure 8.17

The madrassa on the upper level of the Fountain of Abd al-Rahman Kathuda in Cairo, which was to have been converted into a coffee house. (*Photo: The Aga Khan Award for Architecture.*)

Figure 8.18

The Ksar Haddada at Tatahouine in Tunisia, a communal Berber grain storehouse which has been converted into communal housing. It is an unusually interesting example of a new use which was unsuccessfully nominated for an Aga Khan Award for Architecture in 1983. (*Photo: The Aga Khan Award for Architecture; Ismail Serageldin.*)

Figure 8.19

The sixteenth-century Palace of Ibrahim Pasha at Istanbul, converted and extended into a museum and conference centre. (*Photo: The Aga Khan Award for Architecture.*)

decade, came three works which set completely new standards; the first also by Albini, the Museo del Tesoro in Genoa (1954–6), placed tomb-like beneath the small courtyard which separates the apsidal east end of the cathedral from the archbishop's palace. The other two works were the conversion into a museum of the Castello Sforzesco in Milan (1949–56) by the BBPR studio (L. Belgiojoso, E. Peresutti and E. N. Rogers) and Carlo Scarpa's rearrangement of the Museo Abatellis in Palermo (1953–4).

Scarpa's best known museum conversion is the Castelvecchio at Verona (1956–8 and 1962–4). For a long time a barracks, the castle first became a museum in 1926, but suffered from bomb damage in 1945. Scarpa's genius is best illustrated by the placing of the fourteenth-century equestrian statue of Cangrande I Della Scala (to whom Dante dedicated the 'Paradiso') which was removed from the Scala tombs in Santa Maria Antica (itself an act which goes against best conservationist practice) to form the centre piece and all-pervasive presence of the new museum. The statue was set dramatically at an angle high up on the first floor. It was reached by a bridge and protected by a deep roof overhang, both elements being modern additions to the historic building. In the en-suite sculpture galleries on the ground floor, the vertical surfaces of the reveals in the arched openings were lined with single pieces of rough-hewn stone, an effective way of contrasting one plane with another and of emphasizing the arches. The floor stopped short of the walls to leave a gutter-like feature which articulated the joint between horizontal and vertical surface, while the great double steel T-beams supporting the first-floor structure had a gap all round where they penetrated the wall to stress the difference between new and old. For the same reason, new windows filling old openings were set well back, keeping clear of the wall plane.

At the Querini Stampalia, a modest palace on the Campo Santa Maria Formosa in Venice, converted into a library and an institute (1961–3), the same principles applied. Scarpa's work, which was limited to the ground floor and garden, included the provision of a fully glazed wall to the new lecture hall. A short colonnade of Tuscan columns on the garden side enabled him to set the glazing line behind this strong architectural feature. In the entrance hall, part of the floor was raised above flood level and part kept at the old level to allow the water from the canal to penetrate the building. Stepping stones provided a means of access for those who arrived by boat and entered through the iron gates. Like the Castelvecchio, the Querini Stampalia shows Scarpa's way of wedding new work to old, but on a small scale, without the dramatic qualities of the museum, yet richer in detail.

At the city planning level, the 1960 plan for Gubbio by Idolosalire and Leonardo Benevolo became the model for Giancarlo De Carlo's plan for Urbino, adopted in January 1964 and published in 1968, and for Pier Luigi Cervellati's 1969 plan for the historical centre of Bologna. Any conservationist plan at that moment in Italy was bound to draw not only on immediate precedent, like the plan for Gubbio, but also on a legal framework that went back to the general principles laid down in the laws of 1931 and 1938. These stated that later additions to a monument must be respected for their historic or aesthetic value; the use of monuments must be compatible with their structures; all restoration must be the minimum necessary and as simple as possible; and all existing buildings around monuments must be retained. These principles showed a concern for careful restoration and a respect for the minor architecture which plays such an important part in the setting of a monument and in the total fabric of a town. De Carlo's plan for Urbino was based on the premise that the character of the historic town centre must not be altered and that any necessary expansion outside the walls of the town must be designed to make as little impact as possible on the surrounding landscape. An essential part of the plan was the proposal, revolutionary in Italy at the time, to exclude motor traffic from the historic town centre and make it a pedestrian zone served by a minibus system. Urbino's economy would in future depend, not as it had depended, on an agriculture which was becoming obsolete, but on its university of which the student population had risen from 139 in 1936 to 10 000 in 1966, exceeding the resident population in term time and bringing new life to the town; and on a burgeoning tourism which was to expand in an unprecedented manner over the next twenty-five years.

The first step advocated by the plan was the recuperation of the depressed area south of the town centre and Palazzo Ducale, where there were several redundant convents in a semi-derelict condition which were difficult to convert to residential use. Two such convents were converted at an early stage to provide university faculties of law and education, the buildings in each case undergoing radical internal alterations to make the fullest possible use of their courtyards and open spaces.

In the case of the law faculty, completed in 1970, its two courtyards were excavated to provide space for a library, the plastic domes in the library roof structure providing the reading room with daylight, while also lighting the way into the building across the roof at night. Two years later a similar conversion, inspired by the law faculty at Urbino, was carried out by Luciano and Giorgio Giovannini at Macerata. Here a seventeenth-century Barnabite monastery was transformed into several university faculties. The main work consisted of the construction of a three-storey library in the courtyard, stepped in section so as to allow light to penetrate the lower storeys of the surrounding wings, and with a roof garden, across which a steel-framed glazed link provided access to the library from every part of the building. The library was accommodated by both excavating the courtyard and raising the level of the courtyard with the new library roof (even improving the proportions of the courtyard). At Urbino and Macerata large libraries were inserted at the heart of the complex, while lecture halls and common rooms were fitted into the existing fabric of the surrounding wings. The outward appearance remained effectively unaltered, while all the major changes took place in the private world of the courtyard. This is a valid solution with wide implications, for one is here considering the rehabilitation of a whole city block of monastic buildings, a continuous web of cellular structure arranged around courtyards which can absorb large new volumes and accommodate a wide variety of activities.

The 1969 plan for the historic centre of Bologna involved the total regeneration of the existing built fabric, adequately integrated by social services and equipment, for the city was ruled by the Communist Party which had adopted a decentralizing policy by setting up eighteen neighbourhood councils, four of them in the historic centre. The plan proposed two main lines of action: to rehabilitate so as to maintain the local working-class population, and to appropriate and convert some of the city's monuments for community use. Pursuing these two lines of action depended on the municipality's acquiring all land in the centre of Bologna, but the neighbourhood councils turned against their political masters when the many small landowners who owned the real estate in the centre of the city were threatened with expropriation at rural prices. A compromise was reached by which the municipality guaranteed private owners a subsidy for rehabilitation, the amount being regulated according to the means of each owner. In return the owner undertook to protect existing tenants and to control rents.

Within the plan's second line of action, to appropriate and convert some of the city's monuments for community use, an early example was a small palace in the Via Pietralata, which was converted into a community centre for the Malpighi neighbourhood in the western part of the city. A number of other important buildings were either partly converted or planned for conversion as money became available. The convent of St. Leonardo in the Via S. Vitale was to house a neighbourhood centre, a professional institute, a nursery school and, inside the church itself, a theatre. The Baraccano complex in the south-east corner of the city was already operating as a student's union. It too would become a neighbourhood centre with library and nursery school, as well as a theatre, a museum and an exhibition hall. Its large garden would become public. The former Jesuit church, Santa Lucia, in the Via Castiglione was to be converted into a centre for the performing arts for the use of the university, the municipality and the city theatre. The monastic buildings were to house a secondary school and a library. In the adjacent eighteenth-century Collegio San Luigi work was proceeding apace to provide student accommodation and a home for the State Institute of Arts: and there were elaborate plans to convert the disused friary of San Mattia into another neighbourhood centre, museum and elementary school.

In the plan, the historic centre of Bologna was seen as a whole; as an area for housing, university activities, cultural and tourist functions, small trades and business. Any concentration of administrative and bureaucratic institutions, department stores, warehouses or anything requiring large plots and attracting a large amount of traffic was forbidden. The Piazza Maggiore had been a pedestrian zone for some ten years and there were proposals for restricting vehicular traffic in the historic centre by 'feeder' roads off perimeter roads; for developing public transport, which was already free during rush hours; and for extending the pedestrian network by opening up formerly private courtyards, passages and gardens. In recognizing the totality of the historic centre, the plan also recognized Bologna's special character which is derived from its arcades. Still required today under a fifteenth-century law, they give the whole city a sense of unity and a continuity behind which the individual monument fades into relative insignificance.

The Italian contribution to architectural conservation and the re-use of old buildings in the 1960s and 1970s was of decisive importance. The Italian approach to exhibition design found a parallel in other countries in shop design. Exhibition designers in Italy had already extended their activity into the commercial field. Albini had designed a book shop in the Galleria in Milan (1961) and Scarpa a furniture shop for Gavina in Bologna (1962). In London, the South Africa tourist office in Piccadilly, by James Cubitt and Partners (1951), and in Vienna the Reti

Figure 8.20

Horace Jones's Billingsgate Fishmarket in the city of London. Although the conversion by the Richard Rogers Partnership into a dealing floor with ancillary offices was completed in 1986, the building has never been used as such. (*Photo: Richard Bryant.*)

candle shop by Hans Hollein (1965) set new standards in contemporary design at a time when planning authorities and historic buildings commissions were not yet demanding 'heritage-type' shop fronts and street lamps, nor that the new work in a conversion be 'in keeping' or 'in sympathy' with the style of the old.

The dichotomy which developed in later years is best illustrated by two recent conversions in London, of the Billingsgate Fish Market by the Richard Rogers Partnership into a dealing floor, ancillary offices and private restaurant (Figures 8.20, 8.21, 8.22, and 8.23), and of Tobacco Dock by the Terry Farrell Partnership into a two-level shopping mall (Figures 8.24 and 8.25). The architect of the fish market conversion believes that all historic elements of a building should be faithfully restored but that all new components should be clearly and unashamedly modern, using materials and technology of the present day. So the mutilated arcade of Horace Jones's building has been restored to its original classical splendour

Figure 8.21

The dealing floor of the Billingsgate Fishmarket with the new gallery suspended between two rows of existing columns. *(Photo: Richard Bryant.)*

Figure 8.22

The dealing floor of the Billingsgate Fishmarket: detail of the new gallery showing the relationship of the balustrade to the existing columns. *(Photo: Richard Bryant.)*

Figure 8.23

The original arcade of the Billingsgate Fishmarket which has been faithfully restored. The glass and steel balustrade of the new gallery stands separate and in marked contrast. (*Photo: Richard Bryant.*)

Figure 8.24

Tobacco Dock in London Dock, Wapping, built between 1811 and 1814 to the designs of Daniel Asher Alexander: the interior of one of several long, single-storey, parallel bays over a brick-vaulted basement. (*Photo: The Terry Farrell Partnership.*)

Figure 8.25

The same bay after conversion by *The* Terry Farrell Partnership: the brick-vaulted basement has been broken into and the roof above the new opening removed. (*Photo: The Terry Farrell Partnership.*)

while a new gallery of steel and glass hangs from the existing structure with functions between the two kept to an elegant minimum. In pursuit of the principle of visual and physical separation between new and old, the gallery is in the north and south blocks supported on columns that stand against, but do not touch, the existing wall.

In its day Tobacco Dock, built between 1811 and 1814 to the designs of Daniel Asher Alexander, was a pioneering high-tech structure of cast-iron stanchions and struts supporting a timber roof of hybrid construction. It consisted of several long, single-storey, parallel bays over a brick-vaulted basement, the aim having been to provide as large an area of clear floor space as possible. To find a satisfactory means of opening up the basement for commercial use, and of ventilating and dividing the single-storey ground-floor structure which complied with fire regulations, the architect broke into the vaults in seven places so that wide flights of steps could connect the ground floor with the basement. The roof above these openings was removed with the result that a chequerboard has replaced a linear pattern of uninterrupted trusses and long, unbroken vistas of tree-like stanchions and struts. The new public use which brought into play stringent fire regulations, may be considered inappropriate since it resulted in major alterations to the very form of a Grade I listed building. New elements like balustrades and light fittings, however, need not have been designed in such a heavy-handed historicist manner and without any of the sense of economy and elegance of the original.

The influence of Italian models of the 1960s and 1970s was widespread. Examples include Aarno Ruusuvuori's restoration and conversion to a multiplicity of uses of the City Hall block in Helsinki, begun in 1959 and only completed in 1988; Cambridge Seven's conversion of the Lone Star Brewery in San Antonio to a museum of fine art, begun in 1973; Charles Vandenhove's idiosyncratic alterations and additions to the Hotel Torrenthius in Liège of the late 1970s; Esse Fahlman's conversion to a civic museum and library of a group of warehouses at Sundsvall in Sweden, begun in 1980; Ugo Brunoni's conversion to a public library of a fine old school building in Geneva, completed in 1981; and Roland Simounet's brilliant work in converting the seventeenth-century Hotel Salé in Paris into the Picasso Museum in 1988–9. In Italy itself one should mention Franco Minissi's conference centre so imaginatively inserted into the roof space of the vast Ospizio di San Michele in Rome in 1979–80, and then part of which was also adapted in 1981 to house the International Centre for the Study of Preservation and Restoration of Cultural Property (ICCROM); Andrea Bruno's conversion of Filippo Juvarra's Castello di Rivoli near Turin into a museum of modern art begun in 1979; and Guido Canali's conversion of the service quarters, the only surviving part of the Palazzo Farnese in Parma, into the city's art gallery.

THE AMERICAN CONTRIBUTION

The American contribution to what Americans call 'adaptive use' was of a different kind. From the beginning America went in for the conversion of very large industrial complexes which had become redundant. Ghirardelli Square in San Francisco, originally a group of factory and warehouse buildings belonging to the Ghirardelli chocolate factory, was turned into a shopping centre in the late 1960s by converting seven of the old buildings and constructing eight new ones. Also in San Francisco, and at about the same time two vast ice-houses were converted to provide wholesale showroom space for manufacturers of interior furnishings, and the Del Monte cannery underwent a conversion similar to Ghirardelli Square, but not without the complete gutting and reconstruction of the building within its preserved outer walls.

In 1974 the conversion to artists' housing of the enormous quadrilateral Chickering piano factory in Boston was completed; the architects Gelardin Brunerlot having bought the building and taken on the role of developers. The following year saw the completion by the Massachusetts Housing Finance Agency of conversion work to the even larger group of tannery buildings at Peabody. Space was found in three of the factory buildings for 284 flats. In 1976 the conversion into a retail market of the historic Quincy Market in Boston provided an exact parallel to the conversion of the Covent Garden market in London.

Also in 1976 the Tax Reform Act introduced tax concessions on the costs of renovation and reduced existing tax incentives to replace old buildings with new. As a result of the 1981 Economic Recovery Tax Act, private investors were able to obtain a tax credit of up to 25 per cent on the capital cost of converting an old building, provided it was of landmark status. The result was a great upsurge in the adaptation of old buildings to new uses and explains how it became possible, for instance, to restore and convert the vast Union Station in St Louis. Unfortunately, there was also abuse of the tax credit system, which the 1986 Tax Reform Act endeavoured to correct, with the result that by 1987 the number of rehabilitation projects had declined by 35 per cent.

America has also led the way in showing how conservation can be financed by a combination of different sources, with public finance being used to attract private investment. For example, Britain's Urban Development Grant is based on this principle and its success indicates that joint ventures of public and private funds are being accepted as a satisfactory *modus operandi*.

An inspection of the bibliography at the end of this chapter will show that the architectural conservation scene in America, especially in the latter part of the 1970s, was enlivened by a large number of publications, both of a general and specific nature. Many of these were devoted to the problem of converting old buildings to new uses, starting with the pioneering issue of *Architectural Record*, 'Conservation in the Context of Change', (December 1974) and continuing with *Built to Last – A Handbook on Recycling Old Buildings* (1977), *Rescued Buildings: the Art of Living in Former School Houses, Skating Rinks, Fire Stations, Churches, Barns, Summer Camps and Cabooses* (1977), *Buildings Reborn: New Uses, Old Places* (1978), *How to Recycle Buildings* (1979), and *New Life for Old Buildings* (1982). Nor were specific building types neglected. The two volumes of *Re-using Railroad Stations* were published successively in 1974 and 1975, while *Working Places: the Adaptive Use of Industrial Buildings* appeared in 1976. To the subject of economics *Economic Benefits of Preserving Old Buildings* (1976), *Adaptive Use: Development Economics, Process and Profiles* (1978), and *New Profits from Old Buildings* (1979) made significant contributions.

THE BRITISH CONTRIBUTION

To single out the Italian and American contributions is in no way to belittle the British contribution. Cradle of the Industrial Revolution, Britain possesses a legacy of industrial buildings second to none. Water mills, textile mills, maltings, warehouses, docks and railway stations have all been successfully converted to new uses. In Gloucestershire, Niall Phillips Associates have converted the large Ebley Mill near Stroud into council offices (1988), and New Mills at Wotton-under-Edge into the headquarters of a high-tech company (1985). There is the undoubted success of Covent Garden's central market buildings and their conversion to shops, cafes and offices, bringing new life to an area that might otherwise have died. The Royal Agricultural Hall in Islington has found a new use as a home for the interior design industry; the Albert Dock in Liverpool, if not wholly successful as a conversion because of the large variety of uses, has at least been saved from having its dock drained and filled with a multi-storey car park: and, while it is necessary to record the failure to save Battersea Power Station, there is now the promise of success with the conversion of Bankside Power Station into the Tate Gallery's museum of modern art. There remains the challenge, not yet fully confronted, of the Victorian hospital and asylum, of the large Board school, of the army barracks and of the brewery.

Britain's contribution lies also in the work of certain organizations, four of which deserve special mention. The first organization is the National Trust which has been consistently and imaginatively converting its redundant buildings to new uses, as for example, Llanerchaeron Home Farm Buildings now used as training workshops. At Stackpole in Pembrokeshire the stable buildings have been converted into a residential field-study base for schools. At Milldale a field barn has been converted to a simple interpretation centre for the Dales, and there are innumerable holiday cottages and base camps. The second organization is the National Trust for Scotland, whose concern for vernacular architecture led to the Little Houses scheme, a programme of buying, restoring and selling houses which was changed to buying and funding 'restoring purchasers' when inflation and rising building costs reduced the real value of the Trust's revolving fund. The third organization, the Landmark Trust, was founded by John and Christian Smith 'to preserve small buildings, structures or sites of historic interest, architectural merit, or amenity value, and where possible find suitable uses for them; and for protecting and promoting the enjoyment of places of historic interest or natural beauty'. Today the Landmark Trust owns more than 60 properties, most of which have been restored and converted for letting to holiday-makers. They include Keats's house in Rome and Palladio's Villa Pojana in the Veneto, and numbered among its more unusual buildings in Britain are the Gothic Temple at Stowe, the Egyptian House at Penzance and the Water Tower at Appleton in Norfolk.

Kit Martin, founder of the fourth organization, is an enlightened entrepreneur who has successfully tackled redundancy in country houses. He began with the ruined shell of Dingley Hall in Northamptonshire; he continued first with Gunton Hall in Norfolk and then with Cullen on the Banffshire coast, a Scottish baronial house of 1600 greatly enlarged in mid-Victorian times, and successfully converted

Callaly in Northumberland, a fifteenth-century pele-tower buried inside seventeenth- and eighteenth-century extensions and refacings, into twenty houses, cottages and flats. Martin's method is to work with the existing structural divisions of the building, making as few changes as possible and preserving all original features.

CONCLUSIONS

Britain's very special contribution to architectural conservation in general has been its recognition of the wider context: group value, the setting of monument or a historic building and the concept of conservation areas and its corollary, area conservation. The conversion of old buildings to new uses, however appropriate and well executed, should be seen in the wider context of conservation as a whole. A village or a town should first be surveyed, its building stock measured and analysed, records compiled – in short a body of knowledge accumulated, for there can be no theory without knowledge. This research and documentation leads to an assessment of a building or group of buildings. In making this assessment there are a number of questions which have to be answered:

What is the historical value of the building? This question can be answered quite factually: when was it built and for whom?; who have been its owners or users?; who has visited or lived in it?; what historical events have been associated with it? The more important the history, the greater is the historical value.

What is the architectural value of the building? In answering this question taste, fashion or personal opinion can play a part. Once again, a factual approach is best: putting the date; the name of the architect and the style of the building; and examining stylistic influences and noting a new type of plan or the use of a new material.

What is the significance of the building in its wider context of street or town? What role does the building play on the urban stage? Is it a building where the main interest lies in the interior space, like the train shed of a railway station? Is it an isolated building that provides a focal point, as is often the case with churches or public buildings? Is it a building with a principle façade that belongs as much to the exterior space of the street or square in front of it as to the interior accommodation behind, like the palaces of the Campidoglio in Rome? Is it a building that is part of a greater ensemble? Is its significance of urban scale, and does it, like the towers and spires of churches, make a significant contribution to the silhouette of the town as a whole?

What is the essence of the building; its very nature? Answering this question is of particular importance if the building is to be converted to a new use. A church or a railway station are essentially single-space structures: a house or a block or flats is essentially cellular. It is undesirable, in converting a building, to change its nature, to change a single-space structure into a cellular one, or vice versa.

What are the economic and social needs of a particular area, and how can these needs be best met by the introduction of new uses, and by the increase or reduction of existing uses? It would be a disaster for example, if the Weissenhofsiedlung at Stuttgart were ever to be made into a museum quarter like certain well-preserved historic town centres in eastern Europe. Conversely, it would be wrong to deny the possibility of new buildings, or the alteration and extension of existing buildings if the social and economic conditions required this.

Having answered these questions, it becomes possible to put buildings into categories: those that are historically and architecturally so important that they must be preserved and restored with the minimum of alterations, either to their physical form or to their use; those that are good enough and useful enough to keep, but which may be converted to new uses; those that may be converted inside but must not be altered outside.

Taking the category that may be converted freely there are, I believe, three broad principles worth following. First, respect the nature of the building, but judge every case on its own merit. It may sometimes be expedient to sacrifice the interior of a building so as to save the exterior. Second, respect alterations and additions if these have enough intrinsic quality to merit retention. They are, after all, part of the history of the building and it is usually undesirable to restore a building to its original state. But again there can be no rule, and every case should be judged on its own merit. Last, look after your buildings. William Morris understood the problem when he said: 'Stave off decay by daily care'. Any conservation programme should always have written into it a maintenance programme, which would also ensure that the need for conservation on the scale that it exists today will never recur.

BIBLIOGRAPHY

The bibliography which follows is intended to be exhaustive for books published on the subject of re-using old buildings since 1975. It goes beyond re-use, to urban regeneration and contextual architecture, to cover the wider implications of the subject, and it includes housing rehabilitation and loft conversions. The bibliography is arranged under nine headings.

General

Anstett, P., and others (1981). *Alte Bauten neu genützt.* Deutsche Verlags-Anstalt, Stuttgart.

Arnaud, F., and Fabre, X. (1978). *Réutiliser le patrimoine architectural.* Caisse Nationale des Monuments Historiques et des Sites, Paris.

Austin, R. L. (1988). *Adaptive Reuse: Issues and Case Studies in Building Preservation.* Van Nostrand Reinhold, New York & London.

Cantacuzino, S. (1975). *New Uses for Old Buildings.* Architectural Press, London.

Cantacuzino, S. (1989). *Re-architecture: Old Buildings, New Uses.* Thames & Hudson, London.

Cantacuzino, S., and Brandt, S. (1980). *Saving Old Buildings.* Architectural Press, London.

Cunnington, P. (1988). *Change of Use.* Alphabooks, Sherborne.

Diamonstein, B. (1978). *Buildings Reborn: New Uses, Old Places.* Harper and Row, New York & London.

Diamonstein, B. (1986). *Remaking America: New Uses, Old Places.* Crown, New York.

Fischer, A. (1992). *Neue Architektur durch Umnützung alter Gebaüde und Anlagen.* Kramer, Stuttgart.

Higgs, J. (1978). *The Wizard's Eye: Visions of American Resourcefulness* (introduction by Buckminster Fuller). Chronicle Books, San Francisco.

Hoor, D., and Reiners, H. (1990). *Alte Bauten, neues Wohnen: Beispiele und Ideen für die Umnützung.* Callwey, München.

Jacopetti, R., and Van Meter, B. (1977). *Rescued Buildings: The Art of Living in Former Schoolhouses, Skating Rinks, Firestations, Churches, Barns, Summer Camps and Cabooses.* Capra Press, Santa Barbara.

Landmark Trust (1977). *The Landmark Handbook.* The Landmark Trust, Maidenhead.

Latreille, A., and others (1982). *New Uses for Old Buildings in Australia.* Oxford University Press.

Lerch, H. (1993). *Bausubstanz neu genützt: Architektur individueller Arbeitsstatten.* Koch, Leidenfelden-Echterdingen.

Lion, E. (1982). *Building Renovation and Recycling.* J. Wiley, New York & Chichester.

Markus, T. A. (ed.) (1979). *Building Conversion and Rehabilitation: Designing for Change in Building Use.* Newnes-Butterworths, London.

Massachusetts, Department of Community Affairs, Office of Local Assistance (1977). *Built to Last: A Handbook of Recycling Old Buildings.* Preservation Press, Washington (DC).

Meyer-Bohn, W. (1991). *Umbauten: Alternativen zum Neubau.* Deutsche Verlags-Anstalt, Stuttgart.

Montagu of Beaulieu, Lord (Chairman) (1980). *Britain's Historic Buildings: A Policy for their Future Use.* Working Party on Alternative uses of Historic Buildings. British Tourist Authority, London.

Müller-Menckens, G. (1977). *Neues Leben für alte Bauten: über den Continuo in der Architektur.* A. Koch, Stuttgart.

Opher, P., and Samuels, I. (1982). *New Streets for Old Stones: The Practice of Using Old Environments for New Cultural Activities.* Council of Europe, Strasburg.

Rechmann, P. (1995). *Structural Aspects of Building Conservation.* McGraw-Hill, New York & London.

Reiner, L. E. (1979). *How to Recycle Buildings.* McGraw-Hill, New York & London.

Robert, P. (1989). *Reconversions: Adaptations: New Uses for Old Buildings.* Moniteur, Paris.

Robert, P. (1989). *Adaptations: New Uses for Old Buildings.* Princeton University Press, Princeton (NJ) & Guildford.

SAVE Britain's Heritage (1994). *Bargain Buildings: Historic Buildings in Need of a New Lease of Life.* SAVE Britain's Heritage, London.

Schmertz, M. F. (ed.) (1982). *New Life for Old Buildings.* McGraw-Hill, New York & London.

Scottish Civic Trust (1981). *New Uses for Older Buildings in Scotland. A Manual of Practical Encouragement Prepared for the Scottish Development Department.* HMSO, Edinburgh.

Shopsin, W. C. (1986). *Restoring Old Buildings for Contemporary Uses: An American Sourcebook for Architects and Preservationists.* Whitney Library of Design, New York.

Thompson, E. K. (ed.) (1977). *Recycling Buildings: Renovations, Remodellings, Restorations and Reuses.* McGraw-Hill, New York & London.

Economics of conservation

Frey, H. (1977). *Capital Cost Prediction of Building Conservation: Final Report.* Information Services (Publications), Department of Architecture and Building Science, University of Strathclyde, Glasgow.

National Trust for Historic Preservation (1976). *Economic Benefits of Preserving Buildings*. Papers from the Economic Benefits of Preserving Buildings conference (1975) sponsored by the National Trust for Historic Preservation, Seattle. Preservation Press, Washington (DC).

Thorncroft, M. E. T. (1975). *The Economics of Conservation*. Royal Institution of Chartered Surveyors, London.

Urban Land Institute (1978). *Adaptive Use: Development Economics, Process and Profiles*. Washington (DC).

URBED (1979). *Financing the Reuse of Old Buildings: A Practical Guide to Sources of Finance, Materials and Labour for Projects Involving the Provision of Work Space*. URBED (Urban and Economic Development) Ltd, London.

Urban regeneration

Celada, G. (ed.) (1980). *Architettura di riuso città italiane: interventi pubblici*. PEG. Milano.

Harling, A. (ed.) (1987). *Urban Regeneration: International Architectural Exhibition 1987*. Heritage Trust, London.

URBED (1987). *Reusing Redundant Buildings: Case Studies of Good Practice in Urban Regeneration*. HMSO, London.

Witzemann, H.-M., and others (ed.) (1981). *Junge Liebe zu den alten Städten*. W. Kohlhammer, Stuttgart.

Contextual architecture

Fischer, M. A., and others (1975). *Architektur und Denkmalpflege: neue Architektur in historischer Umgebung*. H. Moos, München.

ICOMOS (1986). *Créer dans le créé: l'architecture contemporaine dans les bâtiments anciens, ICOMOS Section Française*. Electa Moniteur, Paris.

Ray, K. (ed.) (1980). *Contextual Architecture: Responding to Existing Style*. McGraw-Hill, New York & London.

Smeallie, P. H., and Smith, P. H. (1990). *New Construction for Older Buildings: A Design Sourcebook for Architects and Preservationists*. J. Wiley, New York & Chichester.

Building types

Barlow, J., and Gann, D. (1993). *Offices Into Flats*. Joseph Rowntree Foundation, York.

Binney, M., and others (1990). *Bright Future: The Reuse of Industrial Buildings*. SAVE Britain's Heritage, London.

Carlson, D. (1991). *Reusing America's Schools: A Guide for Local Officials, Developers, Neighborhood Residents, Planners and Preservationists*. Preservation Press, Washington (DC).

Cobb, H. (1991). *Your Barn House*. Holt, New York.

Corbett-Winder, K. (1990). *The Barn Book*. Century, London.

Croft, V. (1991). *Recycled as Restaurants: Case Studies in Adaptive Reuse*. Whitney Library of Design, New York.

Department of the Environment, Scottish Development Department and Welsh Office (1977). *New Life for Old Churches*. HMSO, London.

Dewe, M. (ed.) (1987). *Adaptation of Buildings to Library Use.* Proceedings of the Seminar held in Budapest, June 3–7 1985. Saur, München & London.

Educational Facilities Laboratories (1976). *The Arts in Found Places.* A Report from Educational Facilities Laboratories and the National Endowment for the Arts, New York.

Eley, P., and Worthington, J. (1984). *Industrial Rehabilitation: The Use of Redundant Buildings for Small Enterprises.* Architectural Press, London.

English Tourist Board (1979). *Self-catering on the Farm: A Practical Guide to Farmhouse, Barn and Other Conversions and Purpose-Built Accommodation for Holiday Lettings.* English Tourist Board, London.

Essex County Planning Department (1980). *The Essex Countryside: Historic Barns: A Planning Appraisal, Supplementary Planning Guidance.* Essex County Planning Department, Chelmsford.

Fracchia, C. A., and Bragstad, J. O. (1977). *Converted Into Houses.* Thames and Hudson, London.

Green, H., and Foley, P. (1986). *Redundant Space: A Productive Asset: Converting Property for Small Business Use.* Harper & Row, London.

Johnson, A. (1988). *Converting Old Buildings: Homes from Barns, Churches, Warehouses, Stations, Mills, etc.* David & Charles, Newton Abbot.

Mack, L. (1993). *Homes by Design: Transforming Uncommon Properties into Stylish Homes.* Cassell, London.

Marsh, P. (1983). *The Refurbishment of Commercial and Industrial Buildings.* Construction Press, London.

Peters, P. (ed.) (1978). *Umbau Alter Bauernhäuser: Städter leben auf dem Lande.* G. Callwey, München.

Powell, K., and de la Hey, C. (1987). *Churches: A Question of Conversion.* SAVE Britain's Heritage, London.

Powell, K., and Watson-Smyth, M. (1988). *Hospitals: A Suitable Case for Treatment.* SAVE Britain's Heritage, London.

SAVE Britain's Heritage (1993). *SAVE Britain's Heritage Presents Deserted Bastions: Historic Naval and Military Architecture.* SAVE Britain's Heritage, London.

Shopsin, W. C. (1994). *Preserving American Mansions and Estates.* McGraw-Hill, New York & London.

Sports Council (1975, 1978). *Sport for All in Converted Buildings* (2 vols). The Sports Council, London.

Thompson, W. P., and others (1977). *Winnipeg's Historic Warehouse Area: Its Revitalisation Through Conservation: A Report Prepared for Heritage Canada and the Manitoba Historical Society.* Heritage Canada, Ottawa.

Umlauf, E., and Schreiner, P. (1990). *Building Design: Improving Commercial Spaces.* PBC International: Hearst Books International, Glen Cove (NY).

URBED (1981). *Recycling Industrial Buildings.* Capital Planning Information, Edinburgh.

Valerio, J. M., and Friedman, D. (1982). *Movie Palaces: Renaissance and Reuse.* Educational Facilities Laboratories, New York.

Watkins, C., and Winter, M. (1988). *Superb Conversions? Farm Diversification: the Building Experience.* Council for the Protection of Rural England, London.

Specific buildings

Amery, C. (1992). *Bracken House.* Wordsearch, London.

Binney, M., and others (1981). *The Colossus of Battersea.* SAVE Britain's Heritage, London.

Binney, M., and others (1982). *Save Broad Street: A New Use for the Station Building*. SAVE Britain's Heritage, London.
Boissière, O. (1990). *Lindblade Tower et Paramount Laundry: Réconversion à Culver City, Californie*. Éditions du Demi-Cercle, Paris.
Bruno, A. (1984). *Il Castello di Rivoli 1734–1984: storia di un recuparo*. Allemandi, Torino.
Building Design Partnership (1975). *Converting Albert Dock for Liverpool Polytechnic: An Exercise in the Re-use of a Major Historic Building*. BDC, Manchester.
Grajewski, T. (1989). *The Building That Would Not Go Away: The Story of How Samuel Morris Rescued the Royal Agricultural Hall*. Royal Agricultural Hall, London
Herdag, K., and others (1976). *The New York Federal Archive Building: A Proposal for a Mixed Re-use*. Columbia University Press, New York.
Hitchmough, W. (1987). *The Michelin Building*. Conran Octopus, London.
Jenger, J. (1986). *Orsay: de la gare au musée: histoire d'un grand projet*. Electa Moniteur, Paris.
Jenger, J., and others (1987). *Gae Aulenti e il Museo d'Orsay*. Electa, Milano.
Jenger, J. (1988). *Orsay: The Metamorphosis of a Monument; from Paris Terminus to National Museum*. Electa Moniteur, Paris.
Polano, S. (1989). *Carlo Scarpa: Palazzo Abatellis: la Galleria della Sicilia, Palermo 1953–54*. Electa, Milano.
Veilletet, P. (1992). *L'entrepôt Laine à Bordeaux: Valode et Pistre*. Éditions du Demi-Cercle, Paris.

Housing rehabilitation

Druenne, D., and others (1976). *Une méthodologie de la réhabilitation de l'habitat ancien: un cas; le quartier Brindeau, Le Havre*. Institut d'Études et de Recherches Architecturales et Urbaines, Paris.
London Research Centre (1992). *Houses into Flats: A Study of Private Sector Conversions in London – Vol.1: Report of Main Findings*. HMSO, London.
National Building Agency, Scottish Division (1980). *Tenement Rehabilitation to Provide Sheltered Housing: A Case Study and General Perspective*. NBA Scotland, Edinburgh.

Loft conversions

Hoor, D., and Reiners, H. (1991). *Der Dachausbau: Planung und Ausführung; neue Wohnräume unterm Dach*. Callwey, München.
Hudson, J. R. (1987). *The Unanticipated City: Loft Conversions in Lower Manhattan*. University of Massachusetts Press, Amhurst.
Huner, O., and Lautsch, J. (1991). *Ausgebaute Dächer*. A. Kohlhammer, Stuttgart.
Zukin, S. (1988). *Loft Living: Culture and Capital in Urban Change*. Radius, London.

Bibliographies

Vance Bibliographies (1986), Monticello, Illinois.
Adaptive Re-use of Railroad Stations.

Adaptive Re-use of Churches.
Housing from Redundant Buildings.
Will, M. T. (1979). *Recycled Buildings: A Bibliography of Adaptive Use: Literature Since 1970,* Vance Bibliographies, Monticello, Illinois.

NOTES

1 Italo Calvino (1992) *Six Memos for the Next Millennium.* Jonathan Cape, London (1992), 1.
2 Calvino, *op. cit.*
3 Raymond Lemaire (1994) 'Quelle doctrine de sauvegarde pour demain?', *ICOMOS Scientific Journal,* Articles of Members, 50–5.
4 Cesare Brandi (1963) *Teoria del Restauro.* Edizioni di Storia e Letteratura: Rome.
5 A phrase used by Mohammed Arkoun in conversation. It gave Maurice Edelman the title 'Éspace de la Tolerance' for an article on the Aga Khan Award for Architecture, *Le Monde,* 10 September 1983.
6 Toni Del Renzio (1952) 'Palazzo Bianco, Genoa', *The Architectural Review,* January, p. 60.

9

GLORIOUS REPAIR, OR THE 'BRITISH WAY OF MAKING GOOD'

James Simpson

INTRODUCTION

Repair is at the very heart of the British approach to building conservation, a subject which, like architectural history, has grown in the last third of the twentieth century into a fully developed discipline. That this should be so is due very largely to the writings of John Ruskin and William Morris, and to their legacy, the Society for the Preservation of Ancient Buildings (SPAB). Even so, the idea of repair has lacked glamour in the throwaway years of the later twentieth century. With the rise of environmentalism, however, repair is again important.

Buildings have been repaired, altered and rebuilt from the earliest of times, to make good damage and decay and in response to changing needs and fashions. In simple societies, in which people built for themselves without the aid of specialist builders, buildings were genuinely vernacular: native, or 'of the country'; in French, *du pays*. Ways of building were developed empirically over many generations and repair and rebuilding were part of a seasonal land-use routine.

In parts of the Hebrides, ways of building and living which would have been recognizable in medieval and prehistoric times survived well into the twentieth century. In most of Britain the development away from subsistence farming towards specialization happened much earlier. As well as classical architecture, the Romans brought new levels of sophistication in building, and much of what they introduced, in ways of building and finishing walls, for example, has remained common practice in Britain and through much of Europe to the present day. There may have been specialist builders in pre- and post-Roman Britain, but it was in the well-ordered burghs of the Middle Ages and through royal, baronial and ecclesiastical patronage that the building trades we recognize today became established. Through the masons, wrights and carpenters in particular, but also through bricklayers, slaters, tilers, plasterers, smiths, plumbers, glaziers, painters

Figure 9.1

The end of truly vernacular house building in Britain. Houses by Loch Duich, Rossshire, photographed by George Washington Wilson, *c.* 1885. Note the 'creel' or wattle construction of the nearest building. (*Photograph by kind permission of the Washington Wilson Collection at the University of Aberdeen.*)

and others, building practice advanced from the level of simple, well-understood common skills to that of a complex activity requiring organization and highly developed specialized crafts, working with particular materials to produce and assemble the various parts of buildings.

The truly vernacular tradition of simple hut or house building survived in rural Britain, side by side with the newer artisan tradition emanating from the burghs, until it was swept away, in all but the remotest corners, by the agricultural improvements, which transformed the rural economy and the appearance of the countryside, between the seventeenth and the nineteenth centuries. Before the development of canals, railways and roads, however, both traditions were limited in their choice of materials and informed by an empirically developed understanding as to how the materials, which were to hand, could best be used. In this way, geology, climate, transport possibilities and other geographical and ethnographical factors determined what might be termed the architectural geography of the United Kingdom.

While the basic walling and roofing materials of traditional buildings, and their forms and details of construction, varied continuously from place to place across the land, there was also consistent development over time. The progression from vernacular to artisan, from self building to craft building, is one example of such development; the introduction of inner linings of joinery and plasterwork, which is the principal characteristic of post-medieval, as opposed to medieval, artisan construction is another.

In the medieval system of building, which has been called 'Type A' traditional construction,[1] nothing is concealed. Ground floors are solid, whether of earth, stone or tile; walls are solid, whether of stone, clay, timber frame or brick, and commonly finished with lime plaster and limewash, often decorated, outside and in; upper floors are of exposed beams and boards, also sometimes decorated; roof structures are similarly exposed and covered from shedding eaves to ridge with thatch or tile of stone or clay; doors are boarded and hung directly into checks; window glass, if any, is set in lead cames into the stone or timber openings. In this simple system, everything is visible and permeable. It was used for buildings of all sorts (churches, barns and houses), and in all essentials it is the system used to this day for simple sheds, outbuildings and garages. Reasonable comfort in domestic buildings of this type was dependent on the maintenance of fires in large fireplaces and the passage of large quantities of air to keep the rooms warm and dry. In vernacular houses the warmth and the smoke from continuously burning central fires helped to keep the rain out, and the thatch dry and free from rot.

Figure 9.2

Unlined 'Type A' construction in a Yorkshire barn, probably of the eighteenth century. (*Photo: James Simpson.*)

Figure 9.3

The eighteenth century Old Manse of Tankerness, Orkney. This sectional photograph of the partly-stripped interior shows the solid 'Type A' construction of the ground floor rooms, with lath and plaster wall linings and ceilings above. (*Photo: James Simpson.*)

Many structures of this type, particularly churches and barns, have survived in their original form. In the case of domestic buildings, however, major technical advances in the sixteenth, seventeenth and eighteenth centuries transformed living standards and resulted in a wholly new post-medieval system of building, which has been called 'Type B' traditional construction.[2] The essential characteristics of this developed system are: a double-skin construction in which the primary structural elements, the walls, floors and roofs are concealed by inner linings, including suspended ground floors, panelling or lath-and-plaster wall linings and ceilings; secondary ventilation through underfloor ventilators, behind linings and through roof voids, to keep concealed timbers dry; the collection and discharging of rainwater, through a system of gutters and pipes, again to keep the fabric dry; reduced and more controllable primary ventilation through sash windows and much smaller fireplaces; and the enhanced importance of the finishing trades, plasterwork and joinery, based on square framing and mortice-and-tenon joints.

It was this most sophisticated system of construction, the origins of some aspects of which may be associated with the thinking of men like Wren and Hooke and the empirical scientists of Charles II's Royal Society, as well as the classical spirit of the 'Age of Enlightenment', which made the Georgian house the most comfortable, beautiful and desirable of dwellings ever devised.[3]

All types of traditional, pre-industrial construction, whether vernacular or artisan, 'Type A' or 'Type B', and whether the basis of everyday building or fine architecture, have certain characteristics in common: they depend on the use of simple, local or easily transported materials, worked and assembled by hand with ordinary tools and a greater or lesser degree of skill; components are generally small and manhandleable, though simple devices, such as rollers and blocks and tackles, have long been available to move surprisingly heavy stones and timbers, for example, when necessary; water shedding is generally achieved by the single or double lapping of small, more or less impervious elements to create a covering which is waterproof yet discontinuous; sheet membranes were not available in the past, and it is of the essence of traditional construction that it is entirely permeable.[4]

The limitations to which traditional construction used to be subject dictated certain common principles, therefore, despite infinite local variation. Both general principles and local practice in the pre-industrial age were developed empirically over hundreds, indeed thousands, of years, and many generations. Though much was recorded in pattern books and treatises in the eighteenth and nineteenth centuries, much went unrecorded and was handed down by oral tradition and from master to apprentice as village wisdom and as the mysteries of craft guilds. Such knowledge, and conscious or unconscious understanding of what worked best and produced the finest results and the easiest to achieve, were based on centuries of trial and error. Nor was it limited to manual skills or to particular ways of doing things: there was an understanding of the building as a whole, of the system of building currently and locally practised, of overall structure, of the way in which materials and components joined and intersected, of heat and cold, of air and moisture movement, and of behaviour over time, maintenance and repair.

Traditional construction was inherently undamaging, economical of energy and non-renewable resources, and sustainable. The skills and materials required for repair and alteration were precisely the same as those required for building and, as long as a building was required, it was kept in repair. At all periods, however, requirements and fashions evolved and traditional buildings were altered, enlarged and updated. As long as materials and components had to be extracted, made, carried and disposed of by hand, the economics of alteration and improvement favoured the re-use of existing fabric whenever and wherever possible. Complete demolition and rebuilding was an expensive option and, even then, it was normal for materials to be re-used for rebuilding, as fertiliser for the fields or fuel for the fire. There was little or no waste.[5]

The rapid development of industry and mechanized transport in the nineteenth century transformed the economics of building and thus building practice itself. The availability of cheap energy, through greatly increased coal extraction, and

Figure 9.4

Melsetter House on Hoy, Orkney. The old house was seamlessly incorporated into a much larger new Arts and Crafts house of superlative quality, by W. R. Lethaby in 1898. (*Photo: John Sanders.*)

of natural resources from the Empire, fuelled massive expansion and a building boom. Though most building remained superficially traditional, the large-scale production and easy transportation of materials led to rapid standardization of practice. The wide distribution of factory-made bricks and tiles, machine-cut stone and timber, Welsh slates, iron and steel and, above all, Portland cement undermined local practice and lent a mechanical, mass-produced air to the architecture, which was none the less impressive for that, and which was in any case a perfect expression of the age which produced it.

However, these developments in building practice combined with greatly increased confidence and prosperity, the growth of the architectural profession and the building industry, led to a greater readiness to destroy, alter radically and restore older buildings. Moreover, as construction became more scientific, codified and based on written technical specifications and calculations, there was a corresponding loss of traditional wisdom and of real understanding of local practice. As a result, though the craftsmanship and systematic construction of Victorian buildings were often outstanding in their own terms, nineteenth-century repairs and other work to older buildings often showed a lack of understanding of the constructional as well as the aesthetic subtleties of the older work, and were, as a result, ill-blended, unnecessarily destructive and damaging.

It was at least partly in response to this that the Arts and Crafts movement and the Society for the Protection of Ancient Buildings were born. In the last quarter

Figure 9.5

Standen, East Grinstead, Sussex. Hollybush Farm House was repaired and extended by Philip Webb in 1891, the parts making a perfectly integrated whole. (*Photo: John Sanders.*)

of the nineteenth century, as a reaction against the brittle eclecticism of much Victorian architectural design, new value began to be attached, not just to medieval churches and similar structures, but also to ordinary, pre-industrial buildings and to the materials and craftsmanship of traditional construction. Old buildings acquired a new aura of beauty and nobility, which was dependent on their narrative qualities and historical legibility and on the authenticity of their materials and surfaces, rather than on the style, purity or other qualities of their design. As a result, the process of honest, conservative repair acquired an intellectual status and a credibility which it had never had before.

If the leaders of the Arts and Crafts movement believed that they could reverse, or even stand aside from, the effects of industrialization however, they were mistaken. Just as houses with central hearths in the pre-improvement vernacular tradition continued to be built in the Western Isles until the late nineteenth

Figure 9.6

Slate-roofed stone Council Houses at Gullane, East Lothian, by Dick Peddie and Walker Todd in 1920. Though largely superseded by the cavity walls and membranes of modern construction, the artisan tradition in house building lingered on until the 1950s. (*Photo: John Sanders.*)

century,[6] so the artisan tradition survived in many parts of rural Britain until about the middle of the twentieth century.[7] That real traditional building was brought to an end by the pressures of social and economic modernism in the twentieth century was, however, inevitable. That the Arts and Crafts movement had an important influence on architectural thinking is unquestionable, and that influence persists, but the virtues of making and mending by hand, and of inventing within parameters prescribed by the traditional use of local materials, could only have marginal relevance in the modern age.

The twentieth century saw global upheavals on an unprecedented scale. The scramble for resources associated with industrialization and the rapid rise of living standards in the West led to national and political rivalry and wars which, in turn, stimulated levels of activity, and social, economic and physical changes on a scale which was quite beyond any previous experience. Developments in science, engineering, transport, electronics, communications and information technology transformed the world, and human perceptions and aspirations everywhere.

As the building trades in Britain receded, their place was taken by a construction industry which was more capital intensive, had less need for labour and craftsmanship, and was much more demanding of energy and resources. In the span of just two or three generations, the coherence and the diversity of the traditional building practices of the various parts of Britain were, in reality, left far behind. The same has happened throughout the developed world, and the pattern is being repeated everywhere, as societies aspire to match the living standards and the levels of consumption established in the West.

At the same time, however, other forces were at work. Just as the consequences of industrialization and mass production in the nineteenth century inspired the Arts and Crafts movement, so the consequences of war and economic development in the twentieth century produced the Conservation movement. With origins in eighteenth-century enlightenment, antiquarianism and romanticism, and formal beginnings with, for example, the first Ancient Monuments Act and the formation of the National Trust in the late nineteenth century, public interest and concern grew steadily as the twentieth century progressed. After 1945 the listing and legal protection of historic buildings was established through a succession of Planning Acts. The growth of interest in buildings and landscapes of cultural value was matched by increasing awareness of human impact on the natural world, and pressure to control pollution, ecological damage and resource consumption. These two, once separate strands, have been gradually woven together to create a conservation movement whose concerns are essentially environmental.

In the early years of the twentieth century, the practice of what is now known as building conservation was strongly influenced by the repair philosophy of the SPAB. Much ordinary repair work before 1950 was carried out unselfconsciously by ordinary tradesmen in the traditional way, and it was this 'honest tradesman's approach', consciously articulated, which the Society sought to perpetuate. Unfortunately, the fundamental changes of the post-war years meant that, increasingly, the ordinary tradesman's attitudes changed, and the materials and skills required for everyday new building progressively ceased to be suitable for the repair of older work. The distinction between new and old construction methods became such that architects and tradesmen alike lost their understanding of traditional building and as a result lost their faith in it. In a complete reversal, the 'honest tradesman's approach' often resulted in bad and inappropriate work, and the repair and 'restoration' of old buildings became a specialized and, in the age of modernism, a rather despised corner of architectural practice.[8] In the 1960s, it was believed by many that few old buildings would survive until the end of the century, and that even those that did would require to be expensively supported by modern means.

In one of the first cracks in post-war planning orthodoxy, the Civic Amenities Act of 1967 established the principle of urban conservation areas. Even so, the Act laid down that it was the *character* of the place which was to be preserved and enhanced, not its fabric. Conservation was said to be about places, not buildings, which, like the proverbial woodman's axe could be progressively renewed, while the place remained the same. The concept of repair could hardly have been further from the legislators' minds. In the same period, the early Civic Trust improvement schemes were about façades and urban spaces, not about the buildings which supported and enclosed them.

More important, ultimately, to the post-war development of building conservation were the Historic Buildings Act 1953, which established the Historic Buildings Councils and their repair grant schemes, and the Inspection of Churches Measure

Figure 9.7

Flats at Hutchesontown, Glasgow, Sir Basil Spence (1960–6). Modern art, architecture and construction in the service of social engineering. 'In the 1960s it was believed by many that few old buildings would survive until the end of the century.' In the event, it was often the new ones which failed: this block was demolished in September 1993. (*Photo: John Sanders.*)

1955 which established the quinquennial system for the repair of English parish churches. These were lifelines for the small number of architects concerned with old buildings at that time, and vital in beginning to re-establish a demand for the materials and skills necessary for repair. The watershed, however, was probably the Town and Country Planning Act 1968 which gave a more positive system of statutory protection to listed historic buildings, and controlled their demolition and alteration more effectively. Much that has been achieved in the development of the practice of building conservation in the last quarter of the century has been underpinned by that legislation.

At the beginning of this period of development, the general standard of repair practice was very low. Modernism in architecture and construction was at the height of its power, the political imperative in new building was for quantity not quality, and little thought was given to the longer term. There were, on the ground, a few tradesmen who had been trained before the war, and traditional standards survived in a few firms, many of them in country areas. Among architects, across the country as a whole, the influence of the SPAB was small, the concepts of conservative or 'like for like' repair were relatively unknown, interiors were little regarded and most interventions were radical.[9] It was common at this time for roofs and interiors to be unnecessarily removed and for repairs and alterations to be crudely undertaken using heavy steel, concrete, Portland cement and common brick. There was little understanding of dry rot, which was thus profoundly feared: roofs were left unrepaired and walls and timbers saturated with chemicals. There was no understanding of the permeability of traditional construction, which was pumped and fitted up with membranes and coated with impervious renders.[10] There was a great deal to learn.

In the course of the 1970s and 1980s, as realization of the importance of the existing building stock slowly permeated society, much *was* learned. Initially, however, perhaps because many architects and contractors came to building conservation from the world of new construction, and perhaps because of the influence of the new profession of architectural historian, neither having sufficient understanding of traditional building practice, mistakes were made. Indeed, even for committed practitioners of long experience, the learning curve was, and still is, steep. There *remains* a great deal to learn.

Contact with the museum and artefact conservation profession was invaluable, as was the contribution of materials and conservation scientists. A whole new profession of conservators grew up to care for the rarer, more valuable, and more fragile parts of buildings. Scientific analysis of the mechanisms of material decay has become more common and sometimes there are science-based solutions to decay problems. Frequently, however, science and technology have been found wanting, and as time has gone on, the empirical wisdom inherent in traditional building practice is held in ever higher regard. Increasingly, traditional craft-based repair methods have been found to work best, even when there is no consensus as to why or how they work. The fact is that traditional buildings are in some respects extremely complex; practice has been developed through centuries of trial and error. With or without science, that process must continue.

The greatest challenge facing society in the twenty-first century will be an environmental challenge: how to manage the building stock and to augment it as necessary with minimum exploitation of non-renewable resources, both in construction and in use, and without polluting or otherwise damaging the environment. This brings building conservation into the front line of vital human activities. The conservation and creation of good buildings are complementary activities which will enrich, as surely as the neglect and destruction of good buildings and the creation of bad ones will impoverish, the lives of future generations. At the core of building conservation is the simple, or not so simple, process of repair.

REPAIR AS THE BASIS OF BUILDING CONSERVATION

What is repair? What indeed is conservation? The following set of interlocking definitions was first written in the belief that the language of the international charters, the ICOMOS Venice Charter and particularly the Burra Charter, was unusable for everyday purposes, and that there could be little hope that it would become common usage, at least in the United Kingdom. The intention was that words should be used with precision, but naturally in terms of everyday use. Probably the most difficult suggestion is that the word *reconstruction* should not be used of physical fabric but only of the abstract concept of design, rather as it is used of a crime or a historical event. A reconstruction should be the result of research or detective work, therefore, not of the physical process of building. Rebuilding and replication seem simpler and more natural for the latter. Restoration and conversion are defined as special forms of alteration, the objectives of which are the re-imposition of design or a change of use respectively.[11]

The terms defined in the list below are those which may be regarded as having precise or technical meanings in the context of building conservation. No definitions have been offered for such general terms as refurbishment, rehabilitation or renovation which mean little more than 'doing up', 'making useful again' or 'making new again' respectively.

- The *design* of a building or artefact is the abstract conception, on the basis of which it was made. A design may exist in the mind or imagination, on paper, or it may be realized and represented in the building or artefact itself. The design of an unaltered building will be pure, that of a much altered building composite, made up of several designs overlaid one upon the other.
- The *fabric* is the physical material of which a building or other artefact is made. Its state at any particular time will be a product of the original design and of everything to which it has been subject in the course of its history, including deliberate alterations based on well considered secondary or subsequent designs, careless changes, the effects over time of weather and use, damage and decay.
- *Archaeology* is the scientific study and interpretation of the past based on the uncovering, recording and retrieval of information from physical evidence. In the context of standing buildings and other structures, such evidence is as likely to be visible or concealed within the superstructure as concealed below ground. It is in the nature of archaeology that it may be destructive.
- *Reconstruction* (used of design, as of a crime or a historical event) is the establishment of what occurred or what was, on the basis of documentary or physical evidence. Reconstruction may be proven or, to a greater or lesser extent, hypothetical, depending on the nature and strength of the evidence.
- *Preservation* of a building or artefact is its survival, whether by historical accident or through a combination of protection and active conservation. Preservation is a state or an objective, not, in a technical sense, an activity.
- *Protection* is the provision of legal restraints or controls on the destruction or damaging of buildings, artefacts, natural features or systems, sites or areas or other things of acknowledged value, with a view to their survival or preservation for the future. Any intervention or work likely to affect the essential qualities of a building or its character, land, or anything which is legally protected would normally require a consent to be obtained through a procedure established by the relevant legislation.

- *Conservation* is action to secure the survival or preservation for the future of buildings, cultural artefacts, natural resources, energy or any other thing of acknowledged value. Where buildings and artefacts are concerned, such action should not involve significant loss of authenticity or essential qualities.
- A *conservator* is a person with professional, technical and usually manual skill in one or more than one aspect of fine or applied art, artefact or building conservation. Conservators are expected to use judgement and to act ethically, with a proper regard for the authenticity and essential qualities of any work of art, artefact or building with which they may be concerned.
- *Intervention* is any action which has a physical effect on the fabric of a building or artefact.
- *Reversibility* is the concept of carrying out work to a building, part of a building or artefact in such a way that it can be reversed at some future time, without any significant damage having been done to the existing fabric.
- *Maintenance* is the routine work necessary to keep the fabric of a building, the moving parts of machinery, grounds, gardens or any other artefact in good order.
- *Repair* is work to the fabric of a building or artefact, to remedy defects, significant decay or damage caused deliberately or by accident, neglect, normal weathering or wear and tear, the objective of which is to return the building or artefact to good order, without alteration or restoration. It is in the nature of repairs that they are irregular and beyond the scope of normal routine maintenance. They may occasionally be required in response to a specific event, such as a storm or accident, and be unpredictable. Usually, however, they should be anticipated and planned.
- *Alteration* is work to the fabric of a building or artefact, which is neither maintenance nor repair, the objective of which is to change or improve its function, or to modify its appearance.
- *Conversion* of a building or artefact is alteration, the objective of which is a change from one use to another.
- *Restoration* of a building, part of a building or artefact, which has decayed, been lost or damaged or is thought to have been inappropriately repaired or altered in the past, is a form of alteration, the objective of which is to make it conform again to its design, or to its appearance at a previous date. Such design or appearance may be based on reconstruction.
- *Rebuilding*, or remaking, is the building, or making again of the fabric of a building, part of a building, or artefact which has been irretrievably damaged or destroyed, on the basis of a recorded or reconstructed design.
- *Replication* is the making of an exact copy, or copies, of a building or artefact.

Conservation, therefore, is action to secure the survival or preservation for the future of buildings, cultural artefacts, natural resources, energy or any other thing of acknowledged value. This is, quite deliberately, a general definition, usable across the whole spectrum of the Conservation movement. The important thing is that conservation is *active*, whereas preservation is passive, denoting little more than survival, but without connotations of aspic. Maintenance and repair are fundamental to conservation, the difference between them being that the former is routine, the latter irregular.

Maintenance is the routine work necessary to keep the fabric of a building, the moving parts of machinery, grounds, gardens, or any other artefact in good order. Repair is irregular work, beyond the scope of maintenance, to remedy defects, significant decay or damage caused deliberately or by accident, neglect, normal weathering or wear and tear, the objective of which is to return a building or artefact to good order. Repairs may be unpredictable, in response to specific events, such as storms or accidents. Normally, however, they should be anticipated or planned.

When a building is in use, performing satisfactorily and earning its keep, maintenance and repair are all that are required to ensure its survival. In real life, however, things are rarely so simple.

REPAIR AND ALTERATION

In order to survive, buildings must be useful and must be wanted. Keeping them useful and keeping them wanted may be a necessary part of conservation. Building conservation, just as much as building creation, is an architectural activity and should have regard to the Vitruvian precepts, Commodity, Firmness and Delight. To be wanted, and to survive, buildings should be not only structurally and constructionally sound, but also practical and efficient to use and pleasurable to look at and be in.

Any building which is in poor repair, awkward or impractical to use, an eyesore or unpleasant to be in, is likely to be more or less at risk. The loss of any building is inherently wasteful; it may entail cultural loss and the loss a potential economic resource, it may have a negative effect on the local economy and on the local environment, the energy embodied in its construction will be lost and the disposal of its materials may have environmental consequences. By comparison, the consequences of conservation are in almost all circumstances likely to be benign.

Building conservation is not only, therefore, about maintenance and repair, and, in practice, repair works are quite likely to be coupled with alterations, the objectives of which may include improved usability in an existing use, conversion to another use, or improved internal or external appearance. The last might include cleaning or redecoration for aesthetic, personal or commercial as well as for purely practical reasons. In certain circumstances, notwithstanding a general presumption against altering a historic building merely to make it conform again to its original design or previous appearance, it might include restoration.

Delight in architecture is a subtle and complex matter, and not just a visual one. Formal design qualities may be significant, but so may the patina of age, associations, legibility and a sense of authenticity. Charters, principles and guidelines are essential, and many cases where decisions are required will be clear cut. There will be others where careful analysis, consideration, the balancing of conflicting pressures, consultation and judgement will be required. It is the skill of a good architect, not merely in a simplistic sense to be a good designer, but to be able to make good judgements about buildings.

A FRAMEWORK FOR MAINTENANCE AND REPAIR

Good buildings are designed to last. To do so, they must have certain universal qualities which make them attractive and comfortable for human use and occupation, they must be reasonably adaptable and above all they must be maintainable. Among the severest criticisms of some of the worst twentieth-century buildings is that they were said to be maintenance-free, but turned out to be unmaintainable and, indeed, irreparable when parts of them failed. Among the merits of traditional buildings is that they are almost infinitely maintainable and repairable with simple materials, tools and skills.

Any building, from the smallest and simplest to the most magnificent, is best and most economically maintained in the long term in a systematic way. The object of any system is to ensure that relevant information is to hand and records kept, and that maintenance and repair work is carried out efficiently and at the right time; neither too soon, which is wasteful, nor too late, which might result in secondary damage and loss, and more work being necessary. Any system of planned maintenance requires to be set up in the first place and will cost money to operate. However, a good system, managed with judgement and skill, will save money over a relatively few years, and will save more the longer it is operated. Continuity of professional advice, as with a family doctor, is highly desirable.

Current best practice requires the maintenance of two documents, a Conservation Manual and a Log Book and the carrying out of quinquennial inspections.[12]

Conservation manuals

It is good practice for the person responsible for any building to establish a conservation manual. This is a permanent, standing document, containing essential information on the building, its history and architecture, materials and construction. It should also contain guidance on routine housekeeping, management and maintenance procedures including safe access arrangements and on the constraints which would be likely to be applied to any proposed repair or alteration work. Its content should be reviewed quinquennially, but only amended with good reason. A conservation manual might contain the following:

- An introduction, setting out its purpose. which is to inform the building owner or manager about the building and its historical or architectural importance, if any, and to describe a housekeeping and management régime which would keep it on the 'plateau of good repair' and protect it from avoidable damage or decay.

- A written description of the building and its various parts, accompanied, if possible, by plans, sections and elevations and drawings or diagrams of services installations. Architectural drawings are important records which are costly to prepare and likely to be of use in the future. They should be valued and carefully and securely kept.
- Specific guidance on management, housekeeping and maintenance procedures, cross referenced to the descriptive section.
- A record of any known toxic or hazardous materials in the construction with an assessment of the risks associated with them if disturbed or left undisturbed and guidance as to their handling or treatment, as well as other essential health and safety information.
- A conclusion, summarizing the contents of the manual and giving emphasis to the principal recommendations.

While for large, complex or important buildings, conservation manuals should be professionally prepared, it might be within the capacity of a capable and conscientious householder, perhaps with some informed advice, to prepare his or her own.

Log books

A log book should be kept for every building in which current information, such as the names, addresses and telephone numbers, including emergency numbers, of key people, including tradesmen and maintenance contractors, should be listed. It should also contain concise instructions on maintenance and inspection routines and on action in an emergency. Routine diary entries should be made recording the carrying out of inspections etc. and noting the results. A record of any work carried out on the building should be entered. Completed log books should not be destroyed, but should be kept as part of the permanent record of the building. The health and safety file, which many building owners have a legal obligation to maintain, might with advantage be incorporated into the Log Book.

Quinquennial inspections

Important buildings should be professionally inspected quinquennially, or at some other appropriate interval. The inspector should preferably know the building and have some continuing responsibility for its care and conservation. Before undertaking the inspection, the inspector should be familiar with the contents of the Conservation Manual, the Log Book and the previous quinquennial inspection Report. The purpose of the inspection is to establish, on the basis of visual inspection from reasonably accessible points:

- the general condition of all parts of the building;
- the progress of repairs carried out since the previous inspection;
- the need for further repairs and, where appropriate, other works;

- the urgency and importance of such works, under four degrees of priority;
- the order of cost of such works.

The inspection should be carried out in a logical sequence. A complex building might be broken down into wings or sections. Individual inspectors will have their own methods, but, in the interest of consistency, it is suggested that the outside of a building should normally be inspected before the inside, and that the sequence should generally be from top to bottom, starting from the same point on plan at each level and working clockwise from volume to volume and within each volume or room. The materials and construction of each element should be described in as concise a way as possible and, separately, their general condition and any defects recorded. Fairly extensive 'snapshot' photographic coverage, particularly of defects, is likely to be helpful.

Quinquennial inspection reports

Reports should be written up by the inspector as soon as possible after the inspection. The report should contain three main sections, an introduction, a description of the fabric and its condition and recommendations for repairs and, if appropriate, for other works. There might also be appendices.

The descriptive section, which is the record of the inspection should be written first. Following analysis, the recommendations should be drafted and items of work required either identified individually or grouped into packages and arranged in four degrees of urgency or priority categories:

Immediate – Priority 1: work which should be put in hand immediately for public safety or health and safety reasons, to prevent imminent damage or to arrest rapid deterioration. This might include immediate further investigation.

Urgent – Priority 2: work which should be put in hand urgently, within weeks or months, or a year-and-a half-at most. Failure to do so would be likely to result in significant further damage or deterioration and increased cost.

Necessary – Priority 3: work which should be carried out before the next quinquennial inspection, for which there is time to plan, and which might be integrated with other work. This is work which is due, in order to keep the building on a 'plateau of good repair' and to maintain the value and usefulness of the property. Most repair work should come into this category.

Desirable – Priority 4: work which is desirable, if not strictly necessary, but which would improve the functioning or performance of the building or enhance its architectural or aesthetic qualities. Alternatively work which is not due, but is likely to become so before the next quinquennial inspection and which might sensibly be incorporated with other work. Much minor restoration work, such as the reinstatement, where appropriate, of suitable windows, should come into this category.

The recommendations could include further investigation, supplementary technical or specialist reports. They should also include any necessary improvements to the safe access arrangements: hatches, trap doors, ladders, crawl boards, lighting, harness fixing points, etc., which are essential for proper inspections and routine maintenance. In deciding on the priority or degree of urgency of any item of work, or the extent or type of repair which should be recommended, or the way in which individual items should be grouped into packages, considerations of access and accessibility are vital. It should be borne in mind that it is wasteful not to allow materials to live out their useful lives and to repair or replace them before it is necessary to do so. On the other hand it is even more wasteful to allow serious consequential decay to develop for want of a simple repair. Decisions on the timing and packaging of repair works require careful judgement. Approximate costs should be set against each package. The introduction is likely to be the last section to be written. It should include general information about the building, a statement of the purpose of the inspection with a note of any limitations or exclusions, an account of works carried out since the previous inspection and a brief summary of the conclusions and recommendations in the report.

REPAIR AND THE PRINCIPLES OF CONSERVATION

There follows a summary of the principles of building conservation.[13] These are not intended to be definitive or strongly prescriptive. No set of principles can be absolute, universal or unchanging. The SPAB Manifesto and the international charters, particularly the Venice and Burra Charters, have been of enormous importance to the development of conservation theory. Any such text, however, reflects the spirit and ideas of the time at which it was written and must, in time, become historical. The summary presented here is an attempt to express the needs of conservation, as it is practised in the United Kingdom in the mid 1990s.

The need for understanding, experience and judgement, based on clear principles
Decisions on building conservation matters require the exercise of judgement, based on an understanding of principles and informed by knowledge, including that of relevant legislation, and experience. Codes of practice, structural codes and calculations and other standard codes and specifications should not be applied unthinkingly. While such specifications and codes should not be ignored, old buildings have stood the test of time and due consideration should be given to empirical knowledge and to an understanding of what evidently works.

The need for work to be prepared, directed and carried out by people of appropriate experience and skill
Any significant work should carefully be prepared and directed by a suitably qualified and experienced professional person. Work should be carried out by conservators, craftsmen or tradesmen with adequate capacity and appropriate knowledge, skill, experience and attitude. Specifications and contracts should be well suited to the nature of the work. In the course of work, it should be clear

where responsibility for decision making lies, there should be good communication and administration and the professional in charge should give appropriate time and attention to the overseeing of the work.

The need for work to be preceded by research and investigation and to be properly monitored and recorded
Any significant work should be preceded by appropriate recording, documentary research and physical investigation. Such work should be carefully monitored in progress and documented on completion. Reports should be written within an agreed period, deposited in a suitable archive and, where appropriate, published.

The need to balance conflicting objectives
The conservation of buildings may be justified on cultural, economic and environmental grounds, usually in combination. There will often be conflicting pressures, between which balances must be struck, and an analysis of the reasons for conservation in a particular case will assist good decision making.

The need for a conservative approach, based on minimum intervention and reversibility, which respects the whole history and authenticity of a building
The conservative approach, in which the level of intervention and disturbance to the fabric of the building are kept to a minimum, is fundamental to good conservation. The most appropriate intervention is usually that which has the least effect on the fabric and which can be most easily reversed with a minimum of damage.

The need for a systematic approach to preventive conservation
Systematic care, based on good maintenance and housekeeping, both saves money and is fundamental to good conservation. Maintenance and housekeeping routines should be planned, and easy safe access to roofs, etc., provided, with a view to the prevention of decay, to the early identification of defects and to the avoidance of consequential damage. For significant buildings in institutional ownership or care, such a routine should be based on a conservation manual, a log book and a quinquennial cycle of professional inspections leading, if necessary, to repairs. In simple form, a systematic approach to conservation and care is applicable to all buildings, however modest. The provision and updating of precautions against fire and other disasters aimed at prevention and damage limitation, and of appropriate insurance cover, should form part of this routine.

The need for repairs to be carefully considered, wherever possible planned, and regarded as part of an ongoing process
Repairs to the fabric of a building to remedy defects or significant decay should, as far as possible, be carried out as nearly as possible at the time and to the extent that they are necessary. Repair is an ongoing process and no repair is ultimate or final. A craft-based, like-for-like approach to repair is normally the most appropriate, and materials and methods should generally be carefully blended to existing work. Where a discreet, non-traditional method enables more existing fabric to remain *in situ* undisturbed, however, it may be preferred.

The need for a holistic approach to buildings, their materials and systems of construction

The construction of a traditional building should be considered as a whole and treated in a holistic way. Its system of construction, materials, structure and patterns of air and moisture movement should be properly understood. New and untried materials and methods should be treated with caution. The introduction of impermeable materials or membranes, materials which might be incompatible on account of their excessive hardness, rigidity or strength, or treatments, the long-term or side effects of which on the building or its occupants might not be fully understood, should generally be avoided. Any changes to a building should be compatible with its overall system of construction.

The need for any restoration to be cautious, conscientious and honest

A presumption against restoration is a characteristic of the British approach to conservation. Restoration, involving alteration of the existing fabric of a building to make it conform again to its original or to a subsequent design, should only be considered where the particular circumstances of the case justify such an intervention and normally only where the extent of the alteration is small in relation to the whole. Materials and methods should, as in repair work, be carefully matched and blended, but there should be no intention to deceive; new work should be discreetly distinguishable and records kept.

The need to regard buildings and materials as finite resources

The continued use of a building and its materials and components *in situ* is inherently desirable. If alteration is necessary, genuinely surplus materials and components should, if possible, be re-used within the building or, otherwise, reclaimed for use elsewhere. The stripping of materials from disused buildings is, however, to be discouraged. If re-use or reclamation should be impossible, recycling or energy recovery is preferable to dumping.

The need for existing and new work to make an honest and well-integrated whole

New work, whether in alteration, addition or new building in an existing settlement or conservation area, and whether the architecture is matching or distinct, should be of appropriate quality, should not draw attention disproportionately and should contribute to the architectural integrity of the altered or extended building or group of buildings as a whole. Such new work should be designed to last, to be maintainable, to weather and age well and to take its place in the context of the building or group of buildings in which it is set.

The need for a building, its contents and setting, to be considered together

A building in its landscape or townscape setting and with its interior however simple with decoration, fixtures, fittings, services, plant, machinery and associated contents may be regarded as a composite work of art and document of history. Proper consideration should always be given to the treatment of a building, in its setting and with its interior and contents, as a whole.

CONCLUSION

Good repair practice is central to good conservation. If life were simpler and less fast moving than it has been in twentieth-century Britain, as indeed it was at certain periods of history, maintenance and repair would be the only action required to enable buildings to survive. The present reality, however, is that other sorts of intervention may be necessary to accommodate change. Alteration of one sort or another, in addition to straightforward repair, must sometimes be inflicted on buildings if they are to continue to be useful and wanted. Conservation, therefore, may entail more than repair.

The profligate modernist years of the 1960s, in which new was good and old was bad are past, and more balanced views prevail. There is recognition that what is new architecture today is history tomorrow and that society needs good buildings of all ages, old and new together, and needs them to be useful, sound and beautiful. There is, moreover, a new understanding that resources are finite, and that buildings are resources in economic, environmental and cultural terms. Destruction is invariably wasteful and may be positively damaging, while the creation and conservation of good buildings is always worthwhile.

Repair may seem a modest and unglamorous activity, but for an architect who is affected by architecture, who respects the labour, the skill and the creativity of

Figure 9.8

The Roseline Building, Sunderland, Tyne and Wear. These stone and brick former warehouses are lonely survivors on the once bustling Sunderland Quayside, across the Wear from the ancient church of Monkwearmouth. Converted with simple materials, labour and skill by Rivergreen Ltd, Jane Darbyshire and David Kendall, Architects, with Simpson and Brown. (*Photo: Alan Mushen.*)

Figure 9.9

339–43 High Street, Kirkcaldy, Fife. An artisan house of the late-sixteenth-century, built with solid plastered walls and painted 'beam and board' ceilings. Plaster ceilings, panelled linings and sash windows were fitted in the late-seventeenth-century. The house was structurally repaired and restored to its *c.*1700 appearance by Simpson and Brown for the Scottish Historic Buildings Trust in 1992–3. (*Photo: James Simpson.*)

others and who senses the continuity of past, present and future, working closely with fine old buildings can be sheer pleasure, and making them good in the British way can, indeed, be glorious.

Notes

1 James Simpson (1992) 'Construction and the Generation of Style', Aspect: *Views from the King's Manor*, Journal of the Alumni Association of the Institute of Advanced Architectural Studies: York, ii, 14–17.

2 Simpson, *op.cit.*

3 The houses and tenements of Georgian Edinburgh show the traditional, pre-industrial construction of a particular place in its most developed form. See Davey *et al.* (1995) *The Care and Conservation of Georgian Houses*, 4th edition, Butterworth.

4 Lead sheet has been the exception. The underside corrosion problems of recent years illustrate the subtle balance of traditional systems, the risks inherent in changing patterns of air and moisture movement, the limitations of science, and trial and error in action.

5 Legal restrictions on dumping and pressure to re-use or reclaim components and to recycle materials, as, for example, in Denmark, provide new incentives for conservative repair and alteration.

6 The Black House at Arnol, a representative example now in the care of Historic Scotland, was built in 1885. The present custodian, a great grand-daughter of the first occupant, was brought up in the house with an open central hearth in the 1950s.

7 Council houses of cob and thatch at Amport in Hampshire, and of stone and slate at Callander in Perthshire, all with coal fires, for example.

8 The continuing failure of large sections of the architectural profession to abandon their obsession with the more superficial aspects of design and to adapt intellectually to the environmental imperative to build half as much twice as well, in addition to maintaining and making best use of the existing building stock, has affected the position and reputation of architects in society.

9 The pioneering work of the National Trust for Scotland's Little Houses Improvement Scheme by architects like Ian G. Lindsay, exemplary as it was at the time, often involved total renewal within the old walls.

10 The development of the 'remedial treatment industry' was a response to loss of understanding of, and of confidence in, the basic principles of traditional construction. Science and 'modernization' seemed to offer solutions to perceived problems.

11 These definitions have been developed for teaching and other purposes over a period of years and I acknowledge the contribution of my teacher and friend, the late Colin McWilliam. They may form the basis of the definitions in the possibly forthcoming British Standard Guide to the Principles of Building Conservation.

12 The procedures recommended here have been developed over a period of years and were drafted in more or less their present form for *The Care of Historic Buildings & Ancient Monuments by Government Departments in Scotland*, Historic Scotland and the Conservation Unit, Department of National Heritage (1994). They may also be incorporated as guidance in the possibly forthcoming British Standard Guide to the Principles of Building Conservation. The methodology for quinquennial inspection is derived from Feilden and Mawson office practice in the 1970s, with acknowledgement to Sir Bernard Feilden and Keith Darby.

13 These principles have been developed over many years and may be incorporated as a summary in the possibly forthcoming British Standard Guide to the Principles of Building Conservation.

10

UNDERSTANDING HISTORIC STRUCTURES

David Yeomans

INTRODUCTION

Much of the published work on historic structures has been by historians of structural design, using techniques of structural analysis to answer historical questions rather than for the tasks of conservation, and it is important to note that the historian and the conservator have different needs even though these may overlap. The historian may want to know how the structure behaved, and was seen to behave, when it was built and may even be interested in the behaviour of structures that no longer survive, or survive in only small numbers. Their interest is principally in unique structures or those at the limit of structural development at the time. In contrast, the conservator may be dealing with the vernacular as much as with the rare building and needs to understand a structure before attempting a repair. While this may include understanding its historic behaviour it must extend to a consideration of the repair itself and the behaviour of the structure during the repair process. The conservator is concerned with whether the structure will continue to carry load and so must involve a quantitative analysis of the performance of individual members as well as the overall behaviour of the structure. Where their interests overlap is that both need to understand building structures as systems which change with time, both need to use the evidence of the building itself, looking to see what deformations have occurred or what signs of distress there may be in the structure, as much as using the formal tools of structural analysis.

Just as there are differences between the conservator and the structural historian, there are also differences between the conservator and the designer of a new building. One distinction here is obvious, the historic building exists, it can be seen and measured, while the new structure exists only on paper. But, while some parts of an existing structure are visible and may have known properties, others, whose properties are equally if not more important, may be inaccessible and cannot be fully investigated without destroying the very thing that we are trying to preserve. Therefore the conservator must sometimes be able to adapt the techniques of structural analysis to cope with these unknowns.

Moreover, because the conservation of existing buildings is a minority activity compared with new design, conservators do not always have the same support of scientific information available to the designer of new structures. The conservator may have limited knowledge of the properties of the materials used in the structure, of the behaviour of the structure itself, or even of the behaviour of the repair methods being used. In extreme cases it may even be necessary to undertake some form of testing. In all this the conservator is limited in the proposals that can be offered if only by the properties of the existing building materials, so that it may not be possible to enjoy the same kind of safety factors to account for uncertainties that one can enjoy in a new structure. But the principal limitation is the need to preserve the historic character of the building: not all structural solutions may be acceptable.

OUTLINE

Because of the importance of work done by historians of structures it is useful to begin by looking at some of their methods, principally to show how they must be adapted to the task. There have sometimes been differences of opinion both about the appropriateness of different methods used and the questions that might be asked. While often asking different questions, the conservator needs to be clear about this. Understanding a structure is not a task that is undertaken in abstract and we need to know why we are asking the questions. It may seem obvious to say that the nature of the task must be clearly defined before selecting the tools for the job, but the tools of structural analysis were not devised with the conservator in mind but for the design of new structures.

The simplest way of looking at the methods which the conservator uses is to consider the life of a structure, its design and construction and the subsequent changes it undergoes including the process of survey and rehabilitation of the building, taking them in that order. That does not imply that the methods considered will be in order of importance. Indeed the importance of any particular technique will depend upon the circumstances of the particular building and the intentions for it.

THE METHODS OF THE HISTORIAN

Historians of structures have used such diverse methods of structural analysis that they might be read as much as experiments in the appropriateness of the different techniques as for answers to questions about particular structures. Robert Mark (1982), for example, used photo-elastic methods to look at the structure of Gothic cathedrals while Jaques Heyman (1966, 249) only considered the principal lines of thrust within such structures. Both were concerned with the overall behaviour of the structure and made major simplifications, assuming the structures to be homogeneous in construction. Neither the photo-elastic techniques used by Robert Mark nor Heyman's analysis of Gothic structures take account of the differences of construction within the structure, that is, no distinction is made between the ashlar facings and the rubble fill.

However, the differences between the approaches of these two are greater than their similarities. Heyman adopted a technique whose starting point was that the behaviour of masonry structures involved no ability to withstand tensile forces. This is largely true of masonry where only limited tensile stresses can be developed so that considerations of stability are more important than the actual stresses and the strength of materials. The question is whether it is possible to draw a line of thrust that lies within the masonry and the techniques used are very simple – all that is required is an understanding of the triangle of forces and the kind of arguments used by Heyman can be most easily followed in his analysis of Clare College Bridge rather than the more complex cathedral structures. This is also an example where the method of analysis can be used to assess the safety of the structure.

In this simple example an arch can be divided into two equal halves (Figure 10.1). An assumption has to be made of the position of the line of thrust at the apex and the springing but from this and the loads on the structure a line of thrust can be drawn. Some trial and error may be needed, especially if the line of thrust touches the edges of the arch ring when it may be necessary to ensure that this corresponds with observed behaviour of the actual structure, for example, open joints which indicate that a hinge has formed at that point. Heyman also notes that more than one solution may be possible from which maximum and minimum values of the supporting forces can be determined. If it is possible to draw such a line of thrust then the structure will be safe.

It is worth quoting Heyman's own words here because they show both how limited and how powerful his approach can be. He says:

> There are strong arguments for saying that an actual state of masonry is not a useful concept. There is always a certain position of the line of thrust in an arch...; however, a small subsequent shift, for example, of an abutment or a small temperature variation, will almost certainly alter the position of the line of thrust.... In such a changing situation what remains constant is not the actual state of the arch, but the framework of principles within which the behaviour of the arch is bound. (Heyman and Padfield 1972)

What is noteworthy about Heyman's technique is that it involves quite simple structural methods. In dealing with simple bridge arches it can be used to estimate the safe live load that a bridge can withstand and in the paper cited he compared the results of his method with that recommended by the (then) Ministry of Transport.

He has also used this method of analysis to show how the design of flying buttresses responded to the different load conditions that occurred both during the construction of a cathedral and after its completion. During construction the buttress has to stand under its own weight alone, while after completion it must

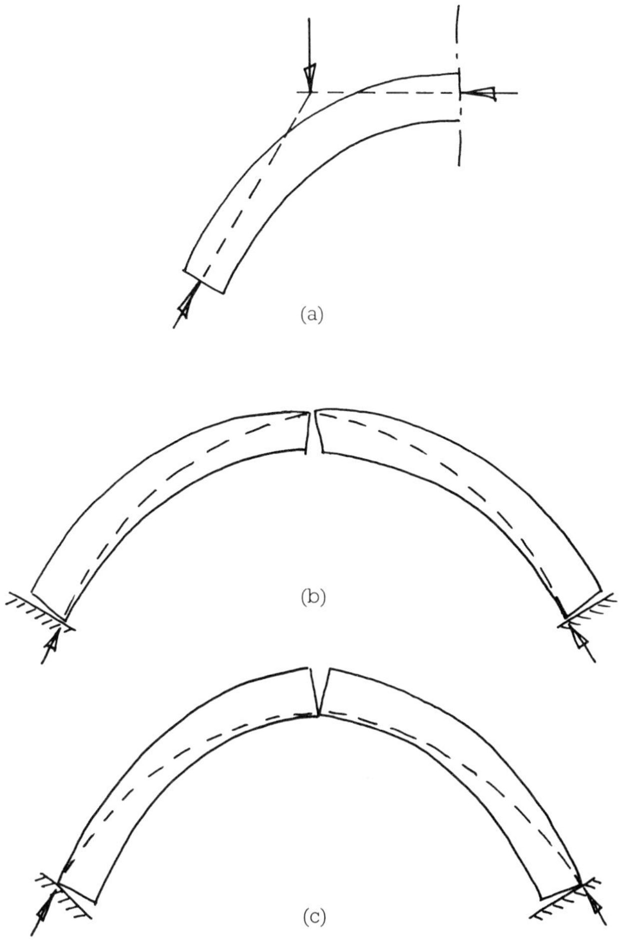

Figure 10.1

Analysis of an arch. Under symmetrical loading, an approximate value for the horizontal thrust can be found from a simple analysis using the triangle of forces (a). However, as can be readily seen from the diagram, the result will depend upon where the forces are assumed to act. The minimum value of the horizontal force can be found by assuming the abutments move out (b) to allow the arch to hinge as shown. The line of thrust must be found by trial and error. Assuming the abutments to move inward producing the hinges shown in (c) gives the maximum value for the horizontal thrust.

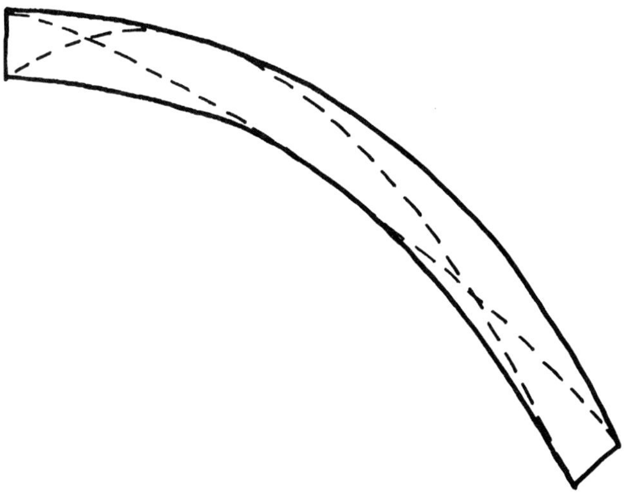

Figure 10.2

Forces in a flying buttress showing the lines of thrust under two loading conditions – one when the buttress is standing under its own weight alone, the other when called upon to resist horizontal thrust. (*From Heyman, 1966.*)

resist the thrust from the vaulting plus any additional wind loads. The diagram of forces (Figure 10.2) shows how the design is adapted to these two different requirements.

In contrast to this simple paper and pencil technique, Mark's use of photo-elastic methods cannot truly represent the behaviour of the actual structures because his models are capable of developing tensile forces under load which the actual structures are not. Nevertheless, his analysis of the cathedral at Chartres, (Mark 1982, 19–21; Borg and Mark 1973), which suggests the development of tensile forces, led him to look for and find cracking in the masonry (Figure 10.3). Such a graphic visualization of the behaviour of the structure, while perhaps less 'correct' as an analytical technique, proved valuable in this instance because it

Figure 10.3

Tracing of the photo-elastic interference pattern produced by simulated wind loading on the buttressing at Chartres Cathedral according to Mark. The lines represent the boundaries of different colours in the original model. Each colour indicates a different level of stress so that closer lines represent areas where there are more rapid changes of stress within the structure. In a column or vault rib, this means larger amounts of bending.

pointed to behaviour which might otherwise have been unrecognized. The graphic view of the way in which the structure works may be attractive for this reason but it is not the kind of technique that can be as simply used as Heyman's methods because it involves making and loading a model and the equipment to look at the resulting pattern of polarized light that shows the different stresses.

DIFFERENCES OF OPINION

Differences between historians go deeper than a simple choice of methodology. There are also different questions being asked by the different scholars, differences in approach that can be seen most clearly in the published encounters between Heyman and Mainstone following the former's paper on the structure of Westminster Hall (Heyman 1967). In this structure questions need to be asked abut the nature of the support provided by the wall which can be tested by direct observation. Different modes of action within a structure may be accompanied by different forces at the supports which may produce observable results. The method of analysis then used by Heyman depended upon the kind of approach used for modern structures: 'If any distribution of forces in a frame can be found in equilibrium with the external loads and not exceeding the yield limits for individual members, then the structure as a whole is safe and there is no need to calculate the "actual" distribution of forces.' (Heyman 1967, 159)

This means that the structure can be simplified by eliminating 'unnecessary' members and the mechanism he assumed did not depend upon load being carried by the curved braces. Mainstone regarded this as an unsatisfactory analysis, wanting to know how the structure was actually working, that is, including the members ignored by Heyman. His view was that 'if it is desired to understand how a structure carries its loads, and not merely to demonstrate that it can do so somehow, it is essential to ensure that the theoretical model is sufficiently representative of the particular structure in the state considered.' (Heyman 1967, 789)

He assumed that there would be some stress in the curved braces and wanted to know how much, the difficulty then being in assessing the stiffness of these members. Heyman's response was that to ask how a structure actually works is an impossible question. One could never know if one had the right answer and the behaviour of the structure today might be quite different from its behaviour when it was built.

Here are different questions. One approach seeks to find a simple mechanism by which the structure may stand. The other asks how the structure is standing and even if the answer cannot be tested by observation of the structure; exploring the possibility may prove interesting. Just as Heyman looked at the different conditions in a flying buttress during and after construction, Mainstone (1969) considered the behaviour of the dome of Santa Maria del Fiore, Florence, during its construction. A question that he needed to ask was whether true dome action could be developed in the structure, that is, whether tensile hoop stresses were possible near the base of the dome. The result of his calculation was that only modest tensile forces would be necessary in the masonry. Had the required

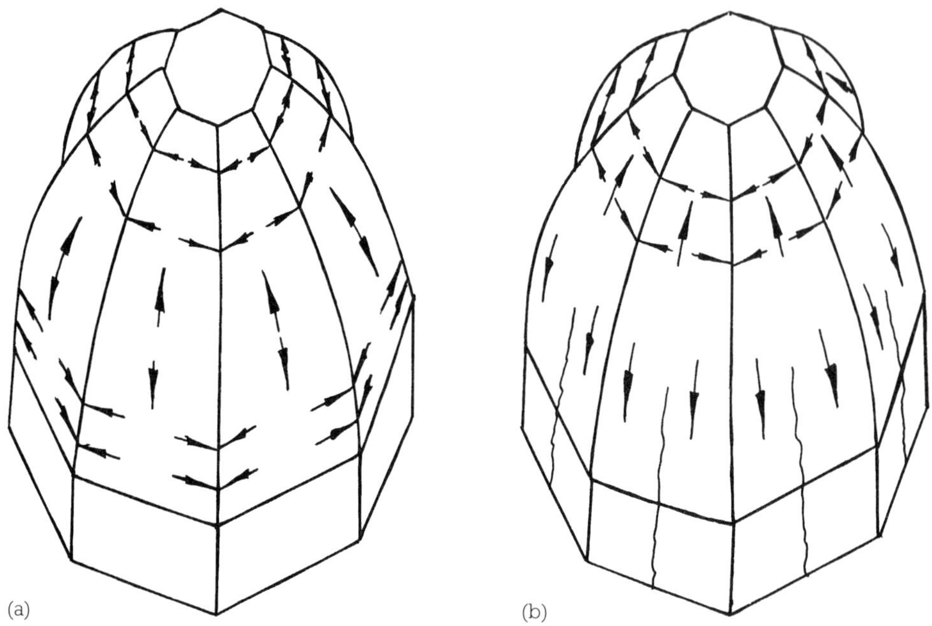

Figure 10.4

Support mechanisms for domes drawn on an outline of Santa Maria del Fiore. If the structure is capable of developing tensile hoop stresses at the bottom then true dome action is possible (a). If tension stresses cannot be developed then cracks will develop splitting the dome into separate arch like segments (b). In either case compressive hoop stresses will be developed at the top of the dome.

stresses been large then tensile cracks would have developed and the support mechanism would have necessarily been quite different (Figure 10.4).

A similar difference in approach can be seen in the discussion of the Clare College Bridge structure, the argument being that a bridge is not simply an arch ring loaded by a completely passive fill within the spandrels. Any distortions of the masonry arch will result in some redistribution of forces within the material of the fill so that this needs to be considered as part of the structural system. Mainstone pointed out that 'if the fills have any stiffness of their own, the slips or rotations will automatically bring about a beneficial redistribution of the loads actually reaching the arch' (Heyman and Padfield 1972, 497).

However, Heyman, given the limited intentions of his analysis, felt it unnecessary to regard the fill as other than a load. This bridge had moved, developing hinges within the arch ring, but he was able to show that a system of forces could still be drawn in which these forces lay always within the masonry of the arch. He did recognize that masonry at the springing of the arch might have a profound affect on the behaviour of the structure, because it would alter the position of the line of thrust at the springing, but he did not consider the fill above the arch to be part of the structure. It will be apparent that the result is for Heyman's analysis to produce a conservative estimate of any load that the bridge will carry.

Although apparently asking different questions and taking quite different views of structural behaviour, both Heyman and Mainstone used mathematical modelling to represent the different structures. Courtenay and Mark (1987) attempted to distinguish between the two views of the structural action of Westminster Hall by carrying out model tests, but this approach presents another set of problems because the model must be constructed in such a way that it accurately replicates the form and behaviour of the structure. Assumptions about a structure may be built into a physical model just as they can into a mathematical model. It might even be argued that a physical model is less satisfactory because of uncertainties about the model's construction.

More recently another analysis (Morris *et al.* 1995) of Westminster Hall roof has used finite element analysis to look at the possible contribution of the decorative tracery within the structure. What engineers have always wanted are methods which are both simple to apply and provide accurate answers, although in practice the methods chosen are usually a compromise between the two. Accuracy is costly in terms of the time taken for the analysis and the engineer is always aware that inevitably the description of the structure will be imperfect. It may be useful to explore more than one load transfer method in the exploration of an historic structure, or assess the effect of assuming different member properties on the overall behaviour of the structure. Finite element analysis offers the opportunity to do this. The structure may be modelled as a number of discrete components with defined properties. The effects of load on the structure may then be determined by computer analysis in terms of both stresses and deflections. The advantage of this is that, while the geometry of the structure remains constant, the properties may be changed.

Comparing the various approaches to this structure the finite element analysis appears to offer the most complete picture because it includes elements of the construction so far ignored by other scholars, although the costs of analysis are much higher. Nevertheless, if we agree with Heyman's view about the limit to our understanding of historic structures we can only regard this as another possible way of looking at the structure of the roof rather than a definitive picture. If we adopt the view that all techniques of structural analysis are guesses, but that some guesses are better than others, then some judgement must be exercised in deciding the kind of answer and the level of precision that is required in selecting the method of analysis to be used.

HISTORICAL QUESTIONS

Although not central to our concerns here it may be useful to look briefly at the use of structural analysis for addressing questions about the history of structural design. The result of Mark's analysis of Chartres was to suggest a possible line of structural development. A failure to provide adequate buttressing resulted in cracking which, when observed, led to a modification of the structure and perhaps a change in structural design. Structural and stylistic reasoning were complementary here and the development of the structure may be seen to follow from some structural logic. Mark (1982) has used this elsewhere in a number of other studies, but the argument that the process of design follows the logic of the structure, as demonstrated by his photo-elastic analysis of other cathedrals, does require some historical data to support it (Addis 1991). Regrettably, we have little direct evidence of the way in which designers of the time viewed the behaviour of structures and, without this, any suggestion that the development of design was related to an understanding of the behaviour of those structures must remain an unproven hypothesis. Part of the problem is that a qualitative understanding of behaviour may not always lead to a correctly proportioned structure (Yeomans 1982).

Understanding the process of structural design may be useful as a means of anticipating the behaviour of particular structures. In early buildings from the period when there were no formal methods of design this means attempting to understand how the structures were seen to behave. Neither is structural understanding a singular skill. There is the understanding of the overall behaviour of the structure, which enables the designer to determine its general proportions and position the members, and there is a detailed understanding of the forces in those members from which their sizes and the details of joints may be determined. The two may not go together. It has been suggested that in the design of roof trusses in the eighteenth century, carpenters and architects had different kinds of understanding (Yeomans 1992b). The practical result is that one may find well designed trusses which are poorly detailed or the converse, poorly designed trusses which are very well constructed.

For more recent structures the formal methods of analysis and structural testing that became available may provide a much surer indication of structural behaviour. Indeed, the history of design methods may be essential to the assessment of some structures for which there is much more information on the stresses that their designers thought acceptable, the way in which they determined these and the loads that they anticipated. Understanding these structures involves the interrogation of contemporary literature as much as the structure itself.[1]

DEFINING THE TASK

If there is not necessarily one correct method for the analysis of a structure and the techniques used depend upon the kind of question asked, the first issue is to define the task clearly. The conservator is using structural analysis to look at a building which has undergone change or is undergoing change. Where change has already occurred, the task is to repair and maintain a structure which is no longer in its original condition. These changes may be a continuing process, such as the result of decay or weathering, or they may be the result of trauma, like a fire. Some changes in structural behaviour may be because of deliberate alteration to the building. Alternatively, the engineer may be consulted because some change is required to the building and the structural consequences need to be considered. Of course, a common situation when changes are required is for the engineer also to need to consider changes that have already occurred. Therefore, analysis might be needed to explain more than one mechanism of behaviour – perhaps the original structural system as well as the present mechanism of load distribution – at the same time establishing the way in which the change from one to another has occurred and hence explaining the resulting movement of the structure. All of this may be different again from the structural system being proposed by repairs or alterations. The ultimate task is to ensure the continuing strength and stability of the structure, bearing in mind the future use of the building, but with the minimum of intervention.

The work of the conservator has some parallels with the work of the doctor, or perhaps it is the veterinary surgeon since the structure cannot speak for itself. Often the task is to recognize symptoms of distress, to diagnose its causes and to prescribe an appropriate remedy. The dilemma facing both conservator and doctor is to avoid harming the patient while carrying out a sufficient examination. To do this a doctor draws upon a knowledge of anatomy and medical science and while the conservator needs an equivalent knowledge it may not always be available. In dealing with an eighteenth-century London town house or medieval timber-framed barn in England one knows what to expect because the building type is familiar. Peculiarities of the individual structure may present some surprises but the general pattern will be known and, just as we know how such building types have been put together, we also know how they behave.

The anatomy of a building derives from the way in which it has been constructed, that is, the materials used, the form in which they have been assembled and the standards of workmanship employed. All these will have some effect on its performance, including its structural performance, and it is essential that the conservator has some understanding of these. In a timber-framed building, for example, the behaviour of individual members will depend upon the species used and their selection. The way that the overall frame behaves will depend upon the jointing techniques used and the care with which the carpenter has assembled them. In Britain there is a high level of knowledge about the construction of historic buildings of most periods, although not always of their structural performance.[2] Long-established traditions have persisted into this century and buildings have been continuously maintained making their construction familiar to both tradesmen and professionals. While older buildings have been the subject of extensive study by archaeologists, from the eighteenth century onward builders' manuals were produced which described good construction practice.

To intrude into existing structures whose construction is not fully understood is like a surgeon proceeding with a less than complete knowledge of anatomy but there are significant types of historic buildings for which there have not been the historical studies upon which the conservator may draw for a knowledge of building anatomy. In these cases the conservator needs to consider acquiring a knowledge of buildings in a way similar to the surgeon, that is, through the study of cadavers, examining as far as possible structures which have become derelict or have collapsed. The selected dismantling of some, the equivalent of a dissection, may provide information and it might also be possible to use such structures as experimental test-beds for either measuring existing properties or developing repair techniques. This applies to both the basic form of construction and to the properties of the materials, including standards of workmanship. Some of this is the kind of information that can be provided by archaeologists and, because the conservator needs to draw upon their studies, the archaeologist should bear in mind the needs of the conservator.

For more recent structures the issue of workmanship and its effect on structural performance extends to manufacturing processes which may also have inherent uncertainties. Today there is a fairly good understanding of the properties of cast iron but many of the early techniques for the working of the material developed at a time when it was not fully understood. This is apparent from the evidence presented at the government enquiry that followed the collapse of the Dee Bridge (Royal Commission 1847). The way in which iron was used, and sometimes misused, in the construction of buildings can be seen from Fairbairn's text on mill construction (Fairbairn 1854). These are just two of many contemporary sources that provide a knowledge of the practice of the period. As James Sutherland (1989) has pointed out, there was a period when ideas about how to design using this material were not always sound. Also the casting process itself is full of uncertainties. There is the possibility of defects within the initial casting, built in stresses occurring during the cooling process observed by Fairbairn and even damage to the completed casting.

During the twentieth century there has been an equivalent learning process in the understanding and construction of reinforced concrete; understanding the way in which the amount of water affected the strength of the mix, the need for adequate cover to the steel and the kinds of aggregates that should be used – all took time. Meanwhile, buildings were being put up which are now recognized as important landmarks in the use of the material and, as such, may be listed buildings. The result is the ultimate irony of a building whose pioneering status makes it, at the same time, both worthy of preservation and a dangerous structure.[3] It is useful to have some knowledge of the development of construction methods when approaching such structures. At the same time it is also useful to know something of the way in which they were designed. Thus, in approaching structures that were put up during periods when the techniques of construction were changing and when new materials and methods were being developed, it is vital to have some knowldedge of this development process.

DESIGN

Historic structures were developed to suit the requirements of the past and within the limitations of the materials and the skills available. That they have survived to the present is a testament to their adequacy even though they may not conform to the expectations of a modern designer. Many were built at a time when there were no means of structural analysis available and their method of construction was refined through a kind of evolutionary process. It is not just that poor designs will not have survived but that methods of construction are

Figure 10.5

The roof structure of St Mary at Hill, London. The dome and vaults were formed of laminated timber ribs resting on trussed girders. Four columns supported the dome. Horizontal thrusts generated by the dome and vaults were transmitted to the walls via compressive forces in the trussed girders. Adjacent buildings and the church tower provided some restraint on three sides but the east wall had no assistance.

likely to have a certain robustness, that is, an ability to cope with a wide range of individual designs. Therefore, we should expect to find low working stresses and for the structure to be inherently stable. The only situations where we should expect excessive stresses, or even signs of distress, is where there have been long-term effects which the builders have not been able to observe and have therefore failed to accommodate by changes in their design, or perhaps in structures which are unusually large or, like the examples just considered, were in some other way pushing at the boundaries of knowledge.[4]

There will occasionally be structures like this which defy attempts at structural analysis, structures which seem to have survived almost in disregard of the laws of statics. Buildings may survive because loads have been carried by elements of construction that would not be regarded as part of the structure and hence in ways that cannot be easily analysed. An example is the roof structure found at St Mary at Hill when it was damaged by fire. Horizontal forces generated within the domed and vaulted roof should have been transmitted through supporting beams as point loads applied to the top of the walls. However, there was no obvious horizontal restraint to prevent the east wall from being pushed outward by this force (Figure 10.5). Therefore it was being contained in some other way, although not one that could easily be quantified. This is not a unique situation but professional responsibility requires engineers to understand how their structures are working and in circumstances like this it may be necessary to add structure where none existed before.

St Mary at Hill had been altered in the early nineteenth-century and perhaps the original Wren roof was better designed. It would certainly not have produced the same degree of thrust. It is so often in later alterations that the original integrity of the structure is lost. Andrew *et al.* (1989) show how the structural integrity of a mid-nineteenth-century (1865) church in Philadelphia was compromised by an extension of 1884 which reduced the buttressing action of the walls which had resisted the outward thrust of collar-braced principal rafters, and again it was restoration after fire damage that was the occasion for improving the structure.

IN-SERVICE BEHAVIOUR

In general, it is time and the processes of deterioration that go with it that raise stresses in the structure and lead to changes that require attention. The most obvious sign of change is movement within the structure and while this may be a sign of poor structural design or the result of some decay, it may be inherent in the materials used. The distinction is important. Mainstone (1988) in his studies of Hagia Sophia, Istanbul, has shown how the plasticity of the mortar in brick masonry has allowed the structure to move with time, the resulting deformations clearly showing the relative strengths of different parts of the structure (Figure 10.6). The most obvious aspect of this is the way in which bulging outward of the dome across the narrow axis, where it must rely upon passive restraint from the buttresses, contrasts with inward movement across the long axis where the semi-domes leant against the main dome with rather more force than the main

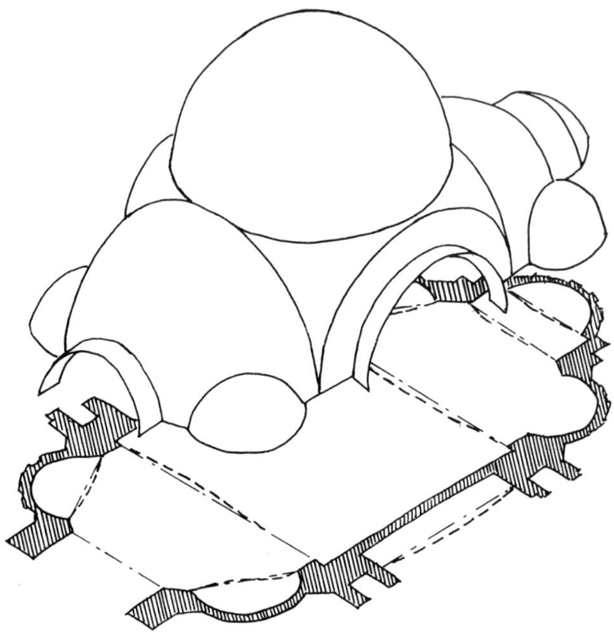

Figure 10.6

Diagram of the structure of Hagia Sophia, with the deformations found by Mainstone plotted on the floor plan.

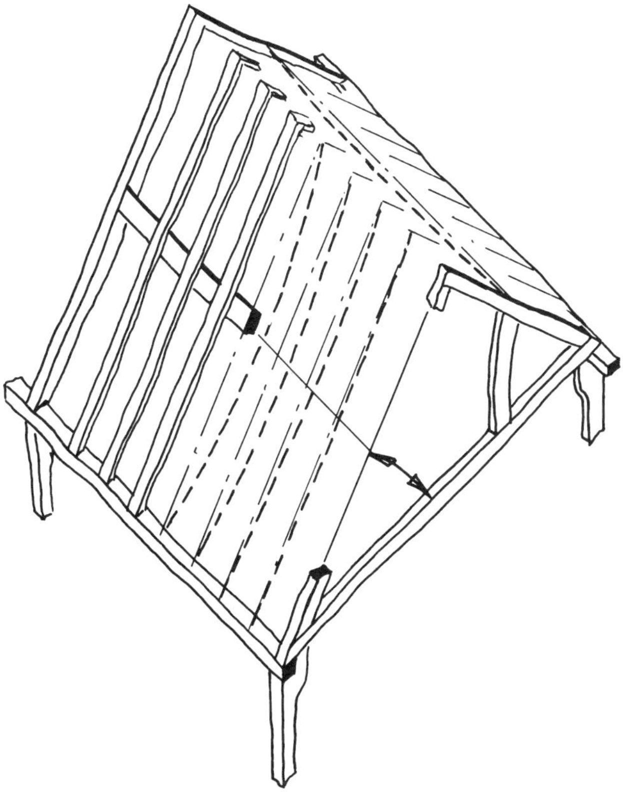

Figure 10.7

An early timber-frame roof showing the deflection of purlins and rafters. The deflection of the purlin is shown (dotted) on the rear slope. The increasing deflection of the rafters towards the centre of the bay is shown on the near slope. In such frames the deflections will be a combination of elastic deformation and long-term creep.

dome's outward thrust. Of course, such diagnosis must make assumptions about the initial geometry of the structure.

Timber is also subject to creep in this way, with the deflection of highly stressed members increasing with time. An example is the deflection of a purlin under the load of the rafters (Figure 10.7). It is possible for deflection in the purlin to increase so much that rafters near the centre of its span are no longer in contact with it providing a very obvious illustration that such movement must involve a redistribution of loads. As this creep takes place, load shifts from the purlin to the rafters and, therefore, from the frame that supports the purlin to the plate that supports the rafters, and such changes may be occurring even when the creep deflection has not become so obvious. The result is that there may be some uncertainly in the load distribution of historic structures, much as Heyman has pointed out. (This particular example presents other issues to which I shall return.)

Historic movement of this kind is an indication of the way in which the structure is working but is not of course a cause for present concern. This may even be so where brittle masonry is displaying all the signs of distress associated with movement. Thus, the engineer is often faced with the task of distinguishing historic from more recent movement, sometimes finding that the principal task is to reassure an anxious client. The starting point must be a reading of the structure and determining its original geometry. There is a useful lesson here in the London church of St John's, Smith Square, where casual observation shows considerable distortion in the walls indicating some settlement of the foundations. However, those looking up will see that the cornice line is straight. The settlement dates from the original construction because building was deliberately slow to allow the settlement to be completed as each year's masonry was built (Yeomans 1988). The conservator may find clues to suggest the extent of historic movement in the geometry of later stages in construction or even in repair work following earlier partial collapses.

Of course, recent movement may be imposed upon historic movement and if recent movement is suspected there may be no alternative to some form of monitoring to determine its progress. In some cases simple tell-tales may be sufficient to show the extent of movement but there may also be occasions when the engineer will have to call in the services of specialist surveyors to measure the extent of movement. Until this is done it may be impossible to understand what is happening to the structure.

When confronted with the problem of movement that has taken place, or perhaps is still continuing, as a result of deterioration of the structure, the conservator is confronted with two tasks; understanding the mechanism which has led to the changes and explaining any resulting changes in the mechanism of load transfer. In some cases the former will be well understood. Timber may decay or be subject to insect attack in adverse conditions, and while diagnosing the extent of the attack may not always be a simple task because of limited access to the damaged areas, at least the properties of the materials and the mechanisms involved are understood. Timber is an example of a material which continued in use as an engineering material into the period when science began to be applied to engineering problems (Barlow 1817), but this is not the case with all materials and methods of construction so that the performance of some is not fully understood. Examples are the long-term behaviour of clay mortars or the performance of thick masonry walls under earthquake loading: structures of concern to the conservator but not to the modern designer.

Of course, all this ignores the most common causes of movement in the building, the effects that produce cyclic movement and which may themselves result in long-term distress. These are very difficult to summarize and equally difficult to diagnose. Changing loads, whether deliberately imposed upon the structure or the result of natural forces like the wind, will produce elastic strains. Changes in moisture content in timber structures will produce dimensional changes which may result in stresses, but it is also possible for exceptionally high loads to exceed the elastic limit of this material and produce permanent deformations that then alter subsequent behaviour. In iron structures it is the high coefficient of thermal expansion that may result in large movement, or large stresses if the member is restrained. Stresses of this kind may increase as corrosion of the material limits what was previously unrestrained movement. And in masonry, vibration on the structure or water action can cause movement of mortar particles within walls that affect their properties and may even result in signs of distress.

Far less predictable effects during the life of a building are the instances of later alterations to the building which may completely change the structure or the loads upon it. A timber-frame building with brick infilling of the frame may respond to subsequent ground movement more like a load-bearing masonry building than a timber frame, and the removal of posts with the consequent shedding of their load to the brickwork is not unknown. In such cases the changes may be obvious but this is not always so. Alterations to the plans of eighteenth-century houses may involve the cutting of doorways in what were trussed partitions destroying their structural integrity. The result may be that, as their trussing action has been lost, loads are now being carried by developing excessive bending moments in their top or bottom chords, or even by the shedding of load to other parts of the structure. Such changes may be quite invisible, hidden behind plaster until opened up in later work, or perhaps to investigate the cause of excessively large deflections.

DIAGNOSIS

With an understanding of the anatomy of the building and the natural processes of change, the mechanism of load transfer and any signs of distress may be correctly diagnosed from a simple observation of the structure. In dealing with buildings that have not been extensively studied, and so whose behaviour is not easily explained, or when faced with inexplicable behaviour in a familiar building type, then more sophisticated methods may be needed possibly involving some exploratory surgery, that is, some opening up of the fabric. Such surgery needs to be as carefully directed and as limited as possible. To achieve this the conservator should be proposing possible explanations of behaviour and examining the structure to confirm or reject these tentative diagnoses. There are therefore two kinds of activity going on in parallel, surveying the structure and carrying out a structural analysis. In practice both of these will, as often as not, be quite informal processes. The complementary nature of the two activities becomes much clearer by looking at instances where more formal methods have been used, that is, cases where perhaps more sophisticated survey methods were needed, and where formal methods of numerical analysis were required.

A couple of examples here may be useful, both involving interaction between timber and masonry structures. Formal and informal methods are not of course discrete types. Informal observation and understanding will be used at the beginning even when the more complex techniques are used later and there can be nothing more important than allowing the building to speak to you at all stages of the process. Neither is there a simple one-way process from survey to analysis because observation of the behaviour of the structure providing a qualitative understanding of the mechanisms involved may be refined by calculation and then tested by further observation.

In the first example a roof had begun to spread and the extent of what was continuing movement was monitored by a survey measuring the outward movement of the wall. A careful level survey carried out immediately adjacent to

Figure 10.8

Different modes of behaviour for collar-braced principal rafters. If the supporting walls can resist the outward thrust of the rafters then the collar will be in compression (a). If there is no horizontal restraint then the feet of the rafters will move outward and the collar will go into tension (b). There will be an associated change in the pattern of bending within the rafters.

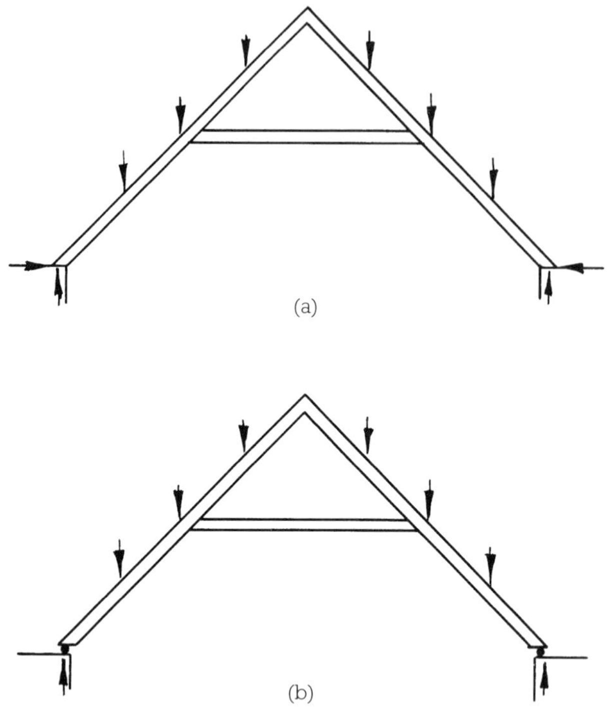

the wall eliminated the possibility of external factors like subsidence to account for the movement of the wall so that the only possibility that seemed plausible was that the wall was being pushed out by the roof. This movement was only possible if the walls had been containing the outward thrust of the roof and had ceased to do so.

Why this should have happened could not be explained but the issue was whether such an explanation was structurally possible. The principals in the roof were collar-braced rafters (Figure 10.8) for which two structural actions were possible: the feet of the rafters could be restrained by the wall, with the collar acting in compression; or if the wall provided no restraint the collars would have to serve this function and be in tension.[5] A numerical test was needed to show whether it was possible for the wall to contain the outward thrust of the principals. If it was, was it so finely balanced that such a change in structural behaviour was credible?

Naturally, a number of assumptions had to be made about the weights of materials used in the structure, and so the result of the calculation was an exercise in determining probable cause. In the event, the calculations suggested that the wall needed to develop a modest tensile stress close to the floor in order to resist the thrust of the roof. The magnitude of this stress did not seem impossible, but if the masonry could not provide this, the roof would spread and the wall would rotate at the same time. A long horizontal crack was found in the plaster at about the height suggested by the calculations. By making some further assumptions

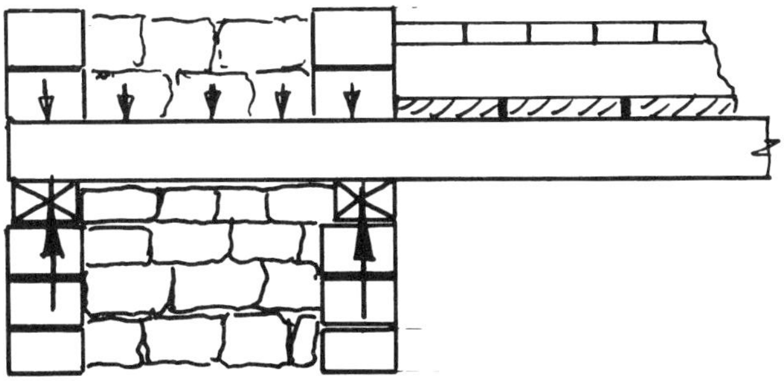

Figure 10.9

Wall construction of buildings in the Kathmandu Valley. Floor joists divide the wall into separate sections. If the core material settles relative to the face veneers then the floor timbers will transmit the weight of the masonry above to the wall plates and so to the veneers.

about the properties of the timber of the principal rafters, it was then possible to assess the amount of movement involved in this shift from one structural action to another and then to use the results of the survey to determine whether movement ceased when this had occurred. Clearly this was not a complete explanation for what had happened because there was still nothing to account for the change in structural behaviour. However, it was sufficient to assess the present state of the roof.

The second example concerns load transfer in a thick brick-masonry wall constructed using clay mortar. The wall was not homogeneous but comprised veneers of good quality brickwork poorly bonded to a core which was of poor quality construction. Signs of distress in the veneer suggested that it might have been carrying excessively high loads. This was possible because at each floor the floor joists were brought through the wall onto plates built into the veneers and, in this way, separated the wall into storey-height lifts (Figure 10.9). A possible explanation for the distress in the veneer was that the core of the wall had settled over time so that it was no longer carrying the load from above. Instead the timbers were collecting the load from the core and transmitting this to the veneer. This could be verified by opening up the structure to see whether there was evidence of such settlement, but before engaging in such surgery it was possible to test to see whether such a mechanism was possible. If loads from above were too large then the timbers would be crushed under this load as the core of brickwork settled. Load would still be taken by the veneer but perhaps not as much. Thus a simple calculation of the compressive stress in the timber as a result of the assumed behaviour was a useful preliminary to the physical investigation.

These are both examples of interaction between different parts of a structure of different materials but analysis of structures of just one material may present the engineer-conservator with the need to undertake calculations which would not be warranted in the design of a new structure of similar form. Consider the simple case of a purlin supporting rafters. The simplest way to design such a structure is to consider the rafter as a rigid support when determining the bending on the rafters. The purlin may then be designed to carry the corresponding load. In fact the purlin is not a rigid support and, in deflecting, will cause more of the load to be carried by the rafters. The resulting redistribution of bending in the rafters is unlikely to increase their maximum stress but it will mean that the stresses in the purlin will be reduced and so the purlin will be 'over-designed' by the simple calculation procedure. In new construction this effect is hardly worth considering but it may be critical in a conservation project where a calculation on this basis might indicate that a purlin which has been performing in a satisfactory manner is overstressed and needs some strengthening.

The effect of the interaction of these two structural components may be calculated[6] but the problem of an historic structure is compounded by the long-term creep behaviour of the timbers which has been discussed above. There can be no universal prescription for analysing such situations; the example is provided to indicate the limits of calculation when dealing with historic structures.

Problems of analysis may occur with a single structural member, as shown recently by Slade and Playle (1995) who were asked to strengthen the floor of the King's Library in the British Museum. They discovered, contrary to the common expectation, that the weakness of cast-iron beams is in the low tensile strength of the iron, in this case failure might first occur by buckling of the compression flange. It was therefore essential to estimate this compression stress as accurately as possible and their analysis of these floor beams provides a nice example of the different approaches that can be used in both structural analysis and in considering structural safety. An obvious method in some instances is to test-load a structure, either as a means of determining the actual load that it can take or, by measuring relative deflections (the way in which a load is distributed between parts of the structure). The first avoids analysis altogether and goes straight to the issue of structural safety. In this case load testing was ruled out, partly because of the consequences of failure. The other 'obvious' approach is to undertake some numerical analysis although, as discussed above, it is not always a simple matter to decide upon the most appropriate model for this purpose. These castings have large holes in the webs (Figure 10.10) and so were modelled as Vierendeel trusses, a structural device not invented until this century and so clearly not that used by John Rastrick when he designed them (Slade and Playle, 1995). Although this was a more accurate model of behaviour than treating it as a solid beam, the authors confessed that this was still only an approximation. Finite element analysis would have been more accurate although more expensive.

They also discussed the other sources of information sometimes available to the engineer-conservator as a means of assessing load-carrying capacity. In the nineteenth century, cast-iron beams were subjected to proof loading and if records are available these may be useful. Unlike timber, where original weaknesses in the material are readily apparent, castings may have weak spots where the iron has not run or where some slag has been included, or there may be built-in stresses as a result of the uneven cooling. Thus, proof loading was a necessary precaution at the time, although the problem for the engineer today is that the test method may not sufficiently replicate the pattern of loading in service and so not produce the same pattern of stresses within the structure.

Another approach is the record of past service. If a structure has withstood a particular loading in the past, and there has been no deterioration in the meantime, then it will normally be able to withstand the same loading again. It is particularly useful to remember this when a structure presents some difficulties

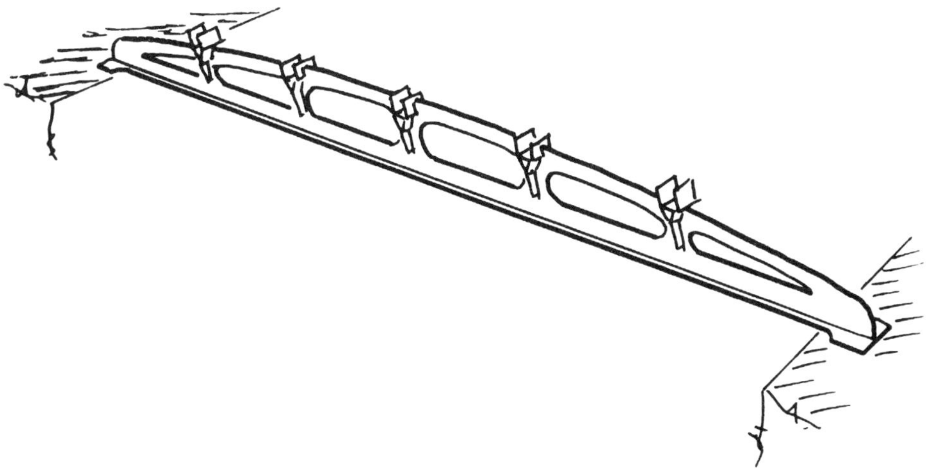

Figure 10.10

Floor beams in the King's Library at the British Museum were cast with large voids between the points at which the major floor timbers were carried. This profile cannot be sensibly analysed as if the web were solid. The cross section of the beam is effectively an inverted 'T' resulting in a slender compression flange.

in analysis and when there is no reason to suppose the loading will change. However, past record of loading will not help if the use of the building involves increased loads or changed loading conditions.

REMEDY

The medical analogy is still useful when considering remedies. The conservator needs to be aware that this is an old building and not a new one so that the standards which are applied to the latter may not be applicable. The building has survived to the present day and needs to be helped to survive into the future and this may be quite different from attempting to bring it back to its youthful vigour or attempting to make it conform to modern standards. The conservator also needs to avoid proposing remedies that are so radical that they leave the patient worse off than before. Amputation at the knee will certainly cure an ingrowing toe-nail or athlete's foot but is hardly to be recommended. At the same time it is not unknown for proposed treatments to historic structures to have something of this character. Langenbach (1989), for example, has discussed the conflict that may arise between the demands of structural safety for traditional structures in earthquake zones and those of conservation. The point is made by taking such an extreme case. If what is valued is the historic fabric then it would be pointless to preserve it from destruction by earthquake if we allow its destruction by engineers, that is, replacing possible loss as a result of an earthquake by certain loss as a result of engineering.

This particular example is also one of those areas which illustrates a need for research. Much of the work on the earthquake resistance of masonry has been concerned with modern masonry structures (Bruneau 1994) which have thin walls laid in brittle mortars. These do not behave in the same way as the thick walls laid in soft lime or clay mortars, or possibly laid dry, that are found in historic buildings.

The methods used to ensure the continued stability of the building must be consistent with the standards of conservation so that the engineer needs to be concerned with the intention of the programme of work, becoming involved in the issue of what is being preserved and why. This is not the place to discuss general issues of conservation standards but there are some aspects which particularly concern the engineer. The first is the value that is to be placed upon the structure itself compared with other aspects of the building. In a medieval English timber barn or a Norwegian loft the building is little more than the structure which is thus of paramount importance. In an eighteenth-century English country house or a French château there may be greater concern for the design and quality of craftsmanship in both internal and external fittings and decoration and the structure may be regarded as of little importance, especially if it is a typical example of its period with no rarity value. However, structural design has its own history and there may well be examples of buildings whose structures, unusual for their day, are a significant contribution to this record. An example was the roof of St Mary at Hill, London, which was an example of the use of nineteenth-century laminated timber and may well have been related to the earlier Wren structure for the same building (Yeomans 1992a).

This means that an assessment of the value of the structure vis-à-vis other aspects of the building needs to be made and the engineer should be prepared to defend his corner. The history of architecture, and of the decorative arts in general, is well established while the history of structural design has received comparatively little attention, particularly those aspects of structure that are not easily visible. What is a minority concern at present may be lost if it is not defended against other interests. Moreover, structures may embody presently unrecognized evidence for historical development which can too easily be lost. A good example here is the recent recognition of the importance of setting-out marks on timber to the understanding of the development of carpentry (Harris 1989).

The engineer also has to recognize that historic structures should not always be expected to meet modern engineering standards. They were developed to suit the requirements of the past and within the limitations of the materials and the skills that were available and may remain viable even though they do not conform to the expectations of the present-day designer. It is therefore important that a structure be assessed in its own terms and whether or not modern standards should apply depends upon its intended use, including the standards of maintenance, which may also be different from those assumed for modern buildings. The most obvious aspects of this affecting the engineer are the design loads to be used and the allowable stresses in materials. These may not appear to be issues of structural understanding but it is important for the conservator to understand how the present-day design figures are derived in order to avoid the unquestioning application of modern standards which may not be appropriate.

The appropriateness of present loading standards for historic buildings has already been questioned in a guidance note from English Heritage (1994) which makes this point, but much more careful consideration needs to be given to the ways in which both these design loads and the values for design stresses are determined. The latter depend upon the properties of the different materials and one example will have to suffice as an illustration. The present design stresses for timbers (British Standards Institution 1991) are provided for commercially available species and have been based upon tests on sections in current use. When dealing with historic timbers, which may be of species not presently available, the engineer needs to refer to tests on small clear specimens, that is, specimens of the timber which are free from any defects, and which was the basis of the previous code (British Standards Institution 1971). Similarly, in assessing the grades of timber in place and obtaining working stresses for different grades the same obsolete code may be used.[7] The advantage of this is that it may be used for exotic timbers for which design values are not available but whose fundamental properties are given by Lavers (1983).

When dealing with standing structures we also have the advantage that the exact position of any knots or other features which lower the safe working stress is known and their effect can be assessed against the stress at that position.[8] It is also possible to assess the effects of high stresses within the timber. Timbers loaded in compression or bending do not fail catastrophically and while the stresses are normally limited to within the region of elastic behaviour, if necessary, the effect of loading above this point can be considered. This might well occur in joints or at bearing surfaces so that what is needed as a basis for deciding whether timbers in an historic structure are under excessive stress is data on the stress/strain relationship which, regrettably, is not easily available.

SUMMARY

At whatever level, whether looking at the overall behaviour of a structure or at the detailed behaviour of a small part, the first task is to allow the structure to speak for itself, to look for the way in which it has responded to loads and the way in which it may have changed over time. While doing this the engineer will be turning the physical object into an abstract model, rather as seen in Figure 10.7 (although all the illustrations here are abstractions). The physical object must be transformed into a structural model in this way so that its behaviour can be understood and eventually quantified. Calculations may be necessary to test the validity of the model being suggested, that is, to see whether it could behave in the way assumed without requiring unreasonably large stresses. What is crucial is to select a model that adequately represents the real structure and is sufficient to answer whatever questions are being asked. In some cases alternative models may be possible and might need to be considered to determine the range of stresses possible but some assumptions will always be necessary. There will be assumptions about the values of the loads and properties of materials, but also about the validity of the model itself, and one should always be aware of the limits these assumptions place upon the answers obtained.

What the engineer will be looking for is structural integrity within the building, a recognizable method by which the load is being transferred to the ground. Sometimes this will have been lost over time with a consequent shedding of load from members that once carried it but are now no longer able to do so. The result is that the structure may appear to be standing by force of habit because loads will have been picked up by other and less obvious members. Either structural members are asked to carry stresses that were not intended, like the tie-beams of trusses coming under excessive bending, or what would normally be regarded as non-loadbearing members have to carry load. In general, one will seek to restore this structural integrity by repair, or occasionally by adding supplementary structure.

Occasionally, because of poor design, there may never have been such structural integrity and in such cases a little more than repair may be required. In these cases the engineer may have to add structure where none existed before. But such instances should be rare. What most structures need is little more than a helping hand to repair the ravages of time.

BIBLIOGRAPHY

Addis, W. (1991). 'Review of Mark 1990', *Construction History,* vii, 111–14.

Andrew, E. N., Osborn, and Newswanger, B. (1989). 'Restoration of the Highway Tabernacle Church', *APT Bulletin,* xxi, 17–25.

Barlow, P. (1817). *Essay on the Strength of Timber and Other Materials,* J. Taylor, London

Bates, W. (1984). *Historical Structural Steelwork Handbook.* BCSA, London.

Borg, A. and Mark, R. (1973). 'Chartres Cathedral: a reinterpretation of its structure', *Art Bulletin,* iv.

British Standards Institution (1971). *The Structural Use of Timber.* CP112: Part 2.

British Standards Institution (1991). *Structural Use of Timber, Code of Practice for Permissible Stress Design, Materials and Workmanship.* BS5268.

Bruneau, M. (1994). 'State-of-the-art report on seismic performance of unreinforced masonry Buildings' *ASCE Journal of Structural Engineering,* cxx (1), 230–51.

Brungraber, R. L. (1985). *Traditional Timber Joinery: A Modern Analysis,* unpublished PhD thesis, Stanford University (CA).

Courtenay, Lynn T., and Mark, R. (1987). 'The Westminster Hall roof: A historiographic and structural study' *Journal of the Society of Architectural Historians,* xlvi, 374–93.

English Heritage (1994). *Office Floor Loadings in Historic Buildings.*

Fairbairn, Sir William (1854). *On the Application of Cast and Wrought Iron to Building Purposes.*

Harris, Richard (1989). 'The grammar of carpentry', *Vernacular Architecture,* xx, 1–8.

Heyman, J. (1966). 'The stone skeleton', *International Journal of Solids and Structures,* ii.

Heyman, J. (1967). 'Westminster Hall roof', *Proceedings of the Institution of Civil Engineers*, xxxvii, 137–62. Discussion, xxxviii, 785–96.
Heyman, J. and Padfield C. J. (1972). 'Two masonry bridges: 1. Clare College Bridge', *Proceedings of the Institution of Civil Engineers* Part 1, lii, 305–30. Discussion; 495–505.
Langenbach, R. (1989). 'Bricks, mortar, and earthquakes: historic preservation *vs.* earthquake safety' *APT Bulletin*, xxi (3/4), 30–43.
Lavers, G. (1983). *The Strength Properties of Timber* (3rd edition), HMSO, London.
Lightfoot, Edgar (1961). *Moment Distribution, A Rapid Method.* Longmans, London
Mainstone, R. J. (1969). 'Brunelleschi's dome of Santa Maria del Fiore and some related structures', *Transactions of the Newcomen Society*, xlii, 107–26.
Mainstone, R. J. (1988). *Hagia Sophia: Architecture, Structure and Liturgy of Justinian's Great Church.* Thames and Hudson, London.
Mark, R. (1982). *Experiments in Gothic Structure.* MIT, Cambridge, Ma.
Mark, R. (1990). *Light, Wind and Structure.* MIT, Cambridge, Ma.
Morris, E. T., Black, R. G., and Tobriner, S. O. (1995). 'Report on the application of finite element analysis to historic structures: Westminster Hall, London', *Journal of the Society of Architectural Historians*, liv (3), 336–47.
Ozelton, C. & Baird, J. (1976). *Timber Designers' Manual.* Granada, St Albans.
Royal Commission (1847). 'Report of the Commissioners Appointed to Inquire into the Application of Iron to Railway Structures'. *Parliamentary Papers*, Cmnd 1123.
Ruddock, T. (1995). 'Repair of two important early Scottish roof structures', *Proceedings of the Institution of Civil Engineers*, Structures and Buildings, cx, 296–307.
Slade, R. E., and Playle, C. (1995). 'The British Museum: upgrading the floor over the King's Library', *The Structural Engineer*, lxxiii (12), 193–9.
Sutherland, R. J. M. (1989). 'The introduction of iron into traditional building' in *Good and Proper Materials: The Fabric of London since the Great Fire* (Hobhouse, H. and Saunders, A., eds), RCHME Conference, RCHME & London Topographical Society, London.
Yeomans, D. (1982). 'Structural design in the eighteenth century: James Essex and the roof of Lincoln Cathedral Chapter House', *Design Studies*, v (1), 41–8.
Yeomans, D. (1988). 'Managing eighteenth century building', *Construction History*, iv, 3–19.
Yeomans, D. (1992a). 'Three timber roofs', *ASCHB Transactions*, xvi, 52–8.
Yeomans, D. (1992b). *The Trussed Roof, Its History and Development*, Scolar, Aldershot.

NOTES

1 The behaviour of iron and steel structures and the design rules used for them have been reviewed by Bates (1984), but there is so far no equivalent review of early concrete.
2 We have no information on the structural performance of historical jointed frames. Some experiments have been carried out on recent frames inspired by a revival of traditional framing techniques in America. See Brungraber (1985).
3 Lion Chambers, Glasgow, is in this position at the time of writing.
4 One common exception where poor design combines with poor materials is where there are poor soil conditions and inadequate foundations.
5 Ruddock (1995) has recently shown how a similar change in structural behaviour occurred in a collar-braced common rafter-roof as a result of decay at the feet of the rafters.

6 See, for example, Lightfoot (1961).

7 A full explanation of how this was derived lies outside the scope of this essay but is explained in Ozelton and Baird (1976).

8 Rules for this can be found in the American Code of Practice, ASTM D245, 'Methods of establishing structural properties of timber', para. 5348.

11

ART HISTORY, ARCHITECTURAL HISTORY AND ARCHAEOLOGY

Richard Morrice

Architectural history could be described as the application of art historical methods to a field of study which has long been also the subject of archaeological research. Little attention has been paid in Britain, however, to a definition of architectural history, apart that is from David Watkin's book *The Rise of Architectural History*, but on reflection, though, a reason might be discernible. As both art history and archaeology have developed they have borrowed more and more from each other's differing approaches to history. There is thus a disinclination among many historians of differing backgrounds to define themselves as architectural historians, though the term remains useful as shorthand. The question arises, therefore, as to what makes up the discipline which deals with buildings as historical documents; is there a distinct academic discipline of architectural history or not?

SOME DEFINITIONS

The answer lies in the twin roots of British architectural history. When, in 1846 Robert Kerr, a young architect and journalist, published *The Newleafe Discourses on the Fine Art Architecture*, he included two characters who represent the twin poles from which architectural history was to develop in these islands.[1] Kerr was concerned that the architectural profession was not paying enough attention to the production of 'an architecture of our period, a distinct style of the nineteenth century', as Professor Donaldson put it at the time. In satire, Robert Kerr introduces a discourse between Mr Heavyith'heel, Fellow of the Institute of British Architects, Mr Newleafe, a thoughtful gentleman, and Mr Verditurus, the Librarian. Among the false prophets introduced are both Mr Smug's son, trained as an architect but who travels and becomes a 'Michael Angelo–Raphael–Palladio–Rome–Florence–Venice–Acropolis–Byzantine–Laocoon–Apollo Belvidere–Titian–da Vinci –Munich–Berlin–Paris sort of Person'; and the antiquarian, the Revd Bluebottle Crape, DD LLD FRS FSA etc., etc., etc., 'the scientific man in rubbish' and 'the utter enemy of art by nature'. Here, broadly drawn, are the popular images of art history and archaeology (little change since the nineteenth century), satirized and lampooned; we may smile but each discipline may still admit a remaining hint of a view of the other coloured by similar prejudices.

Architecture had, of course, been written about for centuries and that writing had often taken a historical turn, as, for instance, in the polemical writings of Renaissance architects about the architecture of antiquity. It was, however, only in the eighteenth century that architectural writing became more disinterested and by the 1840s, when England saw a rash of archaeological society foundations, the growth of field archaeology out of antiquarianism was leading to increasing academic interest in the archaeological potential of architecture. Art history, meanwhile, had seen a less insular development, from the cyclical theories of the seventeenth and eighteenth centuries and the more closely analytical tendencies of the nineteenth century to the cultural history of writers such as Burckhardt and Warburg.[2]

To return to architectural history, it goes without saying that it is a branch of history, 'the past as we know it from the interpretations of historians based on the critical study of the widest possible range of relevant sources, every effort having been made to challenge, and avoid the perpetuation of, myth' (Marwick 1989, 13). The study of the history of architecture and buildings is the central discipline in the study of one of the most important groups of sources available for the study of history. The definition above is by Professor Arthur Marwick, from his primer, intended for students and the general reader, *The Nature of History*. Although the study of documentary sources is generally seen as the most usual area of study for the historian, Marwick (1989, 1) begins his book in a way which underlines the place of buildings as historical documents:

> In almost every city, town, village or country throughout the world the overwhelming majority of buildings currently in existence was built in past times to meet the needs and aspirations of human beings now dead and societies now in greater or lesser degree changed or even defunct. This is most obviously so with respect to great temples and cathedrals, fine palaces and manor houses and castles, city halls, houses of parliament and other public buildings; but it is also true of the most humdrum streets and the meanest housing.

Marwick here not only points out the centrality of the built environment in historical research but also that all buildings are potential documents. Kerr's parody of the art historian and the archaeologist is thus rejected by Marwick as no longer so valid, as both disciplines have come to terms with each other's concerns and learnt from them.

Any definition of architectural history therefore has to note this debt to both art history and archaeology and its origins must therefore be sought in both of these disciplines. For many years art history could have been broadly defined as the history of art, codifying artefacts by style, attributing them to artists and dating them, above all trying to define the aesthetic appeal of the object. In this understanding of the discipline of art history, the concern for the artefact itself overrode its relationship to its cultural background and, in some cases, art historians left the discussion of the place of such objects to others. This led to questioning by some early twentieth-century German art historians, such as Erwin Panofsky and Edgar Wind who saw the cultural and social background as important, against what they saw as formalism (Rees and Borzello 1986, 7).

It is against this background that one should view new directions in art history such as that which has become known as 'The New Art History'. Under the influence of a variety of new historical trends, a reinterpretation has emerged, often referred to by that name. These influences, including psychology, critical theory, feminism and Marxism, have doubtless enriched art history by confronting one of the main problems, the objectivity of historical thought and research, but not with great success and the problem remains, as a glance through much of the literature of the New Art History will show. One particular argument of art historians of this school is that traditional art history tends towards a confirmation of ruling ideologies by the stressing of 'high art and genius' as the most appropriate subject of study. In fact, as Rees and Borzello (1986, 7) have pointed out, traditional art history has not been so exclusively concerned in such matters as their critics might suggest. Nor has the traditional art historian been so unworldly as what has now become the classical definition of New Art History might suggest: 'The art historian's role has always been ... to elucidate the work of art, not least contemporary art, relating it to its social-historical context rather than raising it out of time or "appreciating" it as the disembodied work of genius' (Pointon 1986, 27). Much of the discussion centres around arguments as to whether art, and particularly 'high art', exists; these are regarded as major concerns of New Art Historians of a more radical bent.

Accepting that many of the theoretical techniques and considerations of architectural history derive from art history, and that many distinguished architectural historians would regard themselves as art historians engaged in research on a group of artefacts similar to paintings and sculpture in the way they can hold and display meanings but not, of course, identical, where does the recent flurry of activity leave architectural history? Why is there no New Architectural History?

The answer must lie in the nature of architecture itself. Sir Nikolaus Pevsner made a useful distinction between 'architecture' and 'buildings' when he wrote (1973, 15) that 'A bicycle shed is a building; Lincoln Cathedral is a piece of architecture'; a simple aphorism which bases the definition of architecture in aesthetic appeal and one which suggests that architectural history can be regarded as similar in its methods and purposes to art history. It is for this reason that architectural history has come to be considered as analogous to art history, endeavouring to explain by using similar methods but discussing a separate set of artefacts. Among the results is a marked reluctance by some historians to define themselves as anything other than art historians. The need, however, continues for a convenient shorthand and the term is used in this sense for the rest of this chapter.

This is not to say that an awareness of the twin natures of architecture/building or art history/archaeology is unimportant. While all buildings have interest as historical documents (as Marwick's statement quoted above shows), utility in buildings can take two forms of which the concern for usefulness is by no means the most, or even the least, important. It is well documented that architects whose buildings failed for one reason or another as buildings, may occupy an important place in the history of architecture. More importantly, architectural history has derived from archaeology a concern for all types and kinds of buildings. This may be partly due to the traditional concern of the architectural historian (in the art historical sense) for the codification and authentication of the minor works of architects who have made their living by designing and supervizing the building of a wide variety of types of building but archaeologists, particularly those who excavate, have always been interested in the remains of buildings of all types as documents.

AN ARCHAEOLOGIST'S VIEW

Archaeology derives ultimately from antiquarianism which, in the person of a historian like William Stukeley, lay in a 'capacity to observe and record facts in the open air', '...pure fieldwork and survey carried out in the open air, with a minimum of excavation ... a technique in which today British archaeology is outstanding' (Piggott 1985, 152–3). David Stocker has pointed out how the Revd Prof. Robert Willis in the 1840s used such techniques for the study of buildings.

> Willis, one of the greatest early students of ancient buildings, used the term archaeology to describe his work. With refinements in recording technique and application, his detailed recording, subsequent dissection, phasing and analysis of several English cathedrals remains the best work on these buildings today; and still a procedural model for contemporary workers. Willis knew the value of demonstrating structural and dating arguments through making proper records, and he stands close to the start of the archaeological tradition which requires accurate measurement and drawing before an analysis is undertaken'. (Stocker 1992, 302)

Michael Podro has attempted a definition of the distinction between art historical and archaeological research; he sees art historians as considering 'how the products of art sustain purposes and interests which are both irreducible to the conditions of their emergence as well as inextricable from them'. This he sees as the proper concern of the critical historian rather than the archaeologist who is concerned with fact; in other words, he is stressing that art history represents a different pole from archaeological research as any reference to the extrinsic is impossible not only because writing about art has its own context but also, and more important, that art has no relevant autonomy.

David Stocker has written that the major distinction between archaeological and art historical approaches is 'a disagreement between the art historical tradition and the archaeological over what constitutes authority within their subjects'. His argument concerns the status of the record which, for archaeologists, is of prime importance, but is of less note for architectural historians. For archaeologists, of course, the record is of great import because it has the role of 'showing their working' (as Stocker points out), but that is inevitable in below-ground archaeology because it is largely, by nature, hidden both before and after recording and because, in rescue archaeology in particular, recording often precedes the destruction of that which has been recorded. Architectural historians of an art historical background undoubtedly benefit from the dissection of the history of a building by the exposure and analysis of fabric which allows the confirmation of alterations to plan or aspect, perhaps hidden by later work or claddings. But the workings of the art historian should also be apparent and, in such writings, by description and interpretation, it is made clear. Stocker insists that the authority of a historian like Pevsner is respected not because his 'records of buildings are appropriate and give proper evidence for [his] conclusions', but because 'he was able through the quality of his own experience and genius to describe a building verbally. His accounts of buildings are thus very individual, the nuance of every phrase is significant, and the verbal description makes clear all sorts of connections both within the building and between buildings – capturing more elusive meanings, perhaps, than would be seen in a 1:20 record drawing' (Stocker 1992, 302–3).

Stocker's conclusion is odd; he suggests that the art historian who does not make records and relies on personal authority alone for the authority of the resulting interpretation is in some way different from archaeologists with their reliance on the record. But both surely show their working out of their interpretations and on the relevance of their conclusions rests their authority. It does not suffice to suggest that 'it is an important principle amongst archaeologists that even the most celebrated will not be believed by their colleagues unless they provide detailed records of their discoveries, so that the site can be reconstructed in alternative ways' (Stocker 1992, 302). Is Stocker really suggesting that traditional art historians do not critically analyse arguments in art historical writing? It is more likely that the answer is to be found in the traditional relationship of the architectural historian of an art historical background to source material. Architectural history has very often been concerned with the interpretation of style, and discussion of buildings which have been demolished can be of crucial importance in such histories; the preservation of buildings, even by record, has not been regarded as quite as important as it is now under the influences of both archaeology and the conservation movement, influences which overlap.

The architectural historian will therefore agree with the archaeologist on the utility of archaeological analysis and recording and will generally restrict full recording to the management of the conservation of complex multi-phase buildings where important evidence for the building's previous history might, or will, otherwise be lost or where very detailed analysis is intended to guide repair strategies. Indeed it can be argued that the influence of archaeological methods on architectural history has generally been due to the increasing concern for the conservation of historic buildings, as well as the interests which historians with an archaeological background have found in various kinds of standing structure, particularly medieval and vernacular buildings, and the need for those excavating sites to have a knowledge of the buildings that once stood there.

Architectural history has, therefore, benefited from the very wide focus and the depth of interest which an archaeological training bestows; the origins of British archaeology in antiquarianism is, after all, not dissimilar to that of some strands of architectural history (such as ecclesiology). English architectural history has some of these same roots, plus an inheritance from the architectural profession itself. Before the Second World War architects were, by and large, the keepers of the history of their profession but this pre-eminence was largely lost when British architecture turned away from historicism towards an architecture which made rejection of the past more or less an article of faith. With the increasing splintering of history as an academic discipline, consideration of the history of architecture developed as a recognized subject following the War, particularly under the influence of continental art history, which was pre-eminent in the most important English institution teaching art history, the Courtauld Institute in London. Archaeology, equally, has increased as a major influence as increasing numbers of archaeologists have turned their attention to buildings, as historians turn to buildings as documents and as the importance of the archaeological analysis and recording of buildings for conservation purposes has grown. Thus architectural history has three main uses in the conservation of historic buildings: selection and protection; interpretation; and recording for preservation and management.

HISTORIANS AND THE LEGISLATION

Since the 1960s, the conservation of historic buildings has been underscored by a government policy which leans towards philosophies tabulated so many years ago by William Morris and the Society for the Protection of Ancient Buildings.[3] In this interpretation, buildings are irreplaceable historical documents and they therefore need protection and interpretation before any work of alteration, extension or demolition is begun. Indeed, interpretation will need to be carried out in order to illuminate that process.

There is some truth – although increasingly the distinction is becoming less and less true – that the distinction between architectural history and archaeology can be seen in the different developments of the ways in which the built remains of past ages in England can be protected have developed: ancient monuments, sites which have archaeological potential, are scheduled; historic buildings are, largely, listed; and historic areas are protected by designation as conservation areas. These systems are largely complementary and, although they overlap for historical reasons – some buildings were scheduled so as to protect them before the listing legislation was introduced and certain classes of building which have undoubted archaeological potential (including both houses and churches) cannot be scheduled – they are distinct as legal processes. Interestingly, the greater protection afforded to scheduled monuments and listed buildings may, in part, be due to the existence of recognized academic disciplines for their study; that historic areas should not fall easily within the purview of such a discipline may explain the more diffuse concern for their conservation.

Both listing and grant-aid policies have moved away, since the end of the 1960s, from those based on notions of architectural quality in the abstract, based exclusively on perceived 'quality' or the idea of a 'high' architecture (as, for instance, in the original 'Instructions for Investigators' given to each of the new Investigators employed by the Ministry of Town and Country Planning in 1947 to produce the lists of historic buildings) to ones which place more emphasis on architecture as evidence of material culture (that is, a more archaeological approach). It appears that, from the outset of the listing process, there were differing views as to the limits of the protection of historic buildings; during the deliberations of the Maclagan Committee, John (later Sir John) Summerson and Walter Godfrey had very different ideas of where the line should be drawn and these reflect different interpretations of historic interest, a more exclusive one from an architect turned architectural historian, a less exclusive one from a romantic late-Arts-and-Crafts architect of antiquarian bent.[4]

The most important difference between the listing and scheduling processes is, of course, in the level of the controls, a difference which is well known. This reflects itself in what is, in scheduled monument controls, a concern for the primacy of the artefact; a concern which is directly borrowed from archaeology. Although scheduling is now seen as more appropriate for sites and buildings which are out of use, it can also be seen that a building scheduled as an ancient monument is one chosen for protection where it is deemed likely that any works to the building concerned may be so damaging as to require the most rigorous control. Scheduling is also an exclusive system of control, no building nowadays being selected unless it is reasonably certain that it is unlikely to be disturbed due to protection by other means (ownership, for example). This is very different from the inclusiveness of the listing process whereby buildings are listed for their 'special architectural or historic interest' against a list of exemplars of listability. Listing, moreover, was introduced as a more flexible system of protection for buildings which remain in use; indeed use of a building effectively renders it unschedulable.

The difference between the effects of scheduling and listing a site can be seen in one of the main intended consequences, the reactions of those controlling development on the buildings protected under those differing legislative systems. Of course, all buildings have some kind of archaeological potential and, of course, that potential may require investigation and/or recording during works of repair or alteration. But listing is carried out by comparison with other buildings, by taking both intrinsic and extrinsic facts into consideration. Architectural interest must largely be decided by historians, by comparison with exemplars and by consideration of the place of a particular building, in the light of social and cultural background, in the history of architecture. The existence of a special category allowing the listing of buildings for historical reasons alone might be thought to indicate a distinction in the listing of buildings which reflects Podro's distinction between archaeology and critical art history, mentioned above, or perhaps between the tenets of the New Art History and traditional art history. In fact, historic interest is very narrowly drawn and architectural interest very widely. The effect is probably much the same but it indicates the tendency of British architectural history towards a more catholic interpretation of architecture.

The control of works of alteration, extension and demolition requires an educated staff to advise during the controlling process. In the recent past this process has, in local planning authorities and in the relevant government ministry, where the authority for control largely lies, been seen as analogous to the planning process (indeed, by town and country planners, often as part of their profession). Thus, works proposed are appraised for their possible effect on the historic building, with other matters taken into consideration, and advice is offered which might mitigate adverse effects, with the long-stop of refusal of consent for the works if necessary. In this process architectural historians are only involved, as of necessity, in the original listing of the building and, should central government heritage bodies become involved, in major casework. The broad run of control casework can be dealt with by a local planning authority without the involvement of adequate advice by both applicant and controlling authority. This may change under the influence of developments in environmental legislation, particularly from Europe, which suggest that it is for the applicant to prove that proposals for change are acceptable. The post of local authority conservation officer is becoming professionalized with the proper emphasis on training, particularly in the history of the built environment.

The emphasis on interpretation in architectural history is rightly seen to be of increasing importance in the conservation of historic buildings. This is well reflected in the make-up of such bodies as the Society of Architectural Historians of Great Britain which, while having been founded in the 1950s, largely by academic architectural historians, has seen a growing membership of historians working in conservation. Similarly, while there were very few architectural historians working as conservation staff for local planning authorities even in the early 1980s (except in such relatively specialized bodies as the Historic Buildings Division of the Greater London Council and the Directorate of Ancient Monuments and Historic Buildings in the Department of the Environment), it is not that unusual nowadays to find that a conservation officer has a background in architectural history.

While architectural history, therefore, has increased its voice at the control end of the architectural process, the architectural profession has lost much of its former interest in the history of its art. Partly this is historical, due, as mentioned above, to the reaction of architects of the Modern Movement to history in general, and to their reaction to the immediately preceding tenets of the Beaux Arts tradition of architectural teaching, with its emphasis on architectural history of a rather partial kind, in particular. Nowadays, it is still possible for architectural history to be taught in a partial way at architectural school and for architects to leave with little detailed knowledge of either historic buildings or architectural history. This is particularly seen as a problem by the more radical of the traditionally minded wing of the architectural profession but it is also a wider problem as the profession grapples with increasing controls which are being placed on works to historic buildings. Without doubt, historic awareness and related skills in the interpretation of historic buildings are still usually only partly brought to bear on most applications for listed building consent, and such unconcern may be counterproductive and unwittingly put unnecessary difficulties in the way of both architect or surveyor and the applicant.

HISTORICAL ANALYSIS AND INTERPRETATION

The analysis, interpretation and recording of buildings and sites can be thought of as operating on a sliding scale, generally working from the needs for such work when considering minor works to minor buildings on the one hand to major works to major buildings on the other. But it can be more complex as a full programme of recording may be appropriate for important parts of otherwise relatively insignificant building; equally, a much less rigorous survey and analysis is often likely to be sufficient even when considering relatively intensive works to a major building. Consideration should therefore carefully be given to the appropriate kind of analysis or level of recording in each case. It must be emphasized that the analysis and interpretation of historic buildings, though not often stipulated by local planning authorities as part of the planning process, has been able to be required for years by way of further details needed for the adequate assessment by planning authorities of development proposals concerning listed buildings. This has usually taken the form of drawings of buildings or features as existing; as the majority of applications for listed building consent, however, are not made even with the advice of architects or surveyors possessing the ability accurately to survey and draw a building, there must be reservations as to the utility of such records (and thus the stress on site visits for assessment of proposals for alteration). More detailed analysis can, therefore, be insisted on, can be delivered at modest cost, and can be very useful in the consideration of proposals for change; such analysis and evaluation is more likely to be the work of an architectural historian than an archaeologist because of the need for evaluation rather than record but many archaeologists with experience in the recording of standing structures have moved into evaluation by developing experience as the amount of work of this kind increases.[5] Such interpretation and analysis thus becomes an essential part, not only of adequate management of historic buildings but also of the development control process. In particular, local planning authorities, which are really insisting that owners should manage their own buildings for themselves, should be aware that historic appraisal is essential to the management of those buildings and can be insisted on as part of the day-to-day control process. They should therefore take action accordingly, insisting on such appraisal and evaluation wherever appropriate.

Under the listed building control procedures given in the Planning (Listed Buildings and Conservation Areas) Act 1990, local planning authorities have wide powers to request further information so as to inform their deliberations on the merits or demerits of applications for listed building consent.[6] This is an important, and wide, power which local planning authorities should use more widely than they currently do and it is this which allows them to ask for the evaluations of the historic and architectural interest of a listed building. Such evaluations are important, not only in informing the control régime, but also because they can produce important evidence for alteration/extension/repair programmes. It is worth noting that listed building consent is required for works which would affect the special interest of a building as one listed as of architectural or historic interest. This has usually been taken to mean that like-for-like repairs are unlikely to need such consent but it can be suggested, not unreasonably, that a comprehensive repair programme to a major listed building is unlikely not to affect that historic interest. Thus, it would also appear reasonable that a local planning authority can in some cases not only require applications for listed building consent for such repair programmes but also a comprehensive archaeological evaluation which should inform both decisions by the authority in control terms but also, and usefully, the repair programme. Of course, it must be borne in mind that a local planning authority should always be reasonable in the use of its powers and should only insist on evaluation and recording where it is necessary. In many cases, it may not be needed and such insistence would not only be irrelevant, but in its vexatious effect on applicants would be counterproductive. Careful thought is needed, by fully experienced staff, so that the right balance can be achieved.

ARCHAEOLOGICAL SURVEY AND RECORDING

The architectural historian needs, furthermore, to be aware of the necessity, not only for historical interpretation and analysis, but for archaeological recording in the conservation of some historic buildings; indeed the utility of archaeological recording as a specific tool for the interpretation of historic buildings, not only academically for the elucidation of architectural history, but for the protection of the architectural and historic interest of the building as a part of its continuing management, is a matter which is much interesting the conservation world at this time. Of course, recording has been a part of the planning régime for many years, the Royal Commission on the Historical Monuments of England (RCHME), the main body of architectural record in England, having an important role as a backstop in ensuring recording for academic purposes of buildings threatened in one way or another. The analysis and recording which the RCHME carries out is limited, however, and while the levels used are generally appropriate for academic recording, they are unlikely to be sufficiently detailed to allow conservation management.

Government policy has furthermore, recently and for the first time, allowed the use by local planning authorities of conditions to listed building consents requiring the recording archaeologically of buildings for which consents have been issued. This will doubtless be of great benefit where buildings of particular importance are to be altered or partially or totally demolished but it will not take the place of competent interpretation and analysis.

This change in the regulatory régime has major implications which at the time of writing, two years since Planning Policy Guidance 15 (PPG 15) was issued, have still not become clear. It is possible, as already noted, for the first time, that a local planning authority may insist on interpretation and analysis of a building before works commence which, in a case such as extensive works to an important building of several periods, will be of considerable utility. But full archaeological programmes in advance of works will not be needed in all cases; in most cases the opportunity for analysis and evaluation as part of the pre-application particulars, preparatory to an application being considered by the local planning authority, will be sufficient. Moreover paragraph 3.24 of PPG 15 points out that hidden features of interest can be safeguarded by provision being made for analysis and, if necessary, archaeological recording. It emphasizes the danger to such features during works to a listed building and points out the importance of the local planning authority suggesting this possibility to owners and applicants, as well as suggesting that an appropriate recording condition be attached to the consent concerned.[7]

Full archaeological recording is, of course, of great importance in cases of complex buildings but these cases are not common, both because such buildings are rare and because the opportunities, particularly full repair programmes, are rare too. In fact, the use of conditional recording is likely to continue to be exceptional not only for those reasons but also because the British tradition in historic building conservation lays particular stress on the building or object as the primary record. Conversely, it is possible that the new régime on archaeological recording, mentioned above, may lead to more listed buildings being demolished as local planning authorities cease, in all cases, to seek preservation of listed buildings or parts of listed buildings; the ability now to attach recording conditions gives assurance to authorities that an adequate record of the building will be preserved and, thus, that there is no need to press to the last for the preservation of the building. Of course, the unreasonable attachment of conditions is to be avoided; over-use would almost certainly have a detrimental effect on the recording régime in the long term.

One can therefore see survey, analysis and recording as a matter of degree, rather than of historical approach. Most historians would agree with Jason Wood when he points out (1992, 294) that 'analysis and interpretation [should] develop in tandem with the recording process [;] ... the complexities of phasing within the construction sequence as a whole, throughout the building, can be studied and taken into the interpretation. Drawings illustrating both the suggested development and reconstruction of the building can be prepared, as well as overlays showing fabric surface conditions and the extent of historic interventions, to which the results of new works can be added'. Few building types illustrate the ideal for archaeological recording better than medieval buildings, including houses, castles and churches. Often extremely complex in building history, they demonstrate how, in most cases, even relatively minor works should be accompanied by adequate drawn survey, analysis and interpretation before works begin, and continuing survey and photographic recording while works continue. This has been covered in detail by Jason Wood who has stressed the importance of the integration of the role of the archaeologist into the team carrying out the works to the building and has written a model specification of the eight stages for such recording (Wood 1994, 183–208). Thus, from the start the drawn and photographic record should inform the methodology and decision taking during the lead-in process to works beginning, including production of specifications and schedules of work. Equally, however, interpretation of the fabric depends upon experience not only in an appreciation of what Wood describes as the 'stratigraphy' of a building, but also, as he rightly emphasizes, by 'compare[ing] the results of the archaeological analysis with previous art historical investigations, to see how closely the stylistic development is reflected in the identified construction sequence' (Wood 1994, 193).

There is no doubt that PPG 15 has made archaeological recording of historic buildings as part both of conservation and of alteration and extension works much, much easier. The problem, however, is that local planning authorities which administer the planning system and the professional bodies (apart, of course, from the Institute of Field Archaeologists) which represent the staff who interact within that process have little idea how the responsibilities and duties, following the changes brought in by PPG 15, have increased. Architects and surveyors may have scant concern, in general, for the need for informed recording of historic buildings as part of the conservation process; indeed, if the architectural press is to be believed, historic buildings represent not so much an opportunity as a threat, therefore lessening any inclination to take seriously the rudimentary ability to analyse and evaluate historic buildings. Education, of course, is the answer and the continuing growth in strength of the Association of Conservation Officers and proposals for institutionalization lead to hopes that the control régime, day-to-day largely in the hands of conservation officers, will take archaeological recording as seriously as PPG 15 suggests.[8] Similarly, the Institute of Field Archaeologists is currently discussing standards for archaeological recording; perhaps, likewise the Association of Conservation Officers should be discussing standards for outset survey, analysis and interpretation. Concern on the part of conservation officers is not enough, however, and both the Royal Town Planning Institute and the Royal Institute of British Architects must take note of the alterations in the régime and bring in education in architectural history, interpretation and archaeological recording both at student stage and in terms of continuing professional development.

The role of the architectural historian, and the uses of architectural history, in the analysis of historic buildings is one which is not currently conspicuously well provided for but there are some signs that the position may be changing. Architectural history is increasingly seen as an important part of historic building conservation and the coming opportunities for the training in architectural history of conservation officers can only be beneficial.

BIBLIOGRAPHY

Fernie, E. (1995). *Art History and its Methods: A Critical History*. Phaidon, London.

Marwick, A. (1989). *The Nature of History* (3rd edition). Macmillan, Basingstoke and London.

Pevsner, N. (1972). *Some Architectural Writers of the Nineteenth Century*. Oxford University Press.

Pevsner, N. (1973). *An Outline of European Architecture* (8th edition). Allen Lane, London.

Piggott, S. (1985). *William Stukeley: An Eighteenth Century Antiquary* (2nd edition). Thames and Hudson, New York.

Podro, M. (1982). *The Critical Historians of Art*. Yale University Press, New Haven and London.

Pointon, M. (1986). *History of Art: A Student's Handbook* (2nd edition). Unwin Hyman.
PPG 15 (1994). Department of the Environment, Department of National Heritage, *Planning Policy Guidance: Planning and the Historic Environment.*
Rees, A. L., and Borzello, F. (1986). *The New Art History.* Camden Press, London.
Stocker, D. (1992). 'Broken toes or interdisciplinary ballet: architectural history, archaeology and the recording of buildings', *The Field Archaeologist,* xvi, 302–3.
Walker, D. (1994). 'Listing in Scotland: origins, survey and resurvey', *Transactions of the Ancient Monuments Society,* xxxviii, 31–96.
Wood, J. (1992). 'Integrated recording and analysis for building conservation projects', *The Field Archaeologist,* xvi, 293–4.
Wood, J. (1994). 'Major conservation projects: recording ancient monuments – LUAU's Methodologies and Procedures', in *Building Archaeology: Applications in Practice,* (J. Wood, ed.), 183–208.

NOTES

1 *The Newleafe Discourses* is discussed in some detail in Pevsner 1972, 217–19.

2 This development of art history is well covered in Fernie 1995, 10–21. This volume, the most recent primer on the historiography of art history also has a useful set of definitions, including ones for antiquarianism, art history and archaeology. Revealingly, although architecture is defined, architectural history is not.

3 In the SPAB Manifesto of 1877, Morris wrote that it was important 'to treat our ancient buildings as monuments of a bygone art, created by bygone manners, that modern art cannot meddle with without destroying'.

4 A description of the discussions on what has become known as 'group value' in listing terminology is contained in Walker 1994, 52–3.

5 There has always been a general legal problem with conditions to planning applications which require works to be done because it might be possible to construe such conditions as suggesting that a permission has been granted in return for some kind of payment in kind. Thus DOE Circular 8/87 only mentioned archaeological recording in fairly unspecific terms when discussing 'Conditions which are acceptable in appropriate circumstances' in Appendix VII (p. 65):

> 4. For a period of [14 days] before work commences the [insert name of person or body] shall be given access to the building to enable [him] to [take photographs of the interior] [make measured drawings].
>
> Notes
> 1. This condition must not require the applicant/owner to pay for the photographs or plans; those who wish to make the record must pay. If, however, the applicant/owner has offered to make a record himself, this can be required in terms reflecting the offer made. The time allowed for making a record must be reasonably short and must not unreasonably interfere with the execution of the authorised works.

This was changed when PPG 16 (Archaeology and Planning) was published in 1990 with a suitable wording which overcame the problem about purchase of permissions. PPG 15 followed that lead in 1994 which contains the following advice in paragraph 3.23:

> Local planning authorities should also consider, in all cases of alteration or demolition, whether it would be appropriate to make it a condition of consent that applicants arrange suitable programmes of recording of features that would be destroyed in the course of the works for which consent is being sought. Authorities should not, however, require applicants to finance such programmes in return for the granting of consent. Nor should applicants expect to be granted consent merely because they have arranged suitable programmes.

6 Annex B of PPG 15, dealing with 'Listed Building Control Procedures', gives the following advice on Applications:

> B.3 Applications must be made in triplicate on a form issued by the local authority. Section 10(2) of the Act requires that they include sufficient particulars, including a plan, to identify the building in question and such other plans and drawings as are necessary to describe the works for which consent is sought. For all but the simplest work this should normally mean measured drawings of all floor plans and external or internal elevations affected by the work proposed. There should be two sets of such drawings showing the structure before work and the altered structure or new development to replace it after the proposed work. The inclusion of photographs can be particularly helpful – of all elevations in demolition cases, or of the part of the building affected (interior or exterior) in alteration and extension cases. The Act empowers an authority to seek such particulars as it requires and an authority should certainly seek any particulars necessary to ensure that it has a full understanding of the impact of a proposal on the character of the building in question. An authority should not accept an application for consideration until it has sufficient information to provide such understanding.

7 PPG 15, paragraph 3.24:

> Hidden features of interest are sometimes revealed during works of alteration, especially in older and larger buildings: chimney pieces, fireplaces, early windows and doors, panelling, wattle-and-daub partitions and even wall-paintings may come to light. Applicants for listed building consent should be made aware of this possibility and should seek the advice of the local planning authority when such things are found. If there is any likelihood that hidden features will be revealed, the local planning authority should attach an appropriate condition to the listed building consent to ensure their retention or proper recording, or should require exploratory opening up, with listed building consent as necessary, before considering consent for the main works.

8 Indeed the discussions which are currently leading towards the formulation of suitable vocational qualifications for the construction industry and, in particular, in training in architectural conservation, are taking the needs for appropriate levels of qualification in historical skills – survey, analysis, interpretation and recording – very seriously.

12

HISTORIC LANDSCAPES AND GARDENS AND THEIR CONSERVATION

Peter Goodchild

INTRODUCTION

Since 1980 there has been, in Britain and elsewhere in the world, a marked increase of interest and activity in relation to conserving historic parks, gardens and landscapes. Often these are now referred to collectively as 'cultural landscapes'. In part, this chapter is a more considered response to a question that Sir Bernard Feilden asked me at an international conference in October 1992. In essence the question was 'what are cultural landscapes?' Shortly after this the term was incorporated into the World Heritage Charter.

At the same time a Council of Europe initiative was under way. It was a joint project organized by the Cultural Heritage Committee and the Council for Cultural Co-operation Fund and known officially as Project IV.42 of the Cultural Fund programme. It involved the setting up of a group of specialists on Heritage Landscapes and Sites who were given the tasks of drawing up a recommendation to member states on cultural landscape areas (called *sites culturels* in French) and organizing a series of conferences on their rehabilitation (Cultural Heritage Committee 1994, 2). The setting up of the group began in the later part of 1990, the first meeting was held in May 1991 and in July 1994 a draft recommendation was produced and then released to interested parties. The appendix to the recommendation began with a list of definitions of terms which had been promulgated by the group. The terms in questions were 'landscape', 'cultural landscape areas', 'conservation', 'landscape policies' and 'visual pollution'.

Another piece of work that should be mentioned is the production in April 1990 of a discussion document called *Some principles for the conservation of historic landscapes* (Goodchild 1990). It was originally prepared for use within the professional and postgraduate educational programme at the University of York's Institute of Advanced Architectural Studies, but it was also presented in April 1990 to the ICOMOS UK Committee on Historic Gardens and Landscapes.

It was additionally intended as a contribution to the discussions at the thirteenth Annual Meeting of the Alliance for Historic Landscape Preservation which was organized jointly with US-ICOMOS and held at Seattle, USA, in June 1990. An important element of this document was an attempt to define terms and explain basic concepts. It was based on experience gained at postgraduate level and continuing professional education, as well as a long-standing interest in landscape conservation, and an awareness of a need for a theoretical base, a vocabulary and an ordered system of ideas. The purpose of a theoretical base is not only to facilitate education and instruction, at all levels, on the conservation of the historical dimension of gardens, parks and the wider landscape but also to assist the further development of conservation in practice. This need is still evident and moreover a theoretical base is required that is holistic in its understanding of conservation. For example, the concept of landscape conservation must include both the natural and the cultural dimensions. It must also be understood that conservation has to take place within a context of circumstances.

The exploration and working out of the theoretical aspects of landscape conservation has practical consequences because it will, it is hoped, lead to a clearer and more easily transferable understanding of the subject, but in the meantime advances in the practice of conservation can still be made and are being made.

This paper is a commentary on the present state of landscape and garden conservation in that it discusses the meanings of several basic concepts, it attempts to provide an holistic approach, and it reports on some of the practical measures that have been taken.

A close examination of the term 'cultural landscape' inevitably leads one not only to look again at its component words, 'cultural' and 'landscape', but to think about connected issues such as: 'how does the cultural dimension relate to historical and natural aspects?' 'how do landscapes relate to parks and gardens?' 'what are landscapes?' and 'what is conservation in the context of landscapes, parks and gardens?'. This chapter has provided an opportunity to review all these issues.

Because of the limits of space and the enormous scope of the subject, topics are dealt with in outline but it is hoped that the general principles are clearly and adequately explained. For the same reason the discussion is largely based on the situation in England. However, Britain is amongst the foremost countries in the field of landscape conservation and the principles that operate in England will have wider application. The part of this chapter that deals with World Heritage Sites relates directly to the international scene.

WHAT ARE LANDSCAPES, PARKS AND GARDENS?

Landscapes

In origin, landscapes are the result of the actions of natural processes on naturally occurring elements and organisms. At a certain point in their evolution humans distinguish themselves from the rest of nature. Although they remain part of nature they become a particularly significant force and it is useful to observe the distinction. As a result of their exploitation of the planet, most of its landscapes have to a greater or lesser extent been modified by human activities and are no longer the result of nature without interference. In a physical sense, many landscapes are therefore a mixture of natural phenomena and the uses to which humans have put the land. These include all kinds of rural, urban and industrial uses and all that is connected with them.

Nature and mankind are two of the main factors that determine the character of a landscape. In this context the activities of humans may take the form of the use to which they put the land or the ideas that they formulate with respect to it. The latter may take the form of an imaginative and intellectual engagement with it or the policies and systems that are evolved in connection with practical issues. But there are also other factors which determine the character and appearance of a landscape. Altogether, and in brief, they are its geographical location, natural attributes, the use to which it is put, the managerial environment, the intellectual environment, population densities, condition and the history or course of its evolution. These provide a basis for a systematic approach to the description of landscapes of all kinds and this in turn provides a basis for determining why they are of interest, their comparative significance and the action that can be taken.

Figure 12.1

Farndale in North Yorkshire. Above the farmland is the seemingly natural moorland of the North York Moors. In fact, the moors were created by humans through clearance of woodland, the grazing of sheep and associated management practices, a process which began around 10 000 years ago.

Geographical location not only describes where a landscape is to be found but it places it in a context which may be climatic, ecological, historical, political or administrative. Natural attributes cover the atmosphere, climate, geology and natural soils, geomorphology, hydrological factors, living organisms and their behaviour, and habitats.

The managerial environment refers to such things as the legal, administrative, political, economic and social factors that affect a landscape. These may be local, regional, national or international in origin and the particular combination that operates, or is significant, will vary from one landscape to another. The managerial environment also includes the quantity and quality of the resources that are available for maintenance and management.

Population densities, either of people or other species, are a basic factor that will have an effect on the nature of a landscape. The other species may be animals or plants.

The intellectual and imaginative engagement of humanity with the landscape is a two-way process in that the landscape may be the origin of ideas and stimulate the mind or the mind may project ideas onto the landscape or interpret it in a particular way. By this means concepts of landscape can become a mixture of the physical reality and the workings of the mind. Landscapes can exist in a number of different states. They may be a physical reality; a portrayal in pictorial, cartographic, literary or musical form; a perception in the mind of the observer of a portrayal or of an actuality; or they may exist simply in the imagination. A dynamic relationship can exist between different states of existence and as a consequence they can influence each other.

In connection with actual landscapes it is useful to recognize that an important part of their perceived character is made up of the ideas that are associated with them; for example associations with a particular person, an event, a myth, a legend, religious, spiritual, scientific or artistic ideas. The intellectual environment includes the climate of ideas both within which a landscape was created or modified and within which it is currently interpreted. It can be analysed and described in terms of the different subject areas in which the intellect and imagination operate, such as history, art and aesthetics, science, religion, mythology, anthropology, etc.

'Condition' is a deceptively simple word. In connection with the physical dimension of actual landscapes at least four aspects of condition need to be taken into account. These are the relative age distribution and the health of the different component life forms, including plants, animals and humans; the standard of repair and maintenance of man-made components; the degree of completeness or alteration of individual objects and component areas; and the degree of completeness or alteration of the whole.

Actual landscapes are dynamic because they are formed by living organisms and by continuous natural and man-made processes. As a result, they adjust and change with the life cycles of their components or with the fluctuation of the factors that impinge on them. For a more complete idea of the character of landscape it is therefore necessary to see it in the context of time. A current description refers to only one point in its life and an account of its evolution, or life history, will help to give a much more complete understanding. It will also help with the formulation of proposals for the future.

Like 'condition', 'landscape' is also a deceptive word. Sometimes 'land' can be used as a substitute for 'landscape' but it is not quite the same and the difference highlights the particular meaning of the latter in English. 'Landscape', as a word, was introduced into modern English, probably from the Low Countries, in the sixteenth or early seventeenth century and as a technical term of painters. The first meaning given in the *Oxford English Dictionary* is 'A picture representing natural inland scenery, as distinguished from a sea-picture, a portrait, etc'. From this other meanings may be derived.

Interest in landscape paintings and prints and in the pictorial qualities of real landscapes is an important vein in the history of British culture, particularly since the early eighteenth century and the development of an appreciation of picturesque qualities. As a result there is in the British understanding of landscape an inheritance from this strong pictorially based tradition which means that integral and important parts of the concept are the ideas of appearances and the arrangement, or spatial relationships, of the components that make up the landscape. The concept of land is free of these ideas. It is more focused on those of territory, surface area, ownership and a basic resource for agriculture or some other use.

The visual aspect is a particularly important characteristic when it comes to defining what is meant by landscape. A landscape extends as far as one can see, to the visual horizon or as far as the boundary of what is visible. If one moves to another point the boundary of the view may well change as one does so, perhaps quite dramatically. With this in mind it might be asked how can the outer boundaries of a landscape ever be fixed? The answer is that they can be fixed if the observer keeps to a given route or within a given area. One might, for example, be confined to an area that is in single ownership or which has a distinctive character of one kind or another, such as a garden, a park, or a particular tract of countryside. In these cases the outer boundary of the landscape is made up of the most distant points that can be seen from anywhere within the core area to which the observer is confined. The core area provides the locations from which the landscape as a whole can be seen.

Depending on its nature and extent, the core area may provide within its boundaries the foreground, middle ground and distance for most or only some of the views. It may, however, just provide the foreground and middle ground, or perhaps only the foreground or the immediate foreground of others. In these cases if the views continue, the remainder will be beyond the core area. In any event the observer's experience of the landscape as a whole will be united by the character of the core. With this in mind a landscape may be said to consist of a core area plus adjacent land. Where the adjacent land contributes in a significant way to the character of the core it may be referred to as the setting of the core. If it is self-contained the core will constitute the whole landscape. If the core is large and has a distinctive character and many self-contained or largely self-contained views, it may be the most important part of the landscape even if there is a setting.

When making proposals for the core, it is possible that the adjoining land may be ignored or it may be decided not to make use of it, perhaps because there is no guarantee that it will remain attractive or unchanged, or because there is no opportunity to improve it. In these circumstances it is not difficult to appreciate how, in practice, the idea of a landscape can be reduced to the core area alone, even when there is a setting. The land which is the subject of a particular project is often referred to as the site by landscape architects. The site is the core area for their project. Even if his attention must focus on the site, a landscape architect should always be conscious of its relationship to adjoining land. But because he will spend most of his time dealing with the land within the site it is easy for the site to become synonymous with the landscape, if not in his mind then in the mind of others.

Among archaeologists and landscape historians there currently seem to be at least three different meanings attached to the word 'landscape' in the context of historic landscapes. For some a landscape is an area with a core and a setting, for others it is simply a piece of land, often of limited extent, which has features of archaeological or historical interest, and for some it is the landscape as it existed at some point in history and of which the surviving features are the evidence.

Landscapes (in the sense of an area with a core and a setting) can be of any size. For the purposes of analysis it is useful to think of them as being made up of components which may be areas, objects or details. Objects may be blocks, points or linear. Component areas may be small, medium or large in extent. They may take the form of a particular type of land use or land cover but the use or cover should not be referred to as a type of landscape in itself unless it makes up the whole or the majority of the core area and the setting.

From the above discussion it is possible to propose a primary definition of landscape which is as follows:

> A landscape is a view of scenery in which land or the ground surface, vegetation and other natural elements, including the sky, are usually prominent. It is everything that can be seen, either physically or metaphorically, from a particular point, route or area and it may include a setting as well as a core area. The core area or setting may be of any size depending on the nature of the landscape. A landscape may be open or confined and its components may be analysed in terms of areas, objects and details. The objects may be blocks, linear or individual items occupying a small ground area or a particular point. The arrangement or configuration of components, or their spatial relationships, are an important aspect of the nature of a landscape. Landscapes may exist as a full scale physical reality, as a pictorial, cartographic, literary or musical portrayal, or they may exist in the mind of the observer as a perception or as a self-generated image. These states of existence are interrelated. The subject of landscape includes all the factors and processes by which landscapes are formed and all the associations, meanings and significances with which they are invested by humans.

There is a meaning, which is widely held, according to which a landscape is a place, of any extent, that is in the open air, open to the sky. Perhaps this may be regarded as a simplification or adaptation of the above.[1]

Cultural landscapes

In recent years in landscape conservation circles there has been much discussion about cultural landscapes. The term 'cultural' is used in this chapter in the same sense that it is used in the World Heritage Convention which was adopted by UNESCO in 1972 (UNESCO 1981). In this convention the World Heritage is divided into two parts, the natural and the cultural. In this sense 'cultural' refers to 'the works of man or the combined works of nature and of man'. This definition, however, needs a little augmentation. Human cultures are the product of the physical and mental activities of humankind or of particular populations, communities or groups of humans. Those activities which are characteristic of humankind are of particular significance in determining the nature of a culture.

Originally three categories of cultural heritage were admissible to the World Heritage list, that is, monuments, groups of buildings and sites. Sites were defined as:

> the works of man or the combined works of nature and of man and areas including archaeological sites which are of outstanding universal value from the historical, aesthetic, ethnological or anthropological points of view.

In December 1992, the World Heritage Committee agreed to include the additional category of 'cultural landscape' (ICOMOS 1993, 3–6) and the operational guidelines for the World Heritage Convention were revised accordingly. The revised guidelines state that;

> Cultural landscapes represent the 'combined works of nature and of man' designated in Article 1 of the Convention. They are illustrative of the evolution of human society and settlement over time under the influence of the physical constraints and/or opportunities presented by their natural environment and of successive social, economic and cultural forces, both external and internal.... The term 'cultural landscape' embraces a diversity of manifestations of the interaction between humankind and its natural environment. (ICOMOS 1993, 3–6)

The operational guidelines divide cultural landscapes into three main categories, the first of which is 'the clearly defined landscape designed and created intentionally by man. This embraces garden and parkland landscapes constructed for aesthetic reasons'. The second category is 'organically evolved landscape' which falls into two subcategories, 'relict (or fossil) landscape ... in which the evolutionary process came to an end at some time in the past', and 'a continuing landscape ... which retains an active social role in contemporary society closely associated with the traditional way of life'. The third main category is the 'associative cultural landscape. The inclusion of such landscapes in the World Heritage List is justifiable by virtue of the powerful religious, artistic or cultural associations of the natural element rather than material cultural evidence, which may be insignificant or even absent' (ICOMOS 1993, 3–6).

These categories were drawn up for use in connection with World Heritage Sites which are conceived as exemplars of world-wide significance. Unfortunately

Figure 12.2

A view on Coniston Water. The landscape of the Lake District, like many others, is the product of a mixture of natural and human factors but in it the British feel particularly close to nature. For two centuries or so, its naturalness, its beauty, and the recreation that it provides have drawn people to it and given it a special place in the hearts and minds of the nation. It is an area with very strong cultural associations.

they are not altogether satisfactory for use in the context of a single nation. For example, it is not clear whether intentionally designed landscapes includes those that were intentionally designed for agricultural purposes such as the eighteenth- and nineteenth-century enclosure landscapes of England. This, in turn, raises the question of what is the nature of design in connection with landscapes. Does it only cover places that have been laid out primarily for aesthetic or artistic effect or does it also include places which have been laid out mainly with functional considerations in mind? Is it always a formalized process involving plans on paper for example, or is it the result of any action, no matter how it is organized, that has a considered intention or policy behind it? If it is the latter it might be asked how many agricultural landscapes are not designed? Perhaps most of them are. This question about the nature of design also raises the issue of whether organically evolved landscapes are those that are formed without design or whether they are a mixture, with some parts designed and some not. Or are they places which have developed a part at a time instead of being the result of one overall scheme?

A more important point is raised in connection with continuing landscapes because a significant element of them is an association with a traditional way of life. What is the extent to which one can or should conserve a traditional way of life?

It is not clear whether these World Heritage landscape types are intended as discrete categories or whether they are characteristics that can overlap with each other and co-exist on the same piece of land. At a national, regional or local level one is not necessarily concerned with landscapes that simply belong to one category because often they have a mixture of characteristics. For example, a relict garden, now in the form of grassed earthworks, may be part of a continuing landscape. A garden may also be organically evolved and have significant associations.

In some ways 'cultural landscapes', although it is becoming widely accepted, is not a satisfactory term, because when used in connection with actual landscapes it tends to imply that the landscapes in question are only of interest from a cultural point of view whereas they can still be of interest and perhaps of some importance from that of the natural heritage. The converse may apply to areas that are perceived only in terms of their natural interest. In the interests of propagating an holistic approach to landscape management and conservation it is better to refer to the natural or cultural 'dimension' because the use of this word acknowledges that the other dimension is a factor that must be considered as part of the overall approach. This does not mean that specialists in one dimension should also be specialists in the other but simply that they should recognize each other and work together. Within each of the two main divisions there will be many different specialisms and there is the same need for liaison and cooperation so that balanced assessments and decisions can be made. Organizing effective liaison and coordination between different specialisms and interests is one of the main challenges of environmentalism.

In connection with actual places it is also better, for the same reasons, to use the phrase 'landscapes of cultural interest' rather than 'cultural landscapes'. Perhaps the term 'cultural landscape' is best used in relation to those that exist in the mind, in portrayals and to those actual landscapes that are entirely or largely fashioned according to artistic ideals. Many gardens and parks would come within this latter sort.

Although a distinction can be drawn between the cultural and the natural dimensions they are also closely related because they overlap and interact with each other. For example cultural elements may be composed of, or include, natural ones as in agricultural land, plantations, cultivated varieties of plants, masonry, etc. Natural elements can acquire cultural associations. So the two dimensions often cannot be completely separated from each other. But the relative proportions of each may vary widely from one landscape to another or from one component to another.

In the true wilderness nature may reign supreme but in other areas it is modified or managed by humans to a greater or lesser extent. The degree of modification will depend on the kind and strength of the intervention by humans and also on the ability of the natural system to resist or recuperate. The relative proportions of human influence and nature may remain stable for a period so that a particular equilibrium is reached but this equilibrium may only be attained because for a period the relative proportions are maintained and not because of any inherent stability in a particular equilibrium.

Landscape character areas

The idea of different landscapes consisting of different proportions of human influence and nature enables a couple of basic distinctions to be made. One is between landscapes on which humans have had no or little impact (wildernesses) and those which have been significantly affected. The other is between rural and urban-industrial areas. The issue of where the boundary between rural and urban-industrial should be drawn need not be discussed here except to say that areas of a mixed kind are likely to exist where the two meet or where the one penetrates or is located within the other.

These distinctions are very broad ones and they raise the question of how might landscapes be divided into more specific areas in a way that reflects both their cultural and natural interest. One approach to this might be to employ types of land use or land cover.[2] But these in themselves would not be very satisfactory because although it is possible that a landscape might be entirely or predominantly one type of land use or land cover, it is just as likely that it will be made up of several at the same time.

In practice, because landscape is primarily a visual concept, it is better to concentrate on identifying core areas that have a consistency of character within themselves. Within a country or a region, it is likely that some core areas will be markedly different from each other, but with others and particularly where they are composed of a broadly similar range of components, the differences may be more subtle. But in either case the core areas will be identified on the basis of particular combinations of significant characteristics. It is important, therefore, to decide which characteristics are to be used and to be able to identify the patterns and particular combinations that are formed by them. This approach to the description and identification of landscapes may be called 'characterization' and its purpose in this case is to identify areas that are unified in character and distinctive from their neighbours.

As part of their Countryside Character Programme, the Countryside Commission (CC) is currently developing a technique for breaking down the English landscape into areas of distinctive and cohesive character ('landscape character areas').[3] It is a technique which arises out of their guidelines on landscape assessment which were published in 1993 (CC 1993) and further developed by a pilot study carried out in the south west of England, the results of which were published in 1994 (CC 1994a; CC 1994b). The aim is to deal with England on a regional basis and produce for each region a map showing landscape character areas. These in due course will be linked to the preparation of regional strategies and action plans.

The character areas are primarily descriptive and are based on an analysis and assessment of a broad range of factors including both natural and cultural ones, such as geology and soils, geomorphology and drainage, flora and fauna, habitats, land use, historical features, historical settlement patterns, current population density and scenic characteristics. Two other national government agencies are cooperating to ensure a comprehensive approach. English Nature represents the wildlife and ecological interest and will contribute its map of Natural Areas to the project (English Nature 1994). This map has been produced as part of its Natural Areas programme. English Heritage (EH) is representing the historical dimension in relation to monuments, historic buildings, landscapes and gardens, and will contribute its maps of regional historic settlement patterns (EH 1995a, 17–19). The interests of the Countryside Commission focus on recreation and scenic values in the countryside.

The pilot project for the Countryside Character Programme was carried out in the counties of Cornwall, Devon, Somerset and Avon which represent about 11.6 per cent of the area of England. In this area thirty-eight landscape character areas were identified. The smallest of these was about 59 sq km in size, the largest about 1768 sq km (CC 1994a). The largest is almost thirty times the size of the smallest, but even the smallest is very large compared with most historic sites. The gardens and designed ornamental landscape at Castle Howard in North Yorkshire, which is one of the largest officially registered parks and gardens of special historic interest in England, is in the order of 20 sq km. By comparison the largest English National Park, The Lake District, is 2292 sq km and the smallest, Exmoor, is 693 sq km. The Norfolk Broads which has equivalent status is 303 sq km (CC 1995). The biggest landscape character area in the pilot study area has virtually the same area as the Yorkshire Dales National Park.

These character areas are generally large enough to have a high degree of visual self-containment. They are, however, also intended to be seen in a wider context, that of the landscape of England. Within many of these character areas it will be possible to identify smaller landscapes such as those associated with a particular property such as an estate, a park and garden or a monument.

Parks and gardens

In Britain the concepts of garden and rural landscape overlap significantly with each other and have close historical links. They both also enjoy a very high and generally acknowledged status in the British way of life and culture, and have done so for at least two and a half centuries. Because of these two factors it is easy to think that they are inseparable and must always be dealt with together and yet in some ways they are very different. It is worth clarifying the relationship between the two because if it remains in doubt it gives rise to troublesome ambiguities.

In origin, gardens are special enclosures where horticulture is practiced and where plants are grown to provide food or for medicinal or domestic uses, for pleasure or for scientific, social and even political and religious purposes. Horticulture differs from agriculture in several respects, for example in the range of plants that are grown and in that horticulture is traditionally practiced on a smaller scale and is a more intensive form of cultivation. It is usual to spend a higher degree of care and attention on individual plants and on the conditions in which they grow.

Figure 12.3

View from Backbury Hill, near Hereford. In the mid-seventeenth-century John Evelyn (1620–1705) chose Backbury Hill as the site for an ideal princely garden, his *Elysium Britannicum*. The view over the adjoining vale was one of the important features. The idea was not put into practice but it demonstrates at an early date a desire to link the garden with the countryside.

The commonly recognized branches of horticulture, or gardening, are vegetable growing, floriculture, ornamental gardening, botanical gardening, maintenance, arboriculture and garden design. The subject of gardening is not, however, confined to the kitchen garden, orchard and ornamental gardens because, around a house or other building, the land that is managed for the purposes of amenity or ornament extends beyond the garden proper and also traditions of design have linked gardening to other areas. In the context of the country house, for example, they have linked the ornamental gardens with the deer park, the landscape park and other areas around the house. In urban areas the design of public parks and open spaces is often done within the traditions or current styles of garden design.

In Britain where parks and gardens are numerous, it is usual to talk about them as if they were a unified entity. Because history has extended the original meanings of parks and gardens, it is necessary to include the relevant additional types of place within this concept. As a result 'parks and gardens' has become a convenient umbrella term to cover them all. It is a shorthand way of referring to parks, gardens, designed ornamental landscapes and places of recreation. It has to be recognized that the concept of gardening has moved beyond the boundaries of the garden proper and into the wider landscape where it manifests itself as features that have been intentionally designed or created for ornamental or recreational use.

Figure 12.4

Warwick Castle, Warwickshire. A fine and important example of an eighteenth-century naturalistic landscape. The castle, park and gardens have been composed to form a united whole. The park and garden provide a setting for the castle; the castle is a principal feature of the park and garden.

This extension of the concept of gardening has also been absorbed into the concept of landscape design which in the twentieth century is a recognized discipline. As a discipline it has, however, tended to separate itself from horticulture and traditional domestic ornamental gardening. This tendency has occurred because the profession of landscape architecture has focused more closely on design than on horticulture and because much of its work has not been concerned with private gardens but with the spaces within public, institutional and commercial developments of all kinds where the ideas of grounds, open space, surface treatments are often more appropriate than traditional gardening. Landscape design has been much influenced by its close contacts with architecture and planning. There has also been a considerable increase in public and professional interest in the landscape at large. This has taken place in the context of landscape management and outdoor recreation and it has affected the nature of landscape architecture.

Parks, designed ornamental landscapes and places of recreation have become a kind of middle ground between gardening, on the one hand, and landscape design and management, on the other. In the present time where the general environment and recreation are major public and political issues, it seems right that parks and gardens should be recognized as important components of the environment. In Britain, because of their distinguished history, they are particularly important even if the overall area of the land surface that they cover is small in comparison to rural landscapes. Horticulture, design for ornamental purposes and artistic expression are very significant aspects of parks and gardens and they help to distinguish them from other kinds of landscape where other forms of husbandry and social and economic activity are more important.

LANDSCAPES AND CONSERVATION

Because humans live in the world, they are part of it and it is part of them. The two are inextricably linked. But because of their ability to interact with their surroundings and exploit them, their place in the world is a special one. This ability is at the root of their development. It has enabled them to develop not only in relation to the material and economic aspects of life but also in relation to the spiritual and intellectual. It would seem that their engagement with their surroundings is an essential aspect of their nature. This engagement has taken place, and still takes place, in different ways. It may, for example, be the result of winning a living through husbandry, commerce or industry. It may take the form of religious or artistic experience giving rise in the latter case to poetry, literature, pictorial art or music. It manifests itself through the pleasure and relaxation that it can give. It may also take the form of intellectual enquiry such as scientific or historical investigation.

The curiosity of humans about the world, their intellect, their religious and aesthetic experience and need for periods of relaxation or recreation are all essential to them. Their environment, if it is to remain a healthy one, must provide for all of these, as well as for their material needs. It must also enable the biosphere to remain in a healthy condition, not just because it is a resource for humans but because they are an integral part of it and imbalances may have a detrimental effect on them. In addition, there is the consideration that all forms of life, and not just humans, have a right, as it were, to exist.

The concepts of environment and landscape, although closely linked, are not the same and one might ask how they differ. Accepting the importance of the environment, one might also ask, 'why are landscapes of value'?

A landscape is a part of the overall environment and in this sense whatever values can be attributed to that part can be attached to the landscape. Landscapes are made up of the visible components of the environment and their character is affected by whatever affects the appearance of the environment. In a narrower sense their particular value is that they are a visual resource for mental recreation and relaxation and also for intellectual enquiry.

As part of the British way of life there has been a strong vein of interest in rural landscape and nature since the seventeenth century at least. It has taken several different forms, for example, natural history and scientific interest in nature and wildlife, poetry, landscape and garden design, touring for pleasure and instruction, sketching, painting, incorporation into novels and plays, book illustration, antiquarianism, local and architectural history, music, long-distance walking and other recreations. There is therefore a strong basis for a continuing engagement with the landscape.

Because landscapes are made up of both natural and cultural components, landscape conservation must address them both. Conservation of the cultural heritage can generally be justified in terms of the wise use of resources that are of value to humans, but as has already been pointed out, this on its own is not a completely adequate basis for conserving the natural heritage because it only sees the issue from the particular point of view of humankind. Humans may have assumed the stewardship of the biosphere but they are not its master or creator and ultimately they do not have the right or the knowledge to decide whether other forms of life should survive or perish. There is also the consideration that damaging the biosphere may, in some indirect or unpredictable fashion, change the ecosystem in a way that is detrimental to humans. Resources which they use may be affected by their actions, either indirectly or in a delayed fashion.

In the context of landscape, as in that of the environment in general, a concept of conservation is needed that covers both the natural and the cultural dimensions. Perhaps conservation can be described as a process which includes the protection, care, management and treatment of resources and life forms in all their different kinds, with the purpose of prolonging or perpetuating their existence either because they have, or may prove to have, a use or value which may benefit the present or future generations of humankind or other life forms, or because it is justifiable for other reasons, such as on ethical, moral or religious grounds. Conservation involves taking positive action and this is done through a process that includes the following tasks: identification of the resource, the life forms and the associated issues; preliminary description of them; preliminary assessment; the making of decisions about the nature of the action to be taken and who should take it; formulation of a more detailed proposal for effecting the action; execution of the proposal; monitoring the resource, life forms, issues and the action taken; and reviewing the state of affairs from time to time. Conservation does not take place in isolation but in the context of prevailing circumstances and other factors affecting the resource, life forms or issue in question. An holisitic approach is therefore required.[4]

In this description the word 'resource' can refer to natural or cultural phenomena that are, or may be, used either by mankind or by other life forms. It also recognizes that other life forms have a right to exist whether or not humans make use of them or approve of them. In this way it avoids the charge of being too anthropocentric.

From the point of view of mankind, and some other species, a resource may have a physical, psychological or mental use or value. These uses and values provide a basis for the assessment of resources.

THE HISTORICAL DIMENSION AND ITS CONSERVATION

Landscapes

The cultural dimension of landscapes includes elements that are historical in origin as well as those that are contemporary. Until quite recently the former was given comparatively little attention. It would seem, for example, that in the first half century of its existence the National Trust, which celebrated its centenary in 1995, applied the criterion of historic interest to buildings but not to landscapes, which were covered by the criterion of natural beauty (Hill 1956, 149–50; Lees-Milne 1948, ix–xii) This raises the question of when did interest in the historical dimension of landscapes develop? This dimension has several aspects to it and some of these are long-established as, for example, an ability to appreciate a landscape because of its associations with real or imaginary historical events and characters. In his *Remarks on Several Parts of Italy etc in the Years 1701, 1702 and 1703*, Joseph Addison (1672–1719) records his pleasure in 'seeing my voyage, from Naples to Rome, described by Vergil' (Guthkelch 1914, ii, 128).

There has also been a consciousness of a relationship between an historic monument or building and the landscape in both aesthetic and antiquarian terms, although the latter seems to have been a fairly rudimentary form of historical appreciation. Examples include William Gilpin, who in the late eighteenth century published his observation on the aesthetic aspects in the accounts of his tours.[5] There is also Francis Drake who, in his history of the City of York published in 1736, includes a map to show the routes of Roman roads in the county of York (Drake 1736).

There has also been a long-standing awareness of the traditional elements of the countryside but this perhaps has been more focused on people, their work and artefacts associated with daily life, as for example in Gertrude Jekyll's *Old West Surrey* (Jekyll 1904). But this is not quite the same thing as a scholarly interest in the history and evolution of a landscape as a whole.

If publications can be taken as an indicator of interest, books about aspects of the history of the countryside were being published in the later nineteenth century and the early twentieth. For example, G. Shaw-Lefevre, who was a leading member of the Common's Society, produced his *English Commons and Forests* in 1894. There was also J. C. Cox's *The Royal Forests of England* of 1905, *An Historical Geography of England before 1800*, edited by H. C. Darby, was published in 1936, and in 1938 the first edition of *The Open Fields* by C. S. and C. S. Orwin came out. The latter is a classic study of the medieval open field system of agriculture which remained in operation at Laxton, Nottinghamshire, as a very rare surviving example in this country.

But by far the most important stimulus that awoke general interest in the subject was W. G. Hoskins' *The Making of the English Landscape* which was first published in 1955. Hoskins demonstrated that the landscape itself was an immensely rich document that recorded important aspects of England's social and economic history from the earliest times to the present. He looked not only at rural areas but also at urban and industrial ones. Christopher Taylor, the editor of the 1988 edition of Hoskins's book, and himself a distinguished landscape historian, credits Hoskins with establishing 'landscape history as a new and proper branch of historical study'. He also says that the book 'has inspired two, and perhaps now more, generations of historians, archaeologists, geographers and botanists to follow in their master's footsteps and to explore the mysteries of our country's landscape'. (Hoskins 1988, 7) Amongst the disciplines that Taylor mentions, it is particularly the archaeologists who in recent years have become involved in the study and conservation of the historical dimension of the landscape.

It is interesting to note that in 1952, shortly before the publication of Hoskins's book, the government purchased the manor and fields of Laxton from its private owner in order to protect it. But this was an isolated incident.

An analysis of the motives of some of the main initiatives for the protection of landscape, such as the Commons Society (founded in 1865, now the Open Spaces Society), the National Trust, the Council for the Protection of Rural England (CPRE), the National Parks and the Countryside Commission, shows that their main concern has been with protecting existing aesthetic character and the provision of public access. The CPRE seems to have been conscious of the historical character of the countryside in the early reports that were produced for it before the Second World War but it was not a major consideration. The character of the landscape was interpreted mainly in terms of its natural features, the pattern of land use and its visual qualities. Since the early 1970s it has become increasingly understood that the historical elements of landscapes are of significance in their own right and are not to be judged only or primarily by the aesthetic contribution that they make. This is an important change of attitude and it will help the attempt to save the remainder of a much depleted heritage.

Figure 12.5

The Great Barn at Holkham Hall, Norfolk. The barn was built around 1790 as a key element of the improvement and development of the agricultural areas of the Holkham estate. It was built for Thomas William Coke (1754–1842), 1st Earl of Leicester, known as 'Coke of Norfolk', who was a nationally celebrated agriculturist.

Figure 12.6

Cottesbrooke Hall, Northamptonshire. A rare example of a fully working (private) walled kitchen garden that survived into the 1980s. This was an important and skilled branch of gardening that has now been generally superseded by commercial horticulture and easy importation of vegetables and fruit from around the world.

By way of definition, land or a landscape can be said to be of historic interest when it meets one or more of three conditions. The first is that it provides evidence (either above or below ground) that is of interest for the study and understanding of the history of the use and modification of land by mankind. It may, for instance, have features that are good examples of a particular type or style of land use or of a particular historical period or designer. It may also have features that are of interest to historians of agriculture, horticulture and other forms of husbandry, to historians of art and design, or to historians of technology and social and economic issues.

The second condition is that it has significant associations with particular people, events or ideas that are of historic interest. The third is that it has group value with buildings, structures, monuments or other land and the group value or the ensemble is of historic interest.

The degree of historical significance will vary from place to place and must be determined by assessment. The geographical context within which the assessment is made is one of the important factors. A local context will be quite different from a regional, national or international one.

The concept of historic interest provides a comparatively objective or scientific criterion to assist in the making of decisions about the conservation of land and landscapes.

Figure 12.7

The Royal Crescent, Bath. Carefully located in the landscape, it was built from 1765 to 1775. The open space, immediately in front of it, is an essential element of its majestic character and provides an open foreground to the view out across the valley of the river Avon.

Figure 12.8

Marston Moor near York. An obelisk commemorates the battle of Marston Moor which took place in this area on 2 July 1644. It was an important turning point in the English Civil War and was won by the Parliamentarians. The area is on the official Register of Historic Battlefields published by English Heritage in 1995.

An important early sign of the official recognition of the historical dimension of the landscape was the publication in 1987 by English Heritage of an archaeological management review called *Ancient Monuments in the Countryside* (Darvill 1987). One of the four main sections of the report dealt with man-made landscapes. English Heritage have also undertaken a study of assessment techniques for the historical aspects of landscape and this is due for publication in 1995/96.[6] In the meantime they have also produced a *Register of Historic Battlefields* which was published in 1995 and contains some forty-three sites (EH 1995b). There are, however, no plans for any further registers of historic landscapes.

In 1994 The Department of the Environment and the Department of National Heritage published their Planning Policy Guidance bearing the title *Planning and the Historic Environment* (PPG 15 1994). In it local authorities are reminded that:

> The Town and Country (Development Plan) Regulations 1991 require authorities to have regard to general environmental considerations in preparing their plan policies and proposals. The protection of the historic environment, whether individual listed buildings, conservation areas, parks and gardens, battlefields or the wider historic landscape, is a key aspect of the wider environmental responsibilities, and will need to be taken fully into account in the formulation of authorities planning policies and in development control. (PPG 15 1994, 3)

This piece of guidance would seem to apply to the whole of England and not just to officially designated areas.

Conservation of the historical dimension of the landscape must now also be seen in the broader context of the UK strategy for the environment which was published in 1990 and the UK strategy for sustainable development which was released in 1994 as part of the government's response to the United Nation's conference on the Environment and Development held in Rio de Janeiro in 1992 (CM 1200; CM 2426).

Historic parks and gardens

Historic interest in relation to gardens and parks is also covered by the definition given above. Since the seventeenth century, at least, parks and gardens have enjoyed a special place in the British way of life, and their history is a very distinguished one. One of the main achievements was the development in the first half of the eighteenth century of the naturalistic style of laying out gardens, grounds and parks. It spread across Europe and other parts of the world and was often referred to as the 'English style'. To some it was just a style of design but to others it was an integral part of an enlightened attitude to life and nature which had strong connections with other aspects of life such as literature and painting.

There has also been the development of a very wide range of garden plants through the importation of plants from all over the world and by the selection and breeding of garden plants within our shores. The process happened gradually over the centuries with bursts of activity from time to time. John Parkinson in his *Paradisi in Sole*, published in 1629, described some 830 plants that were suitable for the garden of pleasure. In 1775 Richard Weston listed nearly 4 000 different kinds of garden plants and in 1935 Kingdom Ward estimated that there were at least 12 000 species in cultivation in British gardens (Weston 1775, 90; Kingdom Ward 1935).

A third achievement is based upon the last. This is the kind of ornamental gardening that is based on growing a wide selection of flowering plants and on arranging them in an informal or naturalistic manner. It is the tradition of British flower gardening that was promoted by William Robinson (1838–1935) from the late 1860s on. It was also championed by Gertrude Jekyll (1843–1932) and others. This tradition is still very strong and is known around the world.

Another achievement has been the widespread popularity of gardening as an outdoor activity in Britain. It is one that is enjoyed by people from all sections of society. Many private gardens and parks are open to the public and visiting them is also very popular. Particularly if they are large or numerous in an area, parks and gardens can make an important contribution to the rural and urban landscape in general. Even if they are not open to the public they may be visible from public roads, footpaths and railways for example. Collectively they are a resource not only from the point of view of amenity but as an element of the cultural heritage and as a reserve of cultivated and wild plants.

Historic parks and gardens have been incorporated into the conservation movement at a fairly late stage. Their importance was recognized in the report of

Figure 12.9

A flower border at Great Dixter, Northiam, Sussex. A striking example of a garden that continues and develops the British tradition of flower gardening.

the Committee on Houses of Outstanding Historic or Architectural Interest. The report was published in 1950 and in the introduction it said, 'when we speak of preserving a "house" we also include the preservation of all the land and buildings which can be said to form part of the composition of which the house is the central feature' (Gowers 1950, 1–2). Although parks and gardens, if attached to a house of outstanding interest were able to benefit from the grant-aid provisions of the Historic Buildings and Ancient Monuments Act 1953 this provision did not have much effect in stimulating the development of a broader policy towards them.

It was not until 1965 that a voluntary society for historic parks and gardens was established. This was the Garden History Society which has played an active and important part since then in promoting and developing both the study and conservation of historic gardens. Its membership in 1995 was about 2 000. Other voluntary organizations which are involved at a national level in this aspect of conservation are the Historic Gardens and Landscapes Committee of ICOMOS UK which was formed around 1976, and the Historic Houses Association which was formed in 1973. Since the 1970s these have been joined by county-based organizations known as County Gardens Trusts. An Association of Garden Trusts was formed in 1993.

In 1974 historic parks and gardens received recognition in law for the first time. This was in the context of the Town and Country Amenities Act 1974 by which they were made eligible in their own right to receive grant-aid, providing they were of outstanding interest. They no longer needed to be attached to an outstanding house.

In 1982 the Centre for the Conservation of Historic Parks and Gardens at the University of York's Institute of Advanced Architectural Studies was set up. Its initial funding came from the Countryside Commission and in its early days it concentrated on initiating a survey and inventory of historic gardens and parks, as none existed. But since the establishment of the official Register for England in 1984 it has devoted most of its time to educational activities, although the inventory remains an active project.

The beginnings of a purposeful and special government programme for historic parks and gardens came with the setting up of English Heritage which came into operation in 1984. The origins of the Register go back to work undertaken by the Historic Buildings Council with assistance from the Garden History Society, ICOMOS UK and the Centre at York University. Under the National Heritage Act 1983, English Heritage was given authority to draw up an official Register of Parks and Gardens of Special Historic Interest in England, which is now the basis of their general conservation programme for historic parks and gardens. English Heritage also directly manages some fourteen historic gardens and parks.[7] The National Trust, by comparison, has 162.[8]

The purpose of the Register is to locate and identify parks and gardens that in a national context are of special historic interest and which merit special protection by those who are in a position to protect them in one way or another. However, no additional statutory protective measures have been created for registered sites as its official status is that of an advisory document. Nevertheless there are existing means which can be, and are, used to protect them and more have been instituted. For example, in 1995 a new procedure was introduced by means of a General Development Order by which local authorities must consult English

Heritage about all planning applications affecting the upper grades (Grades I and II*) of registered sites. They must also consult the Garden History Society (GHS) about all applications affecting all grades of registered sites (Grade I, II* and II) (GHS 1995; 6).

At present (in 1995) there are some 1 210 sites on the Register but the number is gradually increasing.[9] The description of each site is kept deliberately short. Its purpose is to note the main existing features of the site in order to identify it. It does not contain an assessment of the site's relative significance other than a note of its grade. Nor does the description give more than a short outline of the site's history.

Between 1984 and 1988 a first edition of the Register was published for each of the forty-six counties of England. During a similar period an official Inventory of historic parks and gardens was also established in Scotland. It currently includes some 275 sites. In 1989 a Register was initiated by Cadw for Wales. At present (September 1995) there are some 129 sites in the published volumes for the two Welsh counties that have been completed.

It has been suggested that a new organization might be established in connection with the conservation of historic parks and gardens. A feasibility study presented its findings in March 1995 (Banks 1995). On reflection there seem to have been three main general issues. One was the need for an organization that will take the lead in promoting the conservation of historic parks and gardens on a national basis and in a pro-active way. Another was the need for an organization that will develop liaison and co-ordination between existing organizations, and the third was the need for an organization that will advise on conservation and repair schemes and whose role will include the putting together of grant-aid packages from the different available sources. Later in 1995 it was decided that these roles could be fulfilled by developing existing organizations, with English Heritage providing the lead.

The system by which historic parks and gardens are protected and conserved in the UK is complex because many different organizations play a part, but there has already been a very beneficial degree of cooperation between them. For example, English Heritage has encouraged the involvement of non-governmental organizations in the preparation of the Register and in conservation case-work. It is to be hoped that the network that is forming will continue to develop and that a cooperative approach will be fostered.

CONCLUSION

Although they are often grouped together and although they have much in common, there are also big differences between parks and gardens, on the one hand, and other kinds of landscape, on the other. Parks and gardens cover a much smaller proportion of the land surface of the nation and come within a tradition that is distinctive by being heavily influenced by horticulture and ornamental design. In England the conservation of historic parks and gardens and that of the historical aspects of the wider landscape are developing along separate lines with the former being linked to historic buildings and the latter to archaeological conservation. This division may well persist but mechanisms are developing by which they can coordinate with each other. Archaeologists are taking more interest in parks and gardens, and garden historians are becoming more appreciative of archaeology.

A mechanism, in the form of the Countryside Commission's Countryside Character Programme, is also developing whereby on a national scale the historical dimension of the landscape is being related to scenic and natural aspects. In this way, a broader understanding of the cultural dimension will, it is hoped, emerge.

BIBLIOGRAPHY

Banks (1995). Elizabeth Banks Associates. *Feasibility Study for an Organization to Assist with the Conservation of Historic Parks and Gardens*, prepared for the Garden History Society.

CC (1993). Countryside Commission. *Landscape Assessment Guide.* CCP 423.

CC (1994a). Countryside Commission. *The New Map of England: A Directory of Regional Landscapes.* CCP 445.

CC (1994b). Countryside Commission, The New Map of England: A Celebration of the South Western Landscape, CCP 444.

CC (1995). Countryside Commission. *Countryside Facts and Figures.* CCP 339.

Command Paper 1200 (1990). *This Common Inheritance. Britain's Environmental Strategy.* HMSO, London.

Command Paper 2426 (1994). *Sustainable Development. The UK Strategy.* HMSO, London.

Cox, J. C. (1905). *The Royal Forests of England.* London.

Cultural Heritage Committee (1994). Cultural Heritage Committee, Group of Specialists on heritage landscapes and sites. *Draft Recommendation of the Committee of Ministers to Member States on the Integrated Conservation of Cultural Landscape Areas as Part of Landscape Policies.* Council of Europe, Strasbourg, July 1994; CC-PAT (93) 80 (rev. 3).

Darby, H. C. (1936). *An Historical Geography of England before 1800.* Cambridge.

Darvill, T. (1987). *Ancient Monuments in the Countryside.* English Heritage Archaeological Report No 5, Historic Buildings and Monuments Commission for England, London.

Drake, F. (1736). *Eboracum or the History and Antiquities of the City of York.* London.

EH (1995a). English Heritage. *Conservation Bulletin*, xxvi (July).

EH (1995b). English Heritage. *Register of Historic Battlefields.*

English Nature (1994). English Nature. *Natural Areas* (July) (a map).

GHS (1995). Garden History Society, *Newsletter* 44 (Summer).

Gilpin, W. (1782). *Observations on the River Wye and several parts of South Wales, etc., relative chiefly to Picturesque Beauty.* London, republished (1973) The Richmond Publishing Co.

Goodchild, P. (1990). *Some Principles for the Conservation of Historic Landscapes.* A draft document for discussion purposes, Institute of Advanced Architectural Studies, University of York, The King's Manor, York.

Gowers, Sir Ernest (Chairman) (1950). *Report of the Committee on Houses of Outstanding Historic or Architectural Interest.* HMSO, London.

Guthkelch, A. C. (1914). *The Miscellaneous Works of Joseph Addison* (2 vols.). London.

Hill, W. T. (1956). *Octavia Hill, Pioneer of the National Trust and Housing Reformer.* London.

Hoskins, W. G. (1988). *The making of the English Landscape.* (C. Taylor, ed.), Hodder and Stoughton, London.
ICOMOS (1993). ICOMOS Landscapes Working Group, *ICOMOS UK Newsletter* (January). This *Newsletter* contains a report on a meeting on the subject of cultural landscapes which was organized jointly by UNESCO World Heritage Centre and ICOMOS and was held at La Petite Pierre, France, on 24–6 October 1992. There is also a report on the 16th Session of the World Heritage Committee held at Santa Fe, USA, on 7–14 December 1992. A copy of the *Newsletter* may be seen at the headquarters of ICOMOS UK, 10 Barley Mow Passage, London W4 4PH.
Jekyll, G. (1904). *Old West Surrey.* Longmans, London.
Kingdom Ward, F. (1935). *The Romance of Gardening.* Jonathan Cape, London.
Lees-Milne, J. (1948). *The National Trust. A Record of Fifty Years' Achievement.* Batsford Ltd, London.
Orwin, C. S., and C. S. (1938). *The Open Fields.* Oxford University Press.
PPG 15 (1994). Department of the Environment, Department of National Heritage. *Planning Policy Guidance: Planning and the Historic Environment,* PPG 15 (September).
Shaw-Lefevre, G. (1894). *English Commons and Forests.* Cassell and Co. Ltd, London.
UNESCO (1981). *The World Heritage Convention.* Paris.
Weston, R. (1775). *English Flora or a Catalogue of the Trees, Shrubs, Plants and Fruits, Natives as well as Exotics.* London.

NOTES

1 These two definitions might be compared with that given in the draft recommendation of 1994 (Cultural Heritage Committee 1994, 12) which describes landscape as the 'Formal expression of the numerous relationships existing in a given period between the individual or a society and a topographically defined territory, the appearance of which is the result of action, over time, of natural and human factors and of a combination of both. Landscape is taken to have a threefold cultural dimension, considering that:
- it is defined and characterized by the way in which a given territory is perceived by an individual or community;
- it testifies to the past and present relationships between individuals and their environment;
- it helps to mould local cultures, sensitivities, practices, beliefs and traditions.'

2 For an explanation of the concept of land cover see CC (1993, 5, 6).
3 Information contained in a letter to Peter Goodchild from David Brooke, Director of the Countryside Character Programme, Countryside Commission, 3 April 1995.
4 This description of conservation may be compared with the definition given in the draft recommendation of 1994 (Cultural Heritage Committee 1994, 12) which defines conservation as 'The dynamic application of appropriate legal, economic and operational measures to preserve specific assets from destruction or deterioration and to safeguard their future'.
5 See for example, Gilpin (1782).
6 Information from English Heritage, September 1995.
7 Information from English Heritage, June 1995.
8 Information from the National Trust, July 1995.
9 Information from English Heritage, June 1995.

13

FINANCING CONSERVATION

David McLaughlin

Financing the conservation of historic buildings has seen significant changes and developments in recent years. These changes have thrown up far-reaching issues, some more contentious than others, which bring into question established principles. As a result, financing conservation is now in a period of great flux. This chapter seeks to identify the major changes, set them in their historic context, and open a debate on the issues they raise.

In the past, financing conservation was much more building centred. Now the emphasis appears to be shifting towards the uncertainties of the market place. Whereas everyone can understand that there is a finite amount of money available for the care of the built heritage which must be used in the most effective way possible, the issue of balancing the conservation needs of the nation's built heritage with its economic viability is a new concept which has yet to be explored in practice. Financing conservation also continues to be concerned with the question of balancing the huge financial demands of the grand project with the demands of the great variety of more modest projects for the care and repair of the majority of the nation's historic buildings.

Perhaps insufficient attention was paid to the overall viability of historic buildings in the past. However, what we must now ensure is that the emphasis does not swing too far in the opposite direction, losing sight of positive aspects of past practice.

Conservation policies are now in a period of change. As a result some anomalies have already been identified which need to be looked at as a matter of urgency if the overall strategy is going to be effective. It is very complex. The needs of historic buildings themselves can be in conflict. What may be good for one set of buildings may not be good for another. One of the problems is trying to establish general principles to apply to individual historic buildings which in turn have their own widely varying individual circumstances; be it condition, use, occupancy, value, ownership or charisma. Charisma? Is it the buildings with charisma – the big name architect, the inherent beauty, the technological feat, the rarity or scarcity value, the spectacular untouched setting of either exterior or interior or both, or the intrinsic and immeasurable value of past occupancy of a great personage – that bestows the magical qualities which naturally draws the attention of conservation campaigners to crusade for their repair at whatever the cost, even where there is absolutely no chance of an economically viable use over the horizon?

The problem for English Heritage, as central government's principal distributor of grant money for the repair of the historic built environment, is that so many different interest groups are scrabbling for a share of a comparatively tiny pot of money. Although the size of the 'tiny pot' has more than doubled in English Heritage's first ten years (from 1984 to 1994), it is not enough to meet current demands.

Recent changes in both central government legislation and guidance have also had a significant impact on financing the care of historic buildings. Notable among these is the impact of Value Added Tax (VAT) on the repair of historic buildings. Conservation bodies seem agreed that the imposition of a 17.5 per cent tax (the VAT rate at the time of writing in 1995) for the repair of historic buildings is wrong and ill-conceived, especially with the knowledge that works of structural alteration which both require and receive listed building consent are zero-rated. It is all the more ironic that if listed building consent were to be granted for the demolition or partial demolition of a listed building, the cost of demolition would be zero-rated. The irony is compounded by the fact that the 're-building' of a listed building with only its foundations or basement remaining along with one or two of its external walls, counts as new construction and is therefore zero-rated for VAT purposes.

The retention of one or two façades may at first appease those anxious to preserve selected vistas within a townscape. But retaining external walls as fragments of an earlier historic structure is a superficial solution in every sense. Earlier uses are lost. The economic base of the neighbourhood may be altered beyond all recognition. The deeper dilemma is that façadism can be seen as an easier route to attract institutional funding as well as having the in-built financial incentive of the new construction being zero-rated for VAT purposes versus the taxed repair of retained historic structures. In addition, larger, more flexible spaces with higher permitted floor loadings in new construction behind retained façades, with clearly and tightly defined building programmes, are perceived by traditional institutional investors as benefits with known returns against financial investment.

To nurture the same listed building into a full, beneficial, and maybe even economic use, with however gentle and conservative repairs, would find the repairs liable to VAT. The anomaly is not lost to conservation campaigners who continue to press for change. In an era where notions like 'value for money' and 'grant leverage' are part of the everyday language of those involved in financing conservation, it is depressing to find that hard-pressed grant budgets are partially diverted into grant-aiding the imposition of VAT on repair works and their associated professional fees. Instead, why not zero rate repair works and associated professional fees on the repair of historic buildings where grants have been offered? This would have the effect of stretching grant budgets further. The grant-offer conditions could include a VAT exemption certificate related to the grant eligible works. Certainly, at the present time, the imposition of VAT on the repair of historic buildings acts as a disincentive for the proper maintenance and repair of historic buildings.

Hopefully the European ministers concerned with the heritage will address this issue at their meeting in Helsinki in 1996.

ECONOMIC VIABILITY

However, an as yet uncharted change in emphasis, but one with profound implications for the future of historic buildings, is the linking of conservation and economic prosperity in *Planning Policy Guidance: Planning and the Historic Environment* (PPG15) published jointly by the Department of the Environment and the Department of National Heritage in September 1994. The subject of the economic viability of historic buildings was raised for the first time in PPG15. The question of economic viability, balancing possible uses against changes, could become the core policy for listed buildings everywhere and for unlisted buildings that make a positive contribution to conservation areas.

PPG15 states the government's view:

> Most historic buildings can still be put to good economic use in, for example, commercial or residential occupation. They are a valuable material resource and can contribute to the prosperity of the economy, provided that they are properly maintained: the avoidable loss of fabric through neglect is a waste of economic as well as environmental resources. In return, economic prosperity can secure the continued viability of conservation areas, and the continued use and maintenance of historic buildings, provided that there is a sufficiently realistic and imaginative approach to their alteration and change of use, to reflect the needs of a rapidly changing world. (PPG15 1994, 2, para. 1.4)

But where is this 'realistic and imaginative approach to . . . alteration and change of use' to come from? Certainly local planning authorities will continue to encourage applicants themselves to adopt 'a sufficiently realistic and imaginative approach' to the re-use of historic buildings. The implication appears to be that local planning authorities will be expected to exercise greater flexibility 'to reflect the needs of a rapidly changing world' when dealing with applications for listed building consent and planning permission for change of use.

Of course 'the continued use and maintenance of historic buildings' must be economically viable if their 'continued use and maintenance' is to happen. But how, and by whom is economic viability to be assessed? PPG15 states:

> In principle the aim should be to identify the optimum viable use that is compatible with the fabric, interior, and setting of the historic building. This may not necessarily be the most profitable use if that would entail more destructive alterations than other viable uses. (PPG 15 1994, 8–9, para. 3.9)

Does this imply the preparation of an 'economic impact analysis' paralleling an environmental impact analysis? For example, what if the proposed change of use leads to increased traffic generation and, consequently, increased pollution levels? Planning applications have already been refused by local planning authorities on the grounds that the traffic generated by the proposals would cause unacceptable pollution levels. Will the cost of pollution be put into the equation and if so, how? Does the consideration of 'the optimum viable use' imply the preparation, analysis and comparison of alternative schemes? If so, who is to pay and how much is this going to draw out the whole process of trying to bring empty historic buildings back into use?

> The Secretaries of State would not expect consent to demolition to be given simply because redevelopment is economically more attractive to the developer than repair and re-use of a historic building, or because the developer acquired the building at a price that reflected the potential for redevelopment rather than the condition and constraints of the existing historic buildings.
>
> (PPG15 1994, 10, para. 3.17)

Will the judgement of a listed building's economic viability push the consideration of its special architectural or historic interest into second place? After all, the building was listed in the first place for being of special architectural or historic interest, not for its economic viability. Conservationists will undoubtedly be carefully watching and monitoring how the question of economic viability is assessed by inspectors at future public inquiries and how, in turn, the Secretary of State responds to the inspectors' reports.

Financing conservation is inevitably bound up with the value of conservation. How is the value of conservation to be measured? Jennifer Page chose the economic context of the heritage as the focus of a valedictory article before leaving her post as Chief Executive of English Heritage in February 1994. Writing in English Heritage's *Conservation Bulletin* she commented:

> Politicians, managers and economists are trying to put heritage concepts into an economic as well as a social and political context – thus providing an intellectually respectable and sustainable basis on which to assess the contribution to wider social and economic objectives gained by spending resources on the heritage. The Council of Europe plans to include in the fourth Conference of European Cultural Ministers, to be held in Helsinki in 1996, the theme of the economic impact of heritage policies.
>
> The wide scope of the heritage makes its economic valuation so much more complex and therefore more prone to failure even than environmental or cultural economics. . . .
>
> English Heritage is leading the way in the UK in thinking about the economics of conservation. Following a series of seminars in 1990, bringing together English Heritage and the Royal Institution of Chartered Surveyors (RICS), the two organizations agreed to work together to establish some economic facts in place of hearsay and prejudice. The first product was joint EH/RICS research on the investment performance of listed office buildings [IPD 1993]. As a result of a second study on the impact of listing on capital values [RICS 1994] – funded by English Heritage, the Royal Institution of Chartered Surveyors and the Department

> of National Heritage – the three organizations are now researching an effective methodology for valuing the wider social and economic impact of conservation policies on the urban environment. (Page 1995, 14–15)

These first two studies focused on the potentially negative effect that the listing of a building in commercial use might have on limiting the options of the building owners, users and investors to alter or demolish it as they seek to maximize property values.

In preparing to research 'an effective methodology for valuing the wider social and economic impact of conservation policies', English Heritage and the Department of National Heritage commissioned a literature review of existing economic appraisal systems, dealing principally with the natural environment and heritage visitor sites (Reading/DTZ 1995).

At a seminar held in June 1995 to consider the literature review, conservation practitioners questioned the limitations of the economic appraisal systems that were being suggested as models for the potential assessment of the concept of the value of conservation (EH 1995). The contrasting outlooks of the researchers and practitioners was highlighted by their different approaches and attitudes to the problems and ethics of valuing the built heritage. The researchers wanted to devise a comparative system from scratch; the pragmatic view of the practitioners was that a framework for a comparative system is already in existence in the form of listed buildings, conservation areas, ancient monuments and World Heritage Sites.

Far from starting from scratch, this comparative system is already well established, using tried and tested criteria. Enshrined in legislation and guidance, publicly accountable and subject to review and consultation, it is already widely used by English Heritage to direct access to its different grant schemes.

GRANT-AID

How have these different grant schemes evolved? It is now over forty years since the introduction of national legislation to grant-aid the repair of historic buildings in England. This was the Historic Buildings and Ancient Monuments Act 1953. The work of the Department of the Environment (DoE) and its predecessors in implementing the Act were guided by the Historic Buildings Council for England (HBC). The thirtieth anniversary of the Act was marked by the National Heritage Act 1983 which transferred the powers of the DoE and HBC to a newly established quango, the Historic Buildings and Monuments Commission for England, better known by is popular title, English Heritage (EH) from 1 April 1984.

The first ten years of English Heritage's work has seen grant offers more than double from a total of £21 400 000 in 1984–85 to £50 700 000 in 1993–94. At the same time the categories of grant schemes have been widened from the earlier grants for individual outstanding secular historic buildings and monuments to include churches, conservation areas, cathedrals and historic gardens.

A committee appointed by the Chancellor of the Exchequer and chaired by Sir Ernest Gowers reported in 1950 on *Houses of Outstanding Historic or Architectural Interest* (Gowers 1950). The report proposed that owners of outstanding historic buildings should be eligible for tax relief and relief from death duties to help offset the cost of repairs and, in return, allow limited public access. Although these proposals were not accepted, the Gowers committee report led to the idea of grant-aiding the repair of outstanding historic buildings that was embraced in the Historic Buildings and Ancient Monuments Act 1953. This Act remains the backbone of legislation enabling English Heritage to grant-aid the repair of outstanding historic buildings and monuments today.

Paradoxically, it was the same legislation that enabled the first Town Scheme in the country to be set up in 1955 which addressed the problems of groups of buildings in historic areas as opposed to an individual outstanding historic building. This was the Bath Terraces Scheme for the cleaning and repair of the front façade stonework of the Circus, a group of thirty terraced houses. The scheme lasted over a period of twenty years. This was the first formal expression of the concept of a partnership between building owners, local authorities and English Heritage's predecessors for jointly funding the repair of historic buildings in historic areas.

Reporting in 1963, the Preservation Policy Group (1970) recommended five historic towns for detailed conservation studies: Bath, Chester, Chichester, King's Lynn and York. One town, King's Lynn, declined to take part. The remaining four towns commissioned studies in partnership with the then Ministry of Housing and Local Government. Published in the late 1960s, the studies led directly to concerted, grant-aided conservation activity in each town from the removal of non-conforming users to encourage the repair and re-use of adjoining historic buildings to straightforward building repairs, the enhancement of streets by coordinating the use of natural traditional paving materials, sympathetic lighting schemes and street furniture, traffic management and pedestrianization, and the removal of a plethora of overhead wires, unauthorized advertisements and redundant traffic signs.

Although local authorities were enabled to grant-aid historic buildings under the Local Authorities (Historic Buildings) Act 1962, the real impetus for the repair of buildings in historic areas came with the introduction of Section 10A grants for the repair of buildings in conservation areas and the wider use of Town Schemes under Section 10B of the Town and Country Planning (Amendment) Act 1972, as amended.

Canterbury began a modest Town Scheme in 1971 covering 416 buildings. In 1975, the Canterbury Town Scheme was increased to cover 1180 buildings and extended again in 1977 to cover all 1480 listed buildings in the City conservation area. For the first three years of its operation the Town Scheme in Canterbury was underspent. While this may at first appear disheartening, the Canterbury experience is borne out time and again in other historic towns. It takes time to establish the viability of any grant scheme, not only in persuading the first building owners to take part but also to convince them to repair their buildings to a much higher standard than they would have been able to without grant-aid. These early projects become the exemplars of both the benefits of the conservation process – repairing the built fabric and bringing empty space back into use – and of excellence in conservation practice.

While most initial grant enquiries were about roof repairs, close inspection by the Conservation Surveyor revealed the extent of fundamental structural problems which might otherwise have gone unnoticed until a building neared collapse. This experience is borne out in other historic areas. Identifying problems and helping building owners understand the need to tackle them is a fundamental role for local authority conservation teams and requires extensive skills and knowledge.

The financing of conservation projects in the past has not been without its problems. Traditionally, building and the repair of buildings has been a seasonal activity, a process carried out in the frost-free months of April to November. At first glance the financial year, 1 April to 31 March, would appear to tie in well with that seasonal process. However, the process of grant-aiding the repair of historic buildings reveals the artificiality and limitations of the financial year. The realization and acceptance by a building owner that their building is in need of repair can be a slow process in itself. Persuading a building owner to properly repair their building is also part of the process. All this takes time. Building owners are understandably reluctant to commission architects or surveyors to prepare the project documentation necessary for the conservation work until they have some assurance that the budget is in place or indeed that a grant offer has been made. But with grant budgets not known or confirmed until after the financial year has started there is an inevitable delay and many projects are unable to start on site until late summer or autumn.

In 1983–84, there were 180 Town Schemes in operation in England with an allocation of £1 800 000 matched on a pound-for-pound basis by the local authority partners. After 1 April 1984 English Heritage became responsible for grants, and in 1993 to 1994 Town Schemes were allocated £4 800 000 and buildings in conservation areas £6 100 000.

CONSERVATION AREA PARTNERSHIPS

During 1993, Town Schemes and grants for buildings in conservation areas began to be phased out by English Heritage in favour of Conservation Area Partnerships, a more targeted and measurable grant system for historic areas.

In March 1994, English Heritage launched a pilot scheme of 16 Conservation Area Partnerships to revitalize areas of neglect and decay in historic towns. In May 1995, English Heritage announced Conservation Area Partnerships for 1995–96 with an additional 111 local authorities across England. They also wrote to all 410 local authorities in England inviting them to submit their applications for further Conservation Area Partnerships in 1996–97.

The whole process of bidding for Conservation Area Partnerships is on a very tight timetable with information on the forthcoming year's partnership schemes theoretically available from English Heritage during April of the preceding financial year. In practice this may differ with all parties understandably tentative. In

1995, for example, preliminary information was sent out to most local authorities in mid-May while the closing date for the submission of preliminary applications remained the last day of June, a period of just six weeks. In addition, English Heritage may also review their guidance of Conservation Area Partnerships, as they did in 1995. It is therefore extremely unwise for local authorities to launch into the preparation of a bid before having received the up-to-the-minute current guidance from English Heritage. This is especially true during these early days of Conservation Area Partnerships as guidance will undoubtedly continue to be developed following the review of the pilot Conservation Area Partnership schemes.

Six weeks is not long to prepare an application for a Conservation Area Partnership given the range of information required by English Heritage who need to assess the preliminary applications and consider their budgetary implications and decide which they can support. Bidding for the resources for a Conservation Area Partnership requires the ability to demonstrate a real commitment on the part of a local authority to conservation. Applications for Conservation Area Partnerships need to give enough information for English Heritage to understand the special architectural or historic interest of the conservation area, the problems it faces, typical repair costs and property values, funding requirements and the local authority's own approach to conservation and other problems.

English Heritage's 'main aim in introducing the new partnership arrangements has been to simplify and clarify the use of English Heritage funding for work in conservation areas and to make it more effective by targeting areas of greatest need. By April 1997, the greater part of English Heritage's budgets for work in conservation areas will be committed to partnership schemes, leaving only a relatively small proportion of the available funding for direct action by means of conservation area ... grants.' (Embree 1995, 8).

English Heritage intend to focus those conservation area grants on Buildings at Risk grants as well as supporting local authorities serving urgent works notices, preparing action plans or establishing new conservation officer posts for the first time. English Heritage's 1989 nation-wide initiative on Buildings at Risk encouraged local authorities to review both the external condition and occupancy of more than a third of the estimated half-million listed buildings in England.

As an example, Bath responded to this initiative, carrying out an augmented buildings at risk survey of 650 listed buildings in 1989, continuing the survey in 1991 and completing it in 1993. Bath had also carried out its own listed building condition survey in 1975. All the surveys have been used to target grants more effectively.

The positive impact of a continuing and effectively targeted campaign of grant-aid in Bath over the fourteen years between 1975 and 1989 is clearly shown by the dramatic increase in the number of buildings in fair and good condition from 1975 to 1989. At the same time the diagram emphasizes the need for a continuing and sustained programme of grant-aid to tackle the repair of the remaining historic buildings in very bad or poor condition in an effective and equitable way. The comparison of the diagrams emphasises the time it takes to establish and maintain effective grant programmes to begin to make an impact in historic areas.

Figure 13.1

Comparison of listed building condition surveys in Bath in 1975 (left) and 1989 (right).

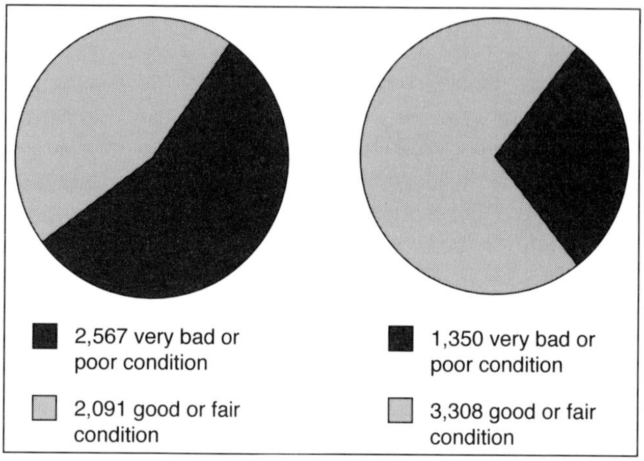

Grant partnerships, their targeting, implementation and conditions are not the only things to have changed over this period. The economy itself has seen severe fluctuations with the falling value of the pound and a very volatile property market leaving many domestic buildings now in negative equity ownership. This means that many building owners are unable to afford to carry out basic maintenance, let alone the comprehensive structural repairs the need for which has become apparent since the purchase of their property. Not everyone has access to the resources necessary for the care of their building. For example, just because the value of the building would enable money to be borrowed against it does not mean that the building owner would be able to repay monies advanced as a loan or mortgage.

Indeed, the cost of repairs, even with the benefit of grant-aid, is not necessarily reflected in an enhanced value of the building. There is a fundamental difference between looking at buildings just as buildings, and looking at buildings as homes. A home is where someone feels safe and secure. It has to be affordable. It has to be suitable to the needs of the occupants.

It therefore seems inconsistent that English Heritage may decline to grant-aid repairs to certain buildings in functional uses such as private homes which may have 'a significant open market value' (EH 1994a) but consider that buildings and monuments with important historic collections or that form part of an estate cannot be judged by market values. In such cases English Heritage review the outgoings of the estate in assessing the need for possible grant-aid (EH 1994a).

Conservation Area Partnerships mark the way for local authorities to take on a true commitment to conservation, literally putting their money where their mouth is when it comes to grant budgets. In addition to encouraging local authorities to adopt a more robust view of all their conservation policies and actions, the delegation of the running of Partnership schemes to local authorities makes a very cost effective way to tackle the repair of large sections of the nation's historic fabric.

There is no doubt that English Heritage's Conservation Area Partnership initiative has the potential to be a great success. Having invited all 410 local authorities in England to submit applications during 1995 for further Conservation Area Partnerships in 1996–97, English Heritage received bids totalling £10 700 000 against a potential budget of £3 350 000.

It would be interesting to know how this compares with bids for grants to outstanding historic buildings and churches in relation to the respective amounts of money allocated to each.

The whole process is a delicate balancing act. English Heritage are justifiably proud that in their first ten years they have managed to increase overall grant budgets from £21 400 000 in 1984–85 to £50 700 000 in 1993–94.

OUTSTANDING HISTORIC BUILDINGS

There is never enough money to do everything and so there has always had to be some form of prioritizing. One of English Heritage's principal concerns at national level has been with the most important buildings. First because of their relative quality, and second because of the sheer size of many of them, the cost of putting them into repair tends to be relatively high. English Heritage apply three tests in considering a grant application for Grade I and Grade II* buildings. The first test is how important the building or monument is. English Heritage's criterion is that a property should be protected as a Scheduled Ancient Monument or as a Grade I or Grade II* listed building. English Heritage also invite owners of Grade II listed buildings or unlisted properties which may not have been recognized during the preparation of earlier lists to apply for grants. English Heritage would then look again at the grade of the building.

The second test applied by English Heritage is in relation to the nature and urgency of the work. Grants are only considered for urgent major repairs and not routine maintenance and minor repairs. Normally grants are aimed at repairs which are essential within two or three years but they may cover other works of repair which, while less urgent, are best carried out at the same time as the urgent repairs. Phased repair programmes can be agreed where extensive repairs are needed to one or more buildings, but while taking into account the extent of the overall repairs, English Heritage will only grant-aid the most urgent work.

The third test applies to financial help. Grant applicants need to supply English Heritage with the nature of their interests; whether it is freehold, leasehold or a tenancy; the date and cost of acquiring the interest in the property; the approximate market value of the property when in good repair; the current use of the building; the proposed use after repairs; the outcome of any application for other grants or loans; and, details of any planning permission or any development in respect of the property or land received in the last five years. This information is then used by English Heritage in their consideration of needs assessment (reviewed in Johnson 1992, 9–10).

The National Audit Office in their report 'Protecting and Managing England's Heritage Property' (1992, 16) note that:

> Although in individual cases English Heritage know the contribution of grant to the total cost of a project, they do not collect information about the total value of private investment 'levered in' by their grant schemes or how this varies between cases. There are problems in doing this – for example in assessing the extent of genuine leverage as grant may stimulate work on neighbouring properties. A practical first step to help assess the range of performance achieved might be to set for each grant scheme a target for the amount of direct private investment to be levered in for each £1 of grant spent. But English Heritage emphasise that sometimes – in cases where buildings at risk have limited economic value or potential – relatively high levels of grant were likely to provide the only means of securing preservation, the main aim of grant.

The success of conservation grant money in 'levering' more money into the effective and equitable repair of the built heritage should be measured. However, it must be remembered that the cost effectiveness of grant schemes is not only measurable in financial terms. Other indicators could include the amount of space brought back into beneficial use, the number of dwellings created, the amount of lettable space created and the number of historic buildings brought into proper repair.

In July 1995 English Heritage announced a review of grant conditions for grants for Scheduled Monuments and Grade I and Grade II* buildings. Some of these, for example access arrangements, address concerns expressed in the 1992 National Audit Office report. Others, for example notification procedures, aim to remove anomalies. Altogether, they mark a change in English Heritage thinking.

The most significant and long-awaited of these revised grant conditions must certainly be the emphasis on maintenance (Hawkins 1995, 16). This is surely the key to the future well-being of a building after extensive repairs. Conservationists will be eagerly awaiting the proposed 'maintenance check-list'. It will be interesting to see how this will work in practice and to what degree the feedback from the results of the maintenance checklist being used might influence a greater emphasis on continuing maintenance as opposed to dramatic major repairs.

But for many large, long-running and, at first sight, insoluble, problematic historic buildings, the intervention of long-awaited dramatic repairs may well be the last chance for their rescue. In recent years English Heritage have increasingly sought to grant-aid the major outstanding (both in historic quality and degree of problem) secular and ecclesiastical historic buildings.

In addition to continuing to grant-aid eligible works of repair, English Heritage are increasingly using grant mechanisms to establish alternative ways to repair and secure the future of historic buildings and structures. Among the most spectacular are the Ribbleshead Viaduct, St Ann's Crescent in Buxton, the Anderton Boat Lift in Cheshire, and Pell Wall Hall in Shropshire.

The Ribbleshead Viaduct is a major engineering structure on the Settle to Carlisle railway line. British Rail, anxious that the cost of repairs to the viaduct

Figure 13.2

The 1875 Anderton Boat Lift, near Northwich, Cheshire, lifted boats the 50 feet from the Weaver Navigation to the Trent and Mersey Canal.

would be between £2 700 000 and £4 300 000 used the repair costs as part of its case for closing the line in 1987. Following a trial repair suggested by an English Heritage engineer and their advice that the repairs of the viaduct would not exceed £2 700 000, English Heritage subsequently offered a grant of £1 000 000 towards the repair of the viaduct helping secure the future of the line.

High Peak Borough Council acquired St Ann's Hotel, The Crescent, Buxton in August 1993 with grant-aid from the National Heritage Memorial Fund. This followed the closure of the hotel in 1989. The remainder of The Crescent is owned by Derbyshire County Council who placed it on the market in 1991. The two authorities prepared a detailed Development Brief in consultation with English Heritage in 1994. English Heritage have made a 100 per cent grant of £1 000 000 to the Borough Council to enable them to undertake comprehensive structural, roof and stonework repairs to the former hotel.

English Heritage have offered £500 000 towards the repair of the boat lift, not enough to repair the boat lift itself but enough to persuade British Waterways into creating a trust to raise the funds to do so.

The Pell Wall Preservation Trust has been offered a grant of £1 000 000 to rescue Sir John Soane's 1822–8 Pell Wall Hall, near Market Drayton, Shropshire. This follows the exemplary and tenacious work of the local authority, North Shropshire District Council, in serving a repairs notice on the former owner in 1985 followed by a compulsory purchase order. Minimum compensation of £1 was agreed by the Secretary of State for the Environment in 1988 after a fire in 1986 and two years of court cases. The building envelope will be fully repaired and the house then sold on with the purchaser required to reinstate the main rooms to Soane's plans.

In 1990 English Heritage were allocated an additional £11 500 000 for three financial years to establish a Cathedrals Grant Scheme. To gather information on the care and management of Cathedral fabric and to assess and predict repair needs over the next decade English Heritage commissioned a survey of all sixty-one Church of England and Roman Catholic Cathedrals in England. By 1994 grants totalling £15 000 000 had been offered to fifty-four of the sixty-one English Cathedrals. As one might expect, roof and stonework repairs predominate.

However, the Roman Catholic Metropolitan Cathedral of Christ the King in Liverpool posed problems with modern materials. English Heritage have offered a grant of up to £1 500 000 over three years towards repairs estimated to cost £6 000 000. Listed Grade II* and one of only seven post-war ecclesiastical buildings to be listed, the completed building was consecrated in May 1967. The decision has been taken to overclad the mosaic-covered concrete ribs with panels of glass reinforced plastic in the hope that in the future the mosaic finish can be reinstated to the concrete ribs.

Grant-aid to churches is set at its highest level yet, at £13 400 000 in the 1994–5 financial year. As ever, there are competing needs for English Heritage funding and a Needs Survey of a selection of churches in five dioceses is in progress, jointly run and funded by English Heritage and the Church of England, in an attempt to fix more accurately the likely future demand for grants. Discussing future strategies between English Heritage and the Church of England, English Heritage set out a framework that they will use to 'approach ecclesiastical buildings of whatever denomination'.

English Heritage will continue to offer high-percentage grants for urgent repairs whenever there is a reasonable chance ecclesiastical use will continue; every effort should be made to assist continuing church ownership and use. English Heritage must sometimes be prepared to accept radical changes to some highly graded buildings so that ecclesiastical use and ownership can continue and will assist church authorities to bring conservation and heritage issues to bear in considering long-term strategies for their buildings. Where ecclesiastical use and ownership is impossible to maintain, English Heritage will seek to collaborate with others to create a strategy for re-use, to protect the essential use of the building in its context (Halsey 1995, 17).

While English Heritage is often the first port of call for applicants, other bodies also provide varying types and amounts of financing for the conservation of historic buildings. While not a comprehensive list, these bodies include local authorities, building preservation trusts, the Architectural Heritage Fund, the Council for Care of Churches, the Society for the Protection of Ancient Buildings, private trusts and foundations – the Gulbenkian Foundation, Manifold Trust, Pilgrim Trust, Getty Foundation, Railway Heritage Trust and the Historic Chapels Trust – the National Heritage Memorial Fund, the European Community Fund for pilot schemes for the architectural heritage and, most recently, the National Lottery.

Other bodies, such as the National Trust, seek to provide a secure future for historic buildings and landscapes through a careful combination of appeals, legacies and grant-aid to enable essential repair works to take place in association with carefully considered access and interpretation schemes while securing an endowment for the future care of the site in perpetuity.

LOTTERY FUNDING

The majority of these well-established agencies operate within tried, tested and accepted criteria in their different approaches to promoting the care of the built environment. The National Lottery is not only one of the newest arrivals on the conservation scene but one with an as yet unknown potential for financing the conservation of historic buildings.

National Lottery tickets first went on sale in November 1994. After deducting prize money, administrative costs and tax, the balance of the money is split between five recipients – charities, heritage, arts, sports and the Millennium Fund. The National Heritage Memorial Fund distributes the heritage share, the Heritage Lottery Fund, with the advice of English Heritage and other publicly-funded advisory bodies.

The National Heritage Memorial Fund was established under the National Heritage Act 1980 and now administers two funds, the Heritage Memorial Fund and the Heritage Lottery Fund. The Heritage Memorial Fund operates 'as a fund of last resort to preserve, acquire and maintain items of outstanding importance to the national heritage which have a strong memorial character' (NHMF 1994a).

> The Heritage Lottery Fund will fund the acquisition, conservation, restoration or enhancement of land, buildings or items of importance to the local, regional or national heritage. Heritage areas include historic buildings; landscape of scenic or scientific interest; museum and gallery collections; manuscripts, books and archives; and industrial, transport and maritime history. Eligible projects include acquisition, repair, conservation or restoration of buildings, land, objects or collections of importance to the heritage; new buildings, or improvements to existing buildings, designed to house museum or archive collections of importance to the heritage; and improving public access to, and understanding and enjoyment of, such buildings, land, objects and collections. Large and small projects will be funded with an emphasis on consolidating the current heritage stock. Applicants must be public sector organizations, charities or voluntary organizations which include the preservation of the heritage or nature conservation among their aims. Private individual or commercial organizations are not eligible. (NHMF 1994b)

Reviewing the Heritage Lottery Fund, English Heritage consider that 'local authorities will show a greater readiness to serve compulsory purchase orders and repairs notices if they know they will not have to pick up all the bills' (EH 1994b, 28). But how will local authorities 'know they will not have to pick up the bills'?

The preparation of a full repairs notice can be a costly and time-consuming process in its own right. If the subsequent compulsory purchase order is challenged by the building owner at public inquiry, will an application for Heritage Lottery funding by the local authority be sufficient to convince an inspector of the course of action the local authority has adopted?

There is a real danger that applying for Heritage Lottery funding becomes a lottery in itself. This is all the more apparent when one studies the Heritage Lottery Fund guidelines which state that if the proposed works 'would normally be eligible or a priority for public grant aid', then;

> they are likely to be a lower priority for the Heritage Lottery Fund. That will not preclude us from assisting with a discrete element, or another phase, of the same project which is not likely to receive public support. We may consider assisting with works which would normally be eligible for public support where we are advised that the scale of the resources needed is such that public funds will be unable to make a meaningful contribution. The important factor for us to determine is whether a worthy project is unlikely to succeed without lottery support. Applicants are strongly encouraged to seek the advice of the relevant historic buildings and monuments agency on these points in advance of making an application. We, in turn, will seek their advice when assessing applications.
> (NHMF n.d., 13)

So the National Heritage Memorial Fund advise applicants to first seek advice from English Heritage (or its counterparts in the rest of the UK) about the possibility of grant-aid. If English Heritage advise the applicant that their proposal is either not eligible or not a priority for grant-aid, the applicant can then apply to the Heritage Lottery Fund. The Heritage Lottery Fund Eligibility Checklist invites applicants to confirm the statement, 'Following investigation we believe that our project is unlikely to succeed without the support of Lottery Funds.'

Should the project subsequently be submitted to the Heritage Lottery Fund, the Fund will seek the advice of English Heritage on the project, and whether it is a 'tangible historic asset'. It seems strange to have to assess the same case twice. Admittedly the National Heritage Memorial Fund pay English Heritage for their advice but why, if there is to be Heritage Lottery funding, is the whole process not handled by English Heritage who already have grant assessment, decision, distribution, partnership and administration systems in place and operating?

When Heritage Lottery funding was first announced it was billed as additional funding for the heritage. It would not take the place of existing grant funding. Applicants for Heritage Lottery funding were told to ensure that their projects were:

> supported by a significant element of partnership funding, and/or contributions in kind, from other sources. Possible sources for the partnership element of joint-funded projects may include not only the private or voluntary sectors or local authorities, but also Government programmes, Non-Departmental Public Bodies, or grants from European Community institutions (e.g. European Regional Development Fund). Partnership funding might be provided in the form of a contribution to capital costs, contributions in kind (e.g. land or voluntary labour), or an undertaking to support the running costs of a scheme after it has been completed.
> (NHMF n.d., 8)

However, the European Union's view that cash from state lotteries counts as public spending because it is generated by the public's money, throws the whole process of whether to bid for Heritage Lottery funding or grant into a dilemma.

In August 1995 the Government therefore stated that total public sector support for projects should not exceed more than half of total costs. This will undoubtedly be a blow to the hopes of many projects and calls into question the nature and purpose of lottery funding.

While there appears to be plenty of scope for uncertainty, Heritage Lottery funding has been agreed for a wide range of projects. In some cases, Heritage Lottery funding is backing significant projects which have also benefited from earlier and continuing English Heritage grant-aid. One example is Highcliffe Castle, Christchurch, Dorset. Gutted by fire in 1966, Highcliffe Castle was compulsorily purchased by Christchurch Borough Council in 1977. In July 1995 the Heritage Lottery fund announced a grant of £2 650 000 in addition to contributions of £1 000 000 from English Heritage and £1 000 000 from Christchurch Borough Council towards the estimated £5 244 000 costs of restoration.

REGULAR MAINTENANCE – THE WAY FORWARD

What of the future? Are there other mechanisms to more effectively grant-aid the conservation of historic buildings? One idea might be to grant-aid maintenance works before major repairs become necessary. Think of the implications of Christopher Wren's maxim, 'Drips come sudden and do great mischief'.

In Belgium, the initiative of Monumentenwacht Vlanderen (Monument Watch Flanders)* is an exciting development in the care of historic building through regular inspections. First established in the Netherlands in 1973, Monument Watch is an independent advisory service which promotes and supports the regular annual maintenance of historic buildings. In Flanders, eight teams of two men cover five regions to carry out regular on-site inspections. These are used to draw up status reports for the owners or managers of historic buildings. Monument Watch does not compete with architects, surveyors or building contractors. It generates considerable work for them by helping building owners understand the condition of their building. The report enables any necessary work to be properly planned and carried out.

All team members strive to balance both theoretical and practical knowledge as well as being able to carry out physically challenging inspections from bosun's chairs or steeplejack's ladders as well as more conventional forms of access. Building owners subscribe £30 per year to 'register' their building with Monument Watch and pay an inspection fee of £9 per person per hour of actual inspection time excluding VAT. The inspection fee represents only 10 per cent of the actual cost of the survey. To encourage surveys, the remainder of the cost is subsidized through the funding of Monument Watch by the Ministry of the Flemish Community, the King Baudouin Foundation and, most interestingly, the National Lottery.

Perhaps the Department of National Heritage will consider promoting the regular maintenance of the nation's built heritage by establishing a Monument Watch in the UK with funding from the National Lottery. This would surely be a fitting and lasting use with wide public benefits in return for public funding. Speaking at the York Institute of Advanced Architectural Studies in February 1995,

*Monumentenwacht Vlanderen, Brederodestraat 21, 1000 Brussels, Belgium (Telephone: 02 – 511 18 40, Fax: 02 – 511 52 21).

Sir Bernard Feilden concluded his contribution to the consultation on the economics of architectural conservation by stating:

> Regular maintenance must be promoted. All buildings must have regular inspections. This is what I want engraved on my tombstone.

The views expressed in this chapter are my own and may not necessarily reflect those of my employer.

BIBIOGRAPHY

EH (1994a). English Heritage, *Historic Buildings and Monuments Grants: Notes for Applicants.* (June).

EH (1994b). English Heritage, 'How to apply for funds', *Heritage Today,* xxviii (December).

EH (1995). English Heritage, 'The value of conservation? A seminar on the economic and social benefit of conservation', held at the Royal Society of Arts, London on 9 June 1995.

Embree, Sally (1995). 'Buildings at risk survey: a success story', English Heritage, *Conservation Bulletin,* xxv (March).

Gowers, Sir Ernest (Chairman) (1950). *Report of the Committee on Houses of Outstanding Historic or Architectural Interest.* HMSO.

Halsey, Richard (1995). 'Strategies for saving our churches', English Heritage, *Conservation Bulletin,* xxv (March).

Hawkins, Judy (1995). 'Review of grant conditions', English Heritage, *Conservation Bulletin,* xxvi (July).

IPD (1993). Investment Property Databank, *The Investment Performance of Listed Buildings.*

Johnson, Stephen (1992). 'Assessing needs for historic buildings and monuments grant applications', English Heritage, *Conservation Bulletin,* xvii (June).

National Audit Office (1992). *Protecting and Managing England's Heritage Property.* (July).

NHMF (1994a). National Heritage Memorial Fund, *Heritage Lottery Fund Fact Sheet.* (November).

NHMF (1994b). National Heritage Memorial Fund, *Summary of key points in Heritage Lottery Fund Guidelines and Application Pack.* (November).

NHMF (n.d.) National Heritage Memorial Fund, *Guidelines for Applicants to the Heritage Lottery Fund.*

Page, Jennifer (1995). 'Can we put a value on the heritage?', English Heritage, *Conservation Bulletin,* xxv (March).

PPG 15 (1994). Department of the Environment, Department of National Heritage, *Planning Policy Guidance: Planning and the Historic Environment.* PPG 15 (September).

Preservation Policy Group (1970). *Report on Historic Towns and Villages.* (May).

Reading/DTZ (1995). University of Reading and DTZ Debenham Thorpe. *Discussion draft: The Value of Conservation? A literature review of the economic and social value of the cultural built heritage.* (June).

RICS (1994). Royal Institution of Chartered Surveyors. *The listing of Buildings: The Effect on Value.*

14

A BIBLIOGRAPHY OF THE WRITINGS OF SIR BERNARD FEILDEN

Keith Parker

This bibliography is in two parts. The first seeks to record Sir Bernard's published writings to early 1995, including books, articles and reviews. While some translations have been included, there are exceptions. In the interests of clarity, articles published in periodicals which appear simultaneously in several languages, such as UNESCO Courier, or Parks, have only been recorded in their English version.

The second part of the bibliography lists unpublished papers and reports delivered to ICCROM and held in its Library. While these are still, properly speaking, unpublished, they are recorded by the publicly available Conservation Information Network, and so have been noted here for completeness. Readers wishing to experience the full range of Sir Bernard's thoughts and writing should bear this in mind, therefore, and consult the Librarian at ICCROM as to their availability.

1 PUBLISHED WORKS

1964

Norwich: a place for people. Norwich Society, Norwich.

1969

'Lessons learnt from the restoration work on York Minster', *Architects' Journal*, cxlix (27), 1765–70.

1970

'Saving York Minster', *International Nickel Magazine*, (4), 19–23.

'Training for restoration', *Architectural Review*, cxlviii (885), 301–2.

1971

'Caring for St Paul's', *RIBA Journal*, lxxviii (11), 490–3.

'York Minster' [reprint from unidentified journal or conference proceedings]. *Akademiai Kiado, Budapest.* 815–19.

1972

'The restoration of York Minster', in *The noble city of York* (Stacpoole, A. *et al.* eds.). Cerialis Press, York. 409–46.

1974

'The restoration of York Minster', *Chartered Surveyor*, cvii (1), 4–9.

'Vibration from motor traffic', *European Heritage*, (2), 44–6.

1975

'The care of cathedrals and churches', *Royal Society of Arts Journal*, cxxiii (5224), 196–215.

'Craftsmen for conservation', *Quantity Surveyor*, xxxi (6), 138–41.

'The storehouse of tradition', *Architectural Association Quarterly*, vii (1), 33–9.

With Derek Linstrum: 'Training for conservation: a European view', *Architectural Review*, clvii (935), 61–4. Reprinted in *Architectural conservation in Europe.* Cantacuzino, S. (ed.). Architectural Press (1975). 127–30.

1976

'Craftsmen for conservation: the skills of craftsmen should be recorded and passed on to the new generation of conservation craftsmen', *Building Technology and Management*, xiv (9), 5–9.

'Restoring three cathedrals: York, Norwich and St Paul's', in *Building early America: proceedings of a symposium held on the occasion of the 250th anniversary of the Carpenters' Company of the city and county of Philadelphia, March 27–29, 1974, Philadelphia* (Peterson C. E., ed.). Chilton Book Co., Radnor. 258–76.

The wonder of York Minster. Cerialis Press, York.

1977

European Centre for training craftsmen in the conservation of the architectural heritage: a feasibility study. Pro Venetia Viva, Strasbourg.

[Translation] *Centre européen de formation d'artisans pour la conservation du patrimoine architectural: étude opérationnelle.* Pro Venetia Viva, Strasbourg.

[Translation] *Centro europeo di formazione degli artigiani per la conservazione del patrimonio architettonico: studio operazionale.* Pro Venetia Viva, Strasbourg.

[Translation] *Europäisches Ausbildungszentrum für Handwerker im Denkmalschutz: Programmstudie.* Pro Venetia Viva, Strasbourg.

'Structural interventions in historic buildings', in *Symposium on structures in historic buildings: provisional papers, Rome, 14–18 Sept., 1977.* ICCROM, Rome. 1–8.

1978

'Exemples de technologie appropriée et de technologie inappropriée en matière de conservation du patrimoine culturel', *Patrimoine culturel de l'humanité, Bulletin*, (12), 10–11.

'Introduzione ai principi e alla pratica della conservazione', *Antiqua*, (9), 5–13.

'The principles and methodology of conservation', *The Scientific Examination of Cultural Property*, (1), 5–12.

1979

'Examples of appropriate and inappropriate technology in conservation of cultural property', *Parks*, iv (2), 21–2.

'The ICCROM for the unification of the principles of restoration', in *Problems of the completion of art objects: proceedings of the Second International Restorer Seminar, Veszprem, 15–27 July 1978.* Institute of Conservation and Methodology of Museums, Budapest. 19–33.

An introduction to conservation of cultural property. UNESCO, Paris.

[Translation] *Uvod u konzerviranje kulturnog nasljeda.* Drustvo konzervatora hrvatske, Zagreb (1981).

[For unpublished French and Spanish translations, see Part 2, Reports and papers at ICCROM.]

'Design of museums for conservation of cultural property', in *Regional seminar on the conservation of cultural materials in humid climates, Canberra, 19–23 February 1979.* Australian Government Publishing Office, Canberra. 66–81.

1980

'Architectural factors affecting the internal environment of historic buildings', in International Institute for Conservation, *Conservation within historic buildings: preprints of the contributions to the Vienna Congress, 7–13 Sept., 1980.* IIC, London. 1–5.

'Presentation of historic buildings', *Parks*, v (1), 11–15.

'Preservation of Islamic Cairo', in *International conference on the conservation of old Cairo.* (n. pub.) Cairo. 14p.

1981

'Conservation of materials and practical applications of scientific research in restoration work: general report', in *Nessun futuro senza passato.* ICOMOS, Rome. 267–302.

'Conservation of the architectural heritage and urban renaissance', *Council of Europe Information Bulletin*, xvii (1), 3–13.

'Earthquakes and historic buildings', in *Nessun futuro senza passato.* ICOMOS, Rome. 481–97.

'A possible ethic for the conservation of timber structures', in *Colloque sur la conservation du bois, Troyes, 16/19 mai/May 1981.* ICOMOS, Paris. 7.

'The principles of conservation', *UNESCO Courier* (March), 27–8.

With Jukka Jokilehto: 'Rehabilitation of historic buildings and quarters', *A Future for our Past*, (17), 4–5.

1982

Book review of Thompson, M. W. 'Ruins: their preservation and display', in *Monumentum*, xxv (2), 155–7.

Conservation of historic buildings. Butterworth Scientific. (Revised edition, Butterworth Architecture, Oxford, 1994).

'The design of museums to preserve the collection', *Al-turath wal-hadhara*, (4), 5–30.

'Is ICCROM relevant to the museologist?', *International Journal of Museum Management and Curatorship*, i (2), 145–8.

'Museum management and natural disasters', *International Journal of Museum Management and Curatorship*, i (3), 231–5.

'Museums: the right place for conservation?', *Museum (UNESCO)*, xxxiv (1), 10–20. (Conversation with Giovanni Scichilone).

'The principles of conservation', in *Conservation of historic stone buildings and monuments: proceedings of the conference held in Washington, D.C., February 2–4, 1981.* National Academy of Sciences, Washington (DC). 22–30.

'Tiilimuurauksen saeilyttaeminen – englantilaisen naekoekanta', *Tiili*, (2), 20–7. [The conservation of brickwork – an Englishman's view].

'A visit to Dennis King's workshop: a personal memoir', in *Crown in glory: a celebration of craftsmanship; studies in stained glass* (Moore, P., ed.), Jarrold, Norwich. 7–11.

1983

'Presentation of historic buildings', *Al-turath wal-hadhara*, (5), 4–13.

'Teaching conservation of wooden structures', in *International symposium on the conservation and restoration of cultural property, the conservation of wooden cultural property, November 1 to 6, 1982, Tokyo and Saitama.* National Research Institute of Cultural Properties, Tokyo. 211–21.

1984

'Working in conservation', *SPAB News*, v (1), 13–14.

1985

With Jukka Jokilehto: 'Air pollution in urban areas', *A Future for our Past*, (25), 10–12.

'Architectural and urban conservation: a review of the state of the art', *Town Planning Review*, lvi (2), 197–221.

'Conservation in the commonwealth', *RIBA Journal*, xcii (5), 71–2.

With Jukka Jokilehto: *Manual for the management of World Cultural Heritage Sites*. ICCROM, Rome.

'Professional interdisciplinary collaboration in conservation: is it possible?' in *Culzean Castle conference: conservation and preservation policies and partnership*. National Trust for Scotland, Edinburgh. 44–61.

'A provisional outline of actions to be considered in order to preserve historic centres and buildings in seismic zones', in *Proceedings of the international course on preventive measures for the protection of cultural property in earthquake-prone regions, 24 June – 5 July 1985, Skopje, Yugoslavia*. ICCROM, Rome. 59.

'Towards a maintenance strategy', in *Building appraisal, maintenance and preservation: proceedings of a 3-day symposium held at the University of Bath, Bath, 10th-12th July 1985*. Bath University, Department of Architecture and Building Engineering, Bath. 5–9.

1986
'Historic monuments: discussion', in Second International Congress on architectural conservation and town planning, Basle, April 1985, *Conservation and tourism*. Heritage Trust, London. 134–40.

'A plea for computerised listing', *Transactions of the Ancient Monuments Society*, xxx, 67–8.

'Report on the Khan Marjan, Baghdad', *Al-turath wal-hadhara*, (8), 22–45; (9), 271–300.

1987
Between two earthquakes: cultural property in seismic zones. ICCROM, Rome; Getty Conservation Institute, Marina del Rey.

[Translation] *Entre dos terremotos: los bienes culturales en zonas sismicas*. ICCROM, Rome; Getty Conservation Institute, Marina del Rey (1991).

'Between two earthquakes: the management of cultural property in seismic zones', in *Old cultures in new worlds: 8th General Assembly and international symposium*. US/ICOMOS, Washington (DC). 582–9.

'Multi-disciplinary professional collaboration: a personal view', *Quaderni dell' Istituto di Storia dell' Architettura*, nuova serie, fasc. 1–10, 551–5.

1988
Book review of Jones, B. G., 'Protecting historic architecture and museum collections from natural disasters', in *Town Planning Review*, lix (2), 230–1.

'Perspectives on ICCROM today', *ICCROM Newsletter*, (14), 2–3.

'A short introduction to conservation of cultural property in Europe', in *Air pollution and conservation, safeguarding our architectural heritage*, (Rosvall, J., and Aleby, S., eds.). Elsevier. 55–76. Also published in a special issue of *Durability of Building Materials*, v (3–4), 239–66.

1989
'A case study of Norwich Cathedral', *APT Bulletin*, xxi (3/4), 23–29.

'Conservation of historic buildings between two earthquakes', in *Proceedings of the 1st international seminar on modern principles in conservation and restoration of urban and rural cultural heritage in seismic-prone regions, Skopje, 17–22 October 1988*. Iziis, Skopje; ICCROM, Rome. 31–8.

'From restoration to maintenance: a case study of Norwich Cathedral', *APT Bulletin*, xxi (3/4), 23–9.

Guidelines for conservation: a technical manual. Indian National Trust for Art and Cultural Heritage, New Delhi.

Book review of Ashurst, J., and Ashurst, N., 'Practical building conservation', in *RIBA Journal*, xcvi (8), 71, 75, 79, 83.

1990
'Civilizing our cities', *Third World Planning Review*, xii (4), 311–21.

'Conservación y turismo', in *Congreso internacional para la conservación del patrimonio cultural*, ICOMOS, Quito. 143–54.

'Integrated conservation: a comparative study', in *Proceedings of an international workshop on integrated urban conservation: an approach towards development, Khargapur, November 23–27, 1990*. Indian National Trust for Art and Cultural Heritage, New Delhi. 2.1–2.43.

1991
With Peter Burman, Jane Fawcett and Lord Kennet (joint editors): *Managing World Heritage sites in Britain*. York University, Centre for Conservation Studies, York.

'The state of Europe's architectural heritage: a personal view', in *Science, technology and European cultural heritage: proceedings of the European symposium, Bologna, Italy, 13–16 June, 1989* (Baer, N. S., Sabbiani, C, and Sors, A. I., eds.). Butterworth-Heinemann, Guildford. 3–16.

1992
'After a fire', in *International symposium on fire protection of historic buildings and towns, Risor, Norway, September 12th-14th, 1990* (Larsen, K. E., and Marstein, N., eds.). Tapir, Flataasen. 85–91.

'Storage: out of sight (out of mind?)', *Museum*, xliv (1), 33–4.

1993
'Conservation and tourism', in *ICOMOS international scientific symposium on cultural tourism*. ICOMOS, Colombo. 59–65.

With Jukka Jokilehto: *Management guidelines for World Cultural Heritage sites*. ICCROM, Rome.

'Training for architectural conservation', in *Training: training of trainers in architectural and urban conservation: an appraisal by the ICOMOS Committee on Training.* Sri Lanka National Committee of ICOMOS, Colombo. 58–67.

[Translation] 'La formation et la conservation architecturale', *ibid.*, 68–79.

1994
Book review of Weaver, M. E., and Matero, F. G., 'Conserving buildings: guide to techniques and materials', in *Studies in Conservation*, xxxix (3), 215.

1995
'Conservation: is there no limit? A review', *Journal of Architectural Conservation*, i (1), 5–7.

'Chesterfield wins a silver medal', in *The economics of architectural conservation.* Institute of Advanced Architectural Studies, York. 99–102.

2 REPORTS AND PAPERS AT ICCROM

1972
Restauro della struttura del Monastero di York.

1973
Heritage crafts trusts.

1974
The care of churches and cathedrals.

1975
Architects and archaeology.
Attitudes to conservation of architecture.
Magdalen College: the bell tower.
St Paul's within the Walls (2 reports)

1976
Choice of stone for conservative repair work.
Maintenance of historic buildings.
Uses of plastics in repair and restoration.

1977
Chesterfield.
Mission report: Venice, 3–4 November, 1977.
Report on mission: Unesco, Paris, 7 and 8 December.
Training for conservation work.
Unesco/UNDP mission to Jerash and Petra.

1978
Report of mission to Asia, 2–23 December 1978.
Report of mission to Japan, 4–17 September 1978.
The role and duties of the cathedral architect.

1979

Design of museums for conservation of cultural property.
[Translation] *Diseño de museos para la conservación de patrimonio cultural, 1980.*
[Translation] *Introducción a la conservación.*
[Translation] *Une introduction à la conservation des biens culturels.*
Maintenance of historic buildings.
[Translation] *La manutenzione degli edifici storici.*
Marble stains and bulges on the outer walls of the Dome of the Rock.
Mission report, SS Giovanni e Paolo, Venezia: the south transept window.
Mission to Canada.
Petra/Jaresh tourism project.
Presentation of historic buildings.
Report on the dome of the Al Aksa.

1980

Changed attitudes in town planning of inner cities: a discussion paper.
[Translation] *Mutamento di atteggiamenti nell'urbanistica dei centri delle città: una base di discussione.*
Earthquakes and historic buildings.
Rehabilitation of historic buildings.
Report on the course organized by ICCROM in the Socialist Republic of Montenegro, Yugoslavia.

1982

Report on the Grand Mosque of Sfax.

1987

With Poul Beckmann: *Konarak: report on the structural condition and the preservation of the stone.*
With Poul Beckmann: *Taj Mahal: report on structural and geotechnical aspects of conservation.*

INDEX

The index contains virtually every occurrence of a name, place, organization or document, with the following exceptions: acknowledgements, bibliographies and endnotes, from which a very small number of items has been selected. I have also a wide selection of subjects which I hope will be useful. All illustrations have been indexed and are indicated with a letter 'i' after the page number, thus: 123i. The several changes to areas of administrative counties in 1964, 1974, 1986, and now in 1996, mean that it is not possible to be entirely consistent, particularly in view of the abolition of county names such as Avon and Clwyd; however, I believe that those I have used will not cause any confusion, often being a reversion to the old county names. Individual people are not given titles or distinctions.

Certain groups of material have been gathered together. All places outside the U.K. have been entered under their several countries. All buildings in Bath, Liverpool, London, and Plymouth have been indexed under those names; otherwise the choice between town or village and building as the key word depends on the most likely location for search. I have also grouped together Acts of Parliament. I have provided extensive cross-referencing particularly to the substance of long-titled official bodies and documents. An asterisk * after a page number indicates that the reference is to Dr Jokilehto's comments in his Appendix on Charters and Recommendations. Ranges of pages do not necessarily indicate a continuous treatment of the subject.